SEDIMENT TRANSPORT MECHANISMS IN COASTAL ENVIRONMENTS AND RIVERS

EUROMECH 310

SEDIMENT TRANSPORT MECHANISMS IN COASTAL ENVIRONMENTS AND RIVERS

EUROMECH 310

13 – 17 September 1993

Editors

M. Bélorgey & R. D. Rajaona

Le Havre, France

J. F. A. Sleath

Cambridge, UK

World Scientific

Singapore • New Jersey • London • Hong Kong

Published by

World Scientific Publishing Co. Pte. Ltd.
5 Toh Tuck Link, Singapore 596224
USA office: 27 Warren Street, Suite 401-402, Hackensack, NJ 07601
UK office: 57 Shelton Street, Covent Garden, London WC2H 9HE

British Library Cataloguing-in-Publication Data
A catalogue record for this book is available from the British Library.

SEDIMENT TRANSPORT MECHANISMS IN COASTAL ENVIRONMENTS AND RIVERS
Euromech 310

ISBN-13 978-981-02-1854-6
ISBN-10 981-02-1854-0

PREFACE

The EUROMECH Colloquium 310 on "Sediment Transport Mechanisms in Coastal Environments and Rivers" is the latest in a series of meetings, in the general area of sediment transport, organised under the auspices of the European Mechanics Committee. It was held in Le Havre from 13th to 17th September 1993 and 36 papers were presented by authors from 14 different countries.

The overall aim of the colloquium was to study basic mechanisms of sediment transport, numerical modelling and new experimental and analytical tools. Papers were presented in eight different sessions. Detailed topics covered ranged from turbulence structure, mechanisms of ripple formation, sediment suspension, numerical modelling and laboratory techniques through to site investigations of rivers and beaches.

The present volume contains a selection of the presented papers. As is usual in these Colloquia, some of the work discussed at the meeting was still at too early a stage for publication. The refereeing process reduced the number of papers still further!

The editors would like to thank all those who have contributed to this volume for their enthusiasm (and patience). They are also most grateful to the referees for their helpful comments. Finally, they would like to acknowledge the excellent work of the staff of the "Laboratoire de Mécanique des Fluides" at the University of Le Havre for their assistance in organising the colloquium, and the financial support of "Région de Haute-Normandie".

M. BÉLORGEY
D. R. RAJAONA
J. F. A. SLEATH

CONTENTS

4. Ripple Dynamics

5. Site Investigations

Turbulence and Transition

STRUCTURE OF TURBULENCE OVER TWO-DIMENSIONAL DUNES

Sean J. Bennett
Department of Earth Sciences, University of Leeds, Leeds, LS2 9JT, UK

and

James L. Best
Department of Earth Sciences, University of Leeds, Leeds, LS2 9JT, UK

ABSTRACT

Detailed measurements of the horizontal and vertical components of turbulence have been obtained over fixed, two-dimensional dunes using laser Doppler anemometry. A grid of over 1800 points allows construction of contour plots of the turbulence data which illustrate: (1) high turbulence intensity of the downstream component of velocity in the zone of flow separation, at reattachment and downstream of reattachment; (2) that highly turbulent vertical velocities occur within, above and just downstream of the separation zone as well as along the free-shear layer; and (3) frequent and energetic quadrant 2 events ('ejections') occur above the zone of flow separation and along the free-shear layer. These data document the strong influence of the shear layer bounding the flow separation zone in promoting eddy shedding and generating upward-directed fluid. Such large-scale turbulence may be critically important, not only in the transport and suspension of sediment above dunes, but also in dune formation and stability.

1. Introduction

Investigation of the relationships between fluid turbulence, sediment transport and bedform dynamics remains a central research area in physical sedimentology and sediment transport theory. Although numerous studies have examined bedform dynamics using a variety of grain sizes (see reviews in Southard and Boguchwal, 1990; Southard, 1991), relatively little is known about the turbulence fields that characterise bedforms and the relationships between turbulence and bedform stability. One problem in investigating bedforms in mobile sediments is the migration and change in form of the bedwave in relatively short periods of time, thus preventing detailed measurements of the entire flow field. One method by which to overcome this problem is to conduct these experiments using bedforms which are either chemically fixed into position or cast from a pre-formed mould (see, for example, Raudkivi, 1966; Nelson and Smith, 1989; Wiberg and Nelson, 1992; Lyn, 1993; Nelson et al., 1993). Whilst information has been generated using fixed ripples and dunes, these studies lack the density of data required to fully document the detailed turbulent flow structure and are of insufficient resolution to quantify the fine-scale fluid dynamic behaviour of the fluid as a function of bedform position. Yet, this information is critical in interpreting the morphological characteristics of bedforms, in explaining the mechanisms of transition between different bedform types and in linking the entrainment, transport and suspension of sediment to specific turbulent fluid events. As part of a larger, on-going investigation examining fluid turbulence, sediment transport and bedform dynamics over a range of mobile and immobile granular sediments and bed conditions, this paper presents detailed measurements of the two-dimensional flow and turbulence structure over fixed dunes in a laboratory channel. These results are discussed in the light of past work on flow and macroturbulence over dunes and the mechanisms of dune formation and stability.

2. Experimental Equipment and Procedure

The experiments were conducted in the Sedimentological Fluid Dynamics Laboratory, Department of Earth Sciences, University of Leeds, using a recirculating flume 10 m long, 0.3 m wide and 0.3 m deep. A total of seven, two-dimensional (i.e. straight crested) concrete dunes were moulded from typical bedform profiles that were generated and measured in a 0.3 mm diameter sand (see Bridge and Best, 1988, their Fig 1); the fixed dunes had a wavelength of 0.63 m, a height of 0.04 m, a slip-face angle of 30° and were coated with 0.22 mm glass spheres, ensuring equivalent grain roughnesses between the mobile bed and fixed bedforms. Flow depth over the dune crest was 0.1 m whilst the mean flow velocity was 0.57 m/s, identical to those flow conditions under which the mobile-bed dunes were generated.

A total of 76 velocity profiles spaced 0.01 m apart were taken over one bedform wavelength starting from the fifth bedform downstream (see points in Fig. 1). Downstream (u) and vertical (v) velocities were measured using a two-component, fiber optic DANTEC laser Doppler anemometer operated in back-scatter mode with a 100 mW Argon-ion laser, a correlation-type DANTEC FVA signal processor, 40 MHz frequency shift and a 400 mm focal length lens. Each profile contained between 20 and 30 points starting from 5 mm above the bed to near the free-water surface, thus yielding a grid of over 1800 sample locations. At each point, vertical and horizontal flow velocities were measured for 60 seconds with data rates varying between 20 and 100 Hz (normally 60 Hz). For the mean flow conditions used in this study, data rates of approximately 60-70 Hz are desirable to ensure full characterisation of the turbulence spectra (see Nezu and Nakagawa, 1993).

These data were processed and analysed for: (1) the spatial structure of turbulent flow as a function of dune position and height above the bed, specifically the root-mean-square of the turbulent velocities defined as:

$$u' = \left[\frac{1}{n}\sum_{i=1}^{n}\left(u_i - \overline{U}\right)^2\right]^{0.5} \tag{1}$$

$$v' = \left[\frac{1}{n}\sum_{i=1}^{n}\left(v_i - \overline{V}\right)^2\right]^{0.5} \tag{2}$$

where u' and v' are the horizontal and vertical rms values, n is the number of observations, u_i and v_i are the instantaneous horizontal and vertical velocities and $\overline{U}$ and $\overline{V}$ are the time-averaged horizontal and vertical velocities at-a-point within the flow; and (2) the spatial discrimination of relatively large turbulent events using quadrant analysis (Lu and Willmarth, 1973; Bogard and Tiederman, 1986) defined as:

$$H = |uv| / u'v' \tag{3}$$

where H is the designated threshold value, u and v are the instantaneous fluctuations about a zero mean, and $u'v'$ is the product of the rms values at-a-point within the flow. The second phase of this study sought to specifically identify the occurrence and relative importance of energetic quadrant 2 events (-u, +v); 'ejections' or 'bursts') at different positions over the dune. The density of sampling over the dune allows for the construction of a uniquely detailed set of contour maps for all turbulence parameters.

3. Results

A vector map of the flow field over the dune profile (Fig. 1) illustrates the strong flow separation cell formed in the dune lee-side, acceleration of flow along the dune back and development of a new boundary layer over the next dune downstream of reattachment. The mean distance from separation to reattachment is approximately 4.5 times the dune height, corresponding well with the work of Engel (1981) who obtained values of between 3.5 and 6. Flow expansion in the lee-side of the bedform leads to an area of flow deceleration whilst time-averaged horizontal flow velocities near the bed within the separation zone are negative (i.e. upstream). This detailed pattern of flow separation and boundary layer growth over a dune agrees well with previous studies of flow over dunes (e.g. Raudkivi, 1963, 1966; Rifai and Smith, 1971; Nelson and Smith, 1989; Mendoza and Shen, 1990; Lyn, 1993; Nelson et al., 1993) and negative steps (e.g. Bradshaw and Wong, 1972; Etheridge and Kemp, 1978, 1979; Nakagawa and Nezu, 1987; Adams and Eaton, 1988).

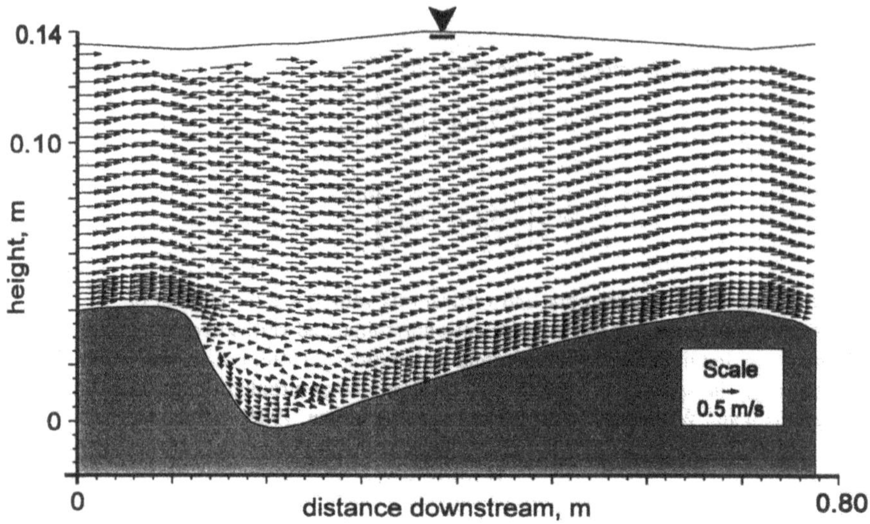

Figure 1. Vector map of flow velocity over a fixed dune comprising over 1800 sample locations. The lowest point in each profile is 5 mm above the bed surface. Vertical exaggeration is ~2.8X.

The spatial distribution of the root-mean-square (rms) values of the downstream component of turbulence over the fixed dune is shown in Fig. 2. The maximum values observed (from 0.2 to 0.3 m/s) occur at and just downstream of reattachment. High turbulent horizontal velocities (0.08 to 0.12 m/s) are also prevalent within the separation zone, downstream of reattachment and along the free-shear layer generated by flow separation.

The rms values of the vertical component (Fig. 3) reveal maximum values (0.06 to 0.09 m/s) within, above and just downstream of the zone of flow separation, having a local maximum downstream of the flow separation cell near reattachment. These high rms values occur in an area where significant erosion and sediment suspension is known to take place (e.g. Iseya and Ikeda, 1986; Nezu and Nakagawa, 1993). A striking feature illustrated in Fig. 3 is the downstream persistence of relatively high vertical rms values (0.04 to 0.06 m/s) along the free-shear layer and its extension into the outer flow region.

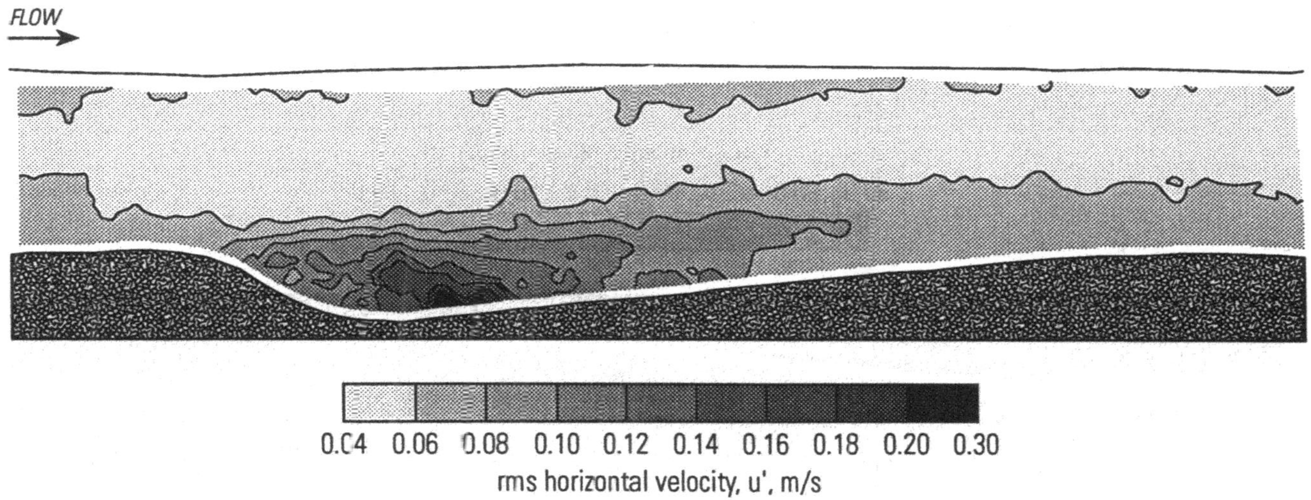

Figure 2. Contour map of the root-mean-square of the horizontal velocity component. Contour interval is 0.02 m/s. Vertical exaggeration is ~1.5X.

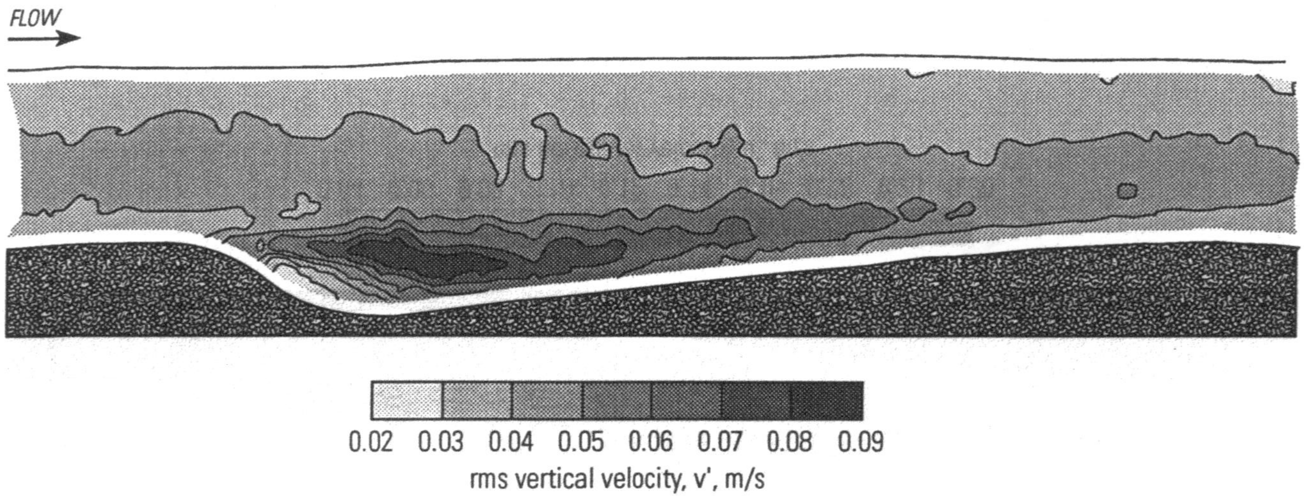

Figure 3. Contour map of the root-mean-square of the vertical velocity component. Contour interval is 0.01 m/s. Vertical exaggeration is ~1.5X.

Quadrant analysis using a threshold value of 4 in Eq. 2, allows examination of the spatial distribution of the frequency of relatively high magnitude quadrant 2 events or ejections (Fig. 4). This threshold value was chosen through a sensitivity analysis using different values for H and corresponds well with values suggested by other authors as being important turbulent events (Lu and Willmarth, 1973; Willmarth and Lu, 1974; Bogard and Tiederman, 1986). The highest frequency of large instantaneous values of uv in quadrant 2 (up to 1.4 Hz) are concentrated just above the zone of flow separation. These turbulent events persist along the free-shear layer well up into the flow, and decrease in frequency downstream over the dune (from 1 Hz to as little as 0.2 Hz). A local maximum of frequent quadrant 2 events also occurs above the upstream dune crest (Fig. 4), representing the continuation of this ejection zone from the previous bedform. These energetic quadrant 2 events also dominate the contribution to the local Reynolds stress along the shear layer.

4. Discussion

These data illustrate several fundamental aspects of flow and turbulence structure over dunes. First, the results show the strong influence of the shear layer bounding the zone of flow separation in promoting eddy shedding and generating upward-directed fluid. It is apparent that vortices may be shed into the outer flow from any position along a dune but very large and frequent vertical movements of fluid ('ejections') occur in this shear layer zone. The frequent, high-magnitude quadrant 2 events are generated before reattachment and along the free-shear layer and are shown to diffuse both downstream and vertically towards the next dune crest. In addition, these data lend support to the flow visualisation studies of Müller and Gyr (1982, 1986) who document the formation of 'hairpin'-shaped vortices and elongated rollers at the dune crest due to flow separation. The zone of strong turbulence generated by Kelvin-Helmholtz instabilities along the free-shear layer is responsible for the vertical ejection of low speed fluid. The generation of this turbulence structure, and its spatial variability, may be critical for dune stability (see below). Furthermore, our results suggest the source of dune-related macroturbulence lies in quadrant 2 events generated along the shear layer and not derived from within the separation zone (i.e. Jackson, 1976; Yalin, 1977, 1992) nor do they share a common origin with boundary layer bursting. This contention is supported by the recent field studies of Lapointe (1992) and Kostaschuk and Church (1993) as well as the experimental work of Nezu and Nakagawa (1989a,b).

Secondly, frequent ejections of fluid from the separation point and along the shear layer provide a source of net upward-directed fluid necessary to initiate and maintain the suspension of sediment. Vertical turbulence anisotropy has commonly been invoked to explain the maintenance of sediment in suspension in turbulent shear flows (e.g. Bagnold, 1966; Leeder, 1983; Wei and Willmarth, 1991; Bridge and Bennett, 1992). Furthermore, several theoretical investigations on the physical processes of

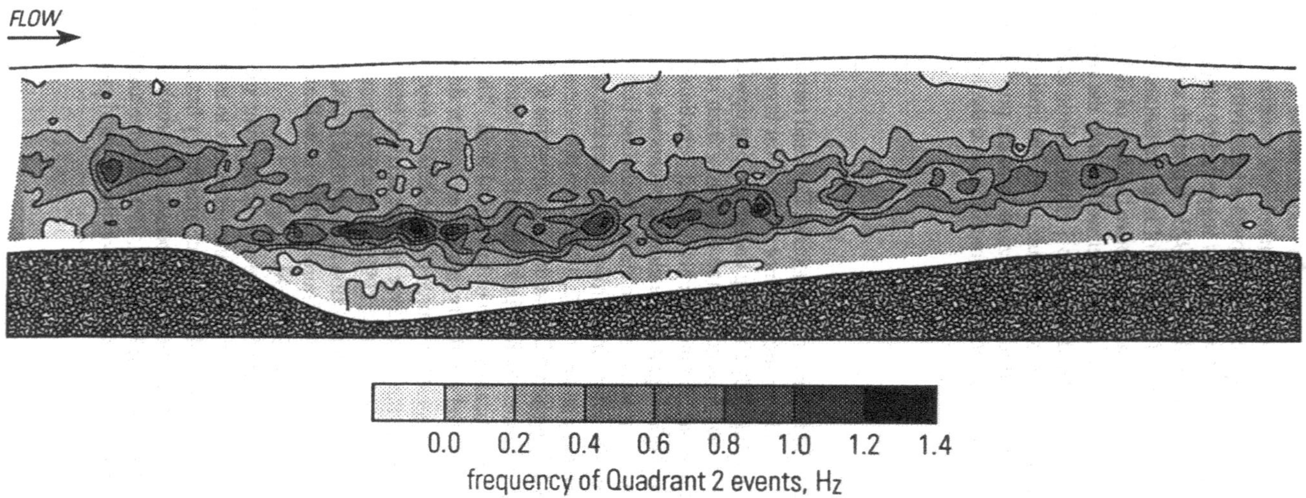

Figure 4. Contour map of the frequency of quadrant 2 events using a threshold value (H) of 4 in equation (3). Contour interval is 0.2 Hz. Maximum values of over 1.2 Hz are located just downstream of dune crest along the separation zone shear layer. Vertical exaggeration is ~1.5X.

suspended sediment transport rely upon the turbulent diffusion and convection of fluid and sediment away from the boundary (see discussion and references in van Rijn, 1984; Bechteler, 1986; Celik and Rodi, 1991; Garcia and Parker, 1991). The frequent vertical ejection of fluid along the shear layer generated downstream of a dune crest provides a physical explanation for both residual fluid stresses and the suspension of sediment. Sediment may be suspended just downstream of the dune crest along the free-shear layer from the flow separation region and entrained and suspended at reattachment. Both sources of sediment, which may reach the surface as 'boils' or 'kolks', are evident in field studies of dune-related macroturbulence (Kostaschuk and Church, 1993). Ongoing work is seeking to establish the relationships between the magnitude of these suspension events and the quantity and size of sediment uplifted into the outer boundary layer.

Thirdly, the generation and stability of dunes have been previously described through linear stability analysis (e.g. Kennedy, 1963, 1969; Hayashi, 1970; Smith, 1970; Richards, 1980; Engelund and Fredsøe, 1982) which predicts the occurrence and stability of bedforms generated by flow over a wavy surface. The common distinction between the hydraulic stability of ripples and dunes has been that ripples are controlled by viscous forces (inner flow variables) and scale with grain size (grain Reynolds number) whereas dunes are controlled by macroturbulence (outer flow variables) and scale with flow depth (e.g. Richards, 1980). The controlling physical mechanisms that cause this bedform transition, however, must lie in the structure of turbulence over the bedform, a structure which cannot be predicted using linear stability analysis. Visual studies suggest the flow structure over ripples is fundamentally different from that over dunes although this contention requires investigation and comparison with the detailed turbulence maps presented here. The occurrence of these frequent, high-magnitude quadrant 2 events may also be linked to the formative processes controlling dune stability. This control is, at present, unclear but may be related to increased bedload and suspended load transport rates induced through the action of these fluid ejections. Finally, the important distinction should be reiterated between the dependence of dune stability on the structure of shear-layer realted turbulence, as suggested herein, and the formation and scaling of dunes as a result of turbulent boundary layer bursting processes (e.g. Jackson, 1976; Yalin, 1977, 1992). Further studies are needed to document the spatial distribution of turbulence and specific turbulent events in the all important bedform transition regions (i.e. ripple→dune, and dune→upper-stage plane bed) to unequivocally assess the role of turbulence in bedform generation and stability.

5. Conclusions

Detailed two-dimensional measurements of turbulence over fixed, straight-crested dunes have allowed construction of unique contour maps which illustrate the strong vertical and horizontal turbulence associated with flow separation and formation of the free-shear layer originating at the dune crest. The occurrence and downstream persistence of this shear layer, with its characteristic frequent and energetic quadrant 2 ejections of fluid, dominate the turbulence structure over dunes and may lie at the heart of explaining the origin and hydraulic stability of these bedforms.

6. Acknowledgements

This research is supported by Natural Environment Research Council grant GR3/8235. We thank John Bridge and Mike Leeder for numerous discussions on the subject and Neil Woodhouse for providing technical support.

7. References

1. E.W. Adams and J.K. Eaton, *J. Fluids Engng.* **110** (1988) 275.
2. R.A. Bagnold, *U.S. Geol. Surv. Prof. Paper* **422-I** (1966).
3. W. Bechteler, ed., *Sediment Transport of Suspended Solids in Open Channels* (A.A. Balkema, Rotterdam, 1986).
4. D.G. Bogard and W.G. Tiederman, *J. Fluid Mech.* **162** (1986) 389.
5. P. Bradshaw and F.Y.F. Wong, *J. Fluid Mech.* **52** (1972) 113.
6. J.S. Bridge and J.L. Best, *Sedimentology* **35** (1988) 753.
7. J.S. Bridge and S.J. Bennett, *Water Resour. Res.* **28** (1992) 337.
8. I. Celik and W. Rodi, *J. Hydraul. Engrng.* **117** (1991) 191.
9. P. Engel, *J. Hydraul. Div. Am. Soc. Civ. Engrs.* **107** (1981) 1133.
10. F. Engelund and J. Fredsøe, *Ann. Rev. Fluid Mech.* **14** (1982) 13.
11. D.W. Etheridge and P.H. Kemp, *J. Fluid Mech.* **86** (1978) 545.
12. D.W. Etheridge and P.H. Kemp, *J. Hydraul. Res.* **17** (1979) 107.
13. M. Garcia and G. Parker, *J. Hydraul. Engrng.* **117** (1991) 414.
14. T. Hayashi *J. Hydraul. Div. Am. Soc. Civ. Engrs.* **96** (1970) 357.
15. S. Iseya and H. Ikeda, in *Proc. of 30th Jap. Confer. on Hydraulics Jap. Soc. Civ. Engrs.* (1986), p. 505 (in Japanese).
16. R.G. Jackson, *J. Fluid Mech.* **77** (1976) 531.
17. J.F. Kennedy, *J. Fluid Mech.* **16** (1963) 512.
18. J.F. Kennedy, *Ann. Rev. Fluid Mech.* **1** (1969) 147.
19. R.A. Kostaschuk and M.A. Church, *Sediment. Geology* **85** (1993) 25.

20. M.F. Lapointe, *Earth Sur. Process. Landforms* **17** (1992) 253.
21. M.R. Leeder, *Sedimentology* **30** (1983) 485.
22. S.S. Lu and W.W. Willmarth, *J. Fluid Mech.* **60** (1973) 481.
23. D.A. Lyn, *J. Hydraul. Engrng.* **119** (1993) 306.
24. C. Mendoza and H.W. Shen, *J. Hydraul. Engrng.* **116** (1990) 459.
25. A. Müller and A. Gyr, in *Mechanics of Sediment Transport*, eds. B.M. Sumer and A. Müller (A.A. Balkema, Rotterdam, 1982), p. 41.
26. A. Müller and A. Gyr, *J. Hydraul. Res.* **24** (1986) 359.
27. H. Nakagawa and I. Nezu, *J. Hydraul. Res.* **25** (1987) 67.
28. I. Nezu and H. Nakagawa, in *Turbulent Shear Flows 6* (Springer-Verlag, 1989a) p. 313.
29. I. Nezu and H. Nakagawa, in *Workshop on Instrumentation for Hydraulic Laboratories* (IAHR/CCIW, Burlington, Canada, 1989b) p. 29.
30. I. Nezu and H. Nakagawa, *Turbulence in Open-Channel Flows* (A.A. Balkema, Rotterdam, 1993).
31. J.M. Nelson, S.R. McLean and S.R. Wolfe, *Water Resour. Res.* **29** (1993) 3935.
32. J.M. Nelson and J.D. Smith, *J. Geophys. Res.* **94** (1989) 8146.
33. A.J. Raudkivi, *J. Hydraul. Div. Am. Soc. Civ. Engrs.* **89** (1963) 15.
34. A.J. Raudkivi, *J. Fluid Mech.* **26** (1966) 507.
35. K.J. Richards, *J. Fluid Mech.* **99** (1980) 597.
36. M.F. Rifai and K.V.H. Smith, *J. Hydraul. Div. Am. Soc. Civ. Engrs.* **97** (1971) 963.
37. J.B. Southard, *Ann. Rev. Earth Planet. Sci.* **19** (1991) 423.
38. J.B. Southard and L.A. Boguchwal, *J. Sediment. Petrol.* **60** (1990) 658.
39. J.D. Smith, *J. Geophys. Res.* **75** (1970) 5928.
40. L.C. van Rijn, *J. Hydraul. Engrng.* **110** (1984) 1613.
41. T. Wei and W.W. Willmarth, *J. Fluid Mech.* **223** (1991) 241.
42. P.L. Wiberg and J.M. Nelson, *J. Geophys. Res.* **97** (1992) 12745.
43. W.W. Willmarth and S.S. Lu, in *Turbulent Diffusion in Environmental Pollution*, eds. F.N. Frenkiel and R.E. Munn (Academic Press, New York, 1974), p. 287.
44. M.S. Yalin, *Mechanics of Sediment Transport*, 2nd. ed. (Pergamon Press, Oxford, 1977).
45. M.S. Yalin, *River Mechanics* (Pergamon Press, Oxford, 1992).

NUMERICAL STUDY ON TRANSITION TO TURBULENCE IN A WAVE BOUNDARY LAYER

HITOSHI TANAKA

Associate Professor, Department of Civil Engineering, Tohoku University, Aoba, Sendai 980, Japan

and

AHMAD SANA

Research Associate, Department of Civil Engineering, Tohoku University, Aoba, Sendai 980, Japan

ABSTRACT

The understanding of transition in a wave boundary layer is very important due to its application in a variety of practical problems. Numerous theoretical and experimental studies have been carried out to explore this complex phenomenon. With the advent of powerful computing systems, numerical models are gaining popularity for the analysis of fluid flow phenomena. Among a large number of existing models to describe the structure of boundary layer, $k-\varepsilon$ model is the most widely used, due to its computational economy and reasonable accuracy. A comparative study of three versions of low Reynolds number $k-\varepsilon$ model, namely; JONES AND LAUNDER [10], HOFFMAN [8] and CHIEN [4] is carried out with special reference to transitional regime in a wave boundary layer on smooth bottom. It is inferred that despite the useful modifications suggested by later researchers, the original model by JONES AND LAUNDER [10] has the most precise prediction of wave boundary layer structure, especially in transition. An abrupt change in friction factor and phase difference is successfully predicted by this model, which is well in agreement with the available experimental data. Finally, the features of transition in oscillatory boundary layer are elaborated.

1. Introduction

In a wave boundary layer, the transition from laminar to turbulent flow is of vital importance, by virtue of its significant effects on bed friction on a sea bottom, wave boundary layer thickness, the change in wave height, drift currents set up by waves and sediment and pollutant transport in coastal environments. Several researchers have, therefore, tried to explore this phenomenon both experimentally and theoretically. The studies based on experiments are very rare and the data to use for verification of theoretical models is scarce. The experimental studies by Kamphuis [16], Jonsson and Carlsen [12], Sleath [21], Hino et al. [7] and Jensen et al. [9] are worth mentioning in this regard. Most of the efforts, though, have been to evolve the theoretical models usually based on temporally constant eddy viscosity. But, as Sleath [21,22] has shown experimentally that this assumption seldom embraces reality. However, time invariant eddy viscosity models proposed by Kajiura [15], Grant and Madsen [6], Tanaka and Shuto [23], Tanaka, Chian and Shuto [24], Myrhaug [17], Asano and Iwagaki [1], Christoffersen and Jonsson [5], have been able to predict some of the features of wave boundary layer with reasonable accuracy and adequate simplicity. But, unfortunately, the assumption of invariant eddy viscosity with respect to time does not solve the problem, because its variation in space also requires considerable attention. It always gets very difficult to judge, as to what extent the accuracy may be sacrificed in favour of simplicity? In all the aforementioned models, the variation in cross stream direction is assumed in three, two or a single layer. For each layer, then, the variation is specified in terms of other boundary layer quantities, separately. As the number of layers increases, so does the complexity of the model.

During the last two decades, as the $k-\varepsilon$ model of Jones and Launder [10,11] gained popularity due to its appreciable accuracy with reasonable computational economy in predicting a wide variety of steady flows, it was tried on oscillatory flow also. First of all, Tanaka [25] and then Sato et al. [19] applied the high Reynolds number version of this model to turbulent wave boundary layer. Later, Asano et al. [2], Aydin and Shuto [3], Justesen [14] and Sana [18] applied the low Reynolds number version, which is, of course, more appropriate in case of oscillatory flow.

Aydin and Shuto [3] used the model incorporating the modifications suggested by Chien [4], while rest of the studies were based on the original model by Jones and Launder [10]. None of these attempts, except that by Sana [18] could predict the abrupt change in the friction factor depicted by the experimental data of Kamphuis [16]. So far, no study has been carried out to appraise the relative performance of different versions of low Reynolds' number $k-\varepsilon$ model. Since, a number of modifications in the original model have been proposed to enhance computing efficiency by reducing the nonlinearity of the governing equations, such type of comparative study would be beneficial.

In the present study, original low Reynolds number $k-\varepsilon$ model by Jones and Launder [10], with the modifications by Hoffman [8] and with those by Chien [4] have been applied to a wave boundary layer on a smooth boundary. The comparison is made on the basis of wave friction factor, phase difference and the time variation of bottom shear stress with reference to the available universal experimental data. It is found that the original model is the most efficient in predicting the transition according to the experiments.

2. Governing Equations

2.1. Original $k-\varepsilon$ Model

Jones and Launder [10] proposed that the eddy viscosity can be expressed in terms of turbulent kinetic energy k , and its dissipation rate ε , thus naming it $k-\varepsilon$ model. The closure problem was solved by using the transport equations of k and ε . Hence, the model equations may be given in dimensionless form as follows;

$$\frac{\partial u^*}{\partial t^*}=\frac{\partial U_w^*}{\partial t^*}+S\frac{\partial}{\partial z^*}\left\{\left(\frac{S}{RE}+C_\mu f_\mu\frac{k^{*2}}{\varepsilon^*}\right)\frac{\partial u^*}{\partial z^*}\right\} \tag{1}$$

$$\frac{\partial k^*}{\partial t^*}=S\frac{\partial}{\partial z^*}\left\{\left(\frac{S}{RE}+f_\mu\frac{C_\mu k^{*2}}{\sigma_k\,\varepsilon^*}\right)\frac{\partial k^*}{\partial z^*}\right\}+SC_\mu f_\mu\frac{k^{*2}}{\varepsilon^*}\left(\frac{\partial u^*}{\partial z^*}\right)^2-S\varepsilon^*-D \tag{2}$$

$$\frac{\partial \varepsilon^*}{\partial t^*}=S\frac{\partial}{\partial z^*}\left\{\left(\frac{S}{RE}+f_\mu\frac{C_\mu k^{*2}}{\sigma_\varepsilon\,\varepsilon^*}\right)\frac{\partial \varepsilon^*}{\partial z^*}\right\}+SC_1 f_\mu C_\mu k^*\left(\frac{\partial u^*}{\partial z^*}\right)^2-Sf_2C_2\frac{\varepsilon^{*2}}{k^*}+E \qquad (3)$$

Where, $RE=\hat{U}_w a_m/\nu$, $a_m=\hat{U}_w/\sigma$ and $S=\hat{U}_w/\sigma z_h$. The quantities with superscript '*' are dimensionless. The parameters used to normalize the quantities are velocity amplitude $\hat{U}_w$, distance from wall to free stream z_h, kinematic viscosity of fluid ν and angular frequency σ. Here u, represents the x-direction velocity. The values of various constants and terms D and E in Eq.2 and Eq.3, respectively, according to original model and the later modifications are presented in Table 1.

TABLE 1. Parameters of Different Low Reynolds Number $k-\varepsilon$ Models.

Parameter	Jones & Launder [10]	Hoffman [8]	Chien [4]
C_1	1.55	1.81	1.35
C_2	2.0	2.0	1.8
C_μ	0.09	0.09	0.09
σ_k	1.0	2.0	1.0
σ_ε	1.3	3.0	1.3
f_2	$1-0.3\exp(-R_T^2)$	$1-0.3\exp(-R_T^2)$	$1-0.222\exp(-R_T^2)$
f_μ	$\exp\left\{-\frac{2.5}{1+R_T/50}\right\}$	$\exp\left\{\frac{-1.75}{1+R_T/50}\right\}$	$1-\exp(-0.0115Bu_\tau^* z^*)$
D	$A\left(\frac{\partial\sqrt{k^*}}{\partial z^*}\right)^2$	$\frac{A}{2z^*}\frac{\partial k^*}{\partial z^*}$	$A\frac{k^*}{z^{*2}}$
E	$A\nu_T^*\left(\frac{\partial^2 u^*}{\partial z^{*2}}\right)^2$	-	$A\frac{\varepsilon^*}{z^{*2}}\exp(-0.5Bu_\tau^* z^*)$

Here, $A=2S^2/RE$, $B=RE/S$, u_τ^* is dimensionless shear velocity, $\nu_T^*=f_\mu C_\mu k^{*2}/\varepsilon^*$ is dimensionless eddy viscosity and $R_T=k^2/\varepsilon\nu$ is turbulence Reynolds Number.

2.2. *Hoffman's Modifications*

The modifications proposed by Hoffman [8] were mainly aimed at reducing the nonlinearity of governing equations and in turn achieving computational economy. According to him the term D as expressed by him was a linear form fully capable of giving correct quadratic behavior of k near the boundary. He also suggested to eliminate the term E in ε equation. Hoffman's modified model showed better agreement with the existing experimental data in Couette flow and fully developed channel flow.

2.3. *Chien's Modifications*

These modifications also proved to be efficient as far as the fully developed channel flow is concerned. Chien [4] compared the results of his modified model and those by the original model and found his modifications giving better agreement with the available data in fully developed channel flow.

2.4. *Boundary Conditions*

In all the above versions of $k-\varepsilon$ model the boundary conditions may be expressed as follows;

$$\text{at } z^*=0 \qquad u^*=k^*=\varepsilon^*=0$$

$$\text{at } z^*=1 \qquad \frac{\partial u^*}{\partial z^*}=\frac{\partial k^*}{\partial z^*}=\frac{\partial \varepsilon^*}{\partial z^*}=0 \tag{4}$$

The dimensionless free stream velocity U_w^* is expressed as follows;

$$U_w^*=\cos(t^*) \tag{5}$$

Here, pressure gradient, which is in fact time derivative of free stream velocity, may be expressed as sine function of t^* in dimensionless form.

3. Numerical Computations

The governing equations were discretized using central differences employing an implicit finite difference scheme of Crank Nicolson type. The constituting equations may be expressed to form a tridiagonal matrix, which, then may be solved using Gauss elimination method.

The cross-stream mesh spacing was made to vary exponentially to allow for better accuracy near the wall. For every value of wave Reynolds' number, the spacing was so adjusted that a sufficient number of mesh points may lie in the viscous sublayer. The convergence limit was set as follows;

$$\left|\frac{X(N)-X(N-1)}{X(N)}\right| \leq 0.00005 \tag{6}$$

Where X is the parameter on which convergence is based and N is the last iteration. Since, iteration procedure was adopted to solve the nonlinear equations for every time step, so, the maximum difference of two consecutive values of velocity, turbulent kinetic energy and its dissipation rate was considered to be the convergence parameter. For reaching the steady state, the convergence was based on friction factor, which covers the overall behavior of wave boundary layer.

4. Results

4.1. Laminar Flow

As is evident from Fig.1 that all the three models predict the velocity profile in laminar condition in an excellent manner. Since, the mesh spacing criterion was same, the points pertaining to the three models lie on one another in such a manner that it is difficult to differentiate among the symbols assigned to them. In this case, three of them show negligible values for turbulent kinetic energy and its dissipation rate, both temporally and spatially, theoretically which is quite satisfactory.

4.2. Transition

Here the definition by Jonsson [13] for the friction factor is adopted, which is given by the following equation;

$$\tau_{0\max} = \frac{\rho}{2} f_w \hat{U}_w^2 \tag{7}$$

Where, $\tau_{0\max}$ is maximum bottom shear stress and ρ is the mass density of the fluid.

Fig.2 shows clearly the capabilities of the three models to predict the friction factor in laminar regime with very good precision. Though, slight variations can be observed as compared with the analytical solution. At $RE \approx 2\times10^5$ an abrupt increase in friction factor is predicted by the model of Jones and Launder [10], while the models of Hoffman [8] and Chien [4] go on depicting laminar behavior. Then Hoffman's model deviates from laminar trend rather mildly at $RE \approx 4\times10^5$. Afterwards, Chien's model shows the abrupt change at $RE \approx 5\times10^5$.It can be observed that Jones and Launder model predicts the jump in friction factor a little earlier, while Chien's one predicts it a little later as compared to the experimental data by Kamphuis [16]. However, the values of friction factor obtained by the former model are closer to those of experiments.

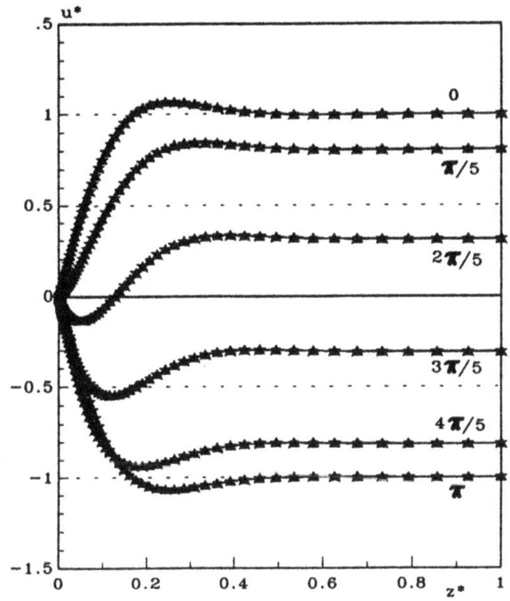

Fig.1 Velocity Profile for Laminar Flow, $RE = 10^4$, Theory; Laminar (solid line), Present computations; Jones and Launder Model (square), Hoffman's Model (triangle), Chien's Model (star).

A consistent abrupt change is evident in the phase difference values (Fig.3). But, it is clear that though Jones and Launder model predicts the friction factor quite efficiently, phase difference values are not in such a good agreement with the experimental data.

To get a more elaborate comparative perusal of the three models, the temporal variation of bottom shear stress normalized by τ_{0max}, is also presented (Fig.4). As the wave Reynolds number increases, the Jones and Launder model depicts a bump. The height of this bump goes on increasing to reach a value that is maximum over the whole cycle, and hence, according to the definition, the friction factor is computed by it. The phase difference, here, is determined by the harmonic analysis, considering the first harmonic component as the governing one. This is why, after $RE \approx 1.5 \times 10^5$ the phase difference drops rapidly, so as to reach $RE \approx 3 \times 10^5$, after which it shows a weak dependance on Reynolds number. In this condition, the values of phase difference are much smaller than those in laminar regime. With slight deviation of the predicted values , this behavior is consistent with the experimental data of Jensen et al. [9]. Especially, for higher Reynolds' numbers the predicted shear stress profile is very close to the experimental curve. Chien's and Hoffman's models can not predict this bump distinctly, despite their ability to give very close values of friction factor and phase difference for $RE = 6.5 \times 10^5$, as shown in Fig.4.

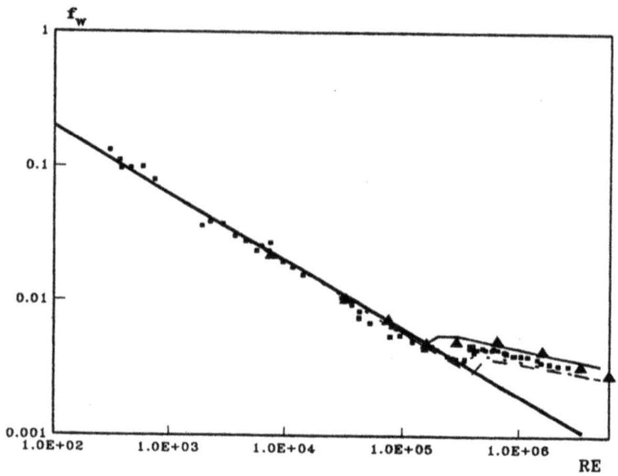

Fig.2 Comparison for Wave Friction Factor, Theory; Laminar (thick solid line), Present computations; Jones and Launder Model (thin solid line), Hoffman's Model (dotted line), Chien's Model (dashed line), Experimental data; Kamphuis, [16] (small square), Hino et al., [7] (big square), Jensen et al., [9] (triangle).

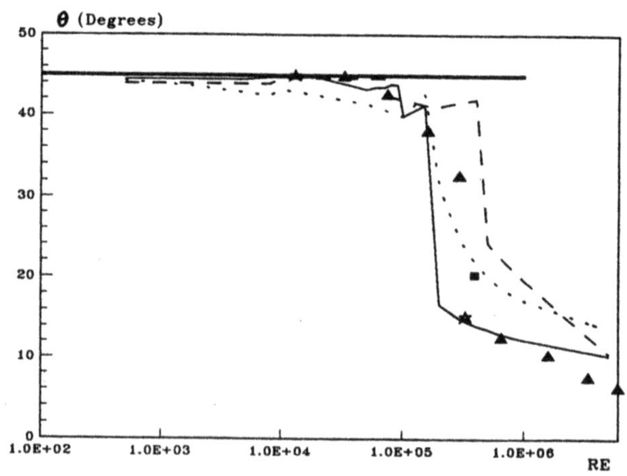

Fig.3 Comparison for Phase Difference, Experimental data; Sawamoto and Sato, [20] (star), Explanation of other lines and symbols is the same as in Fig.2.

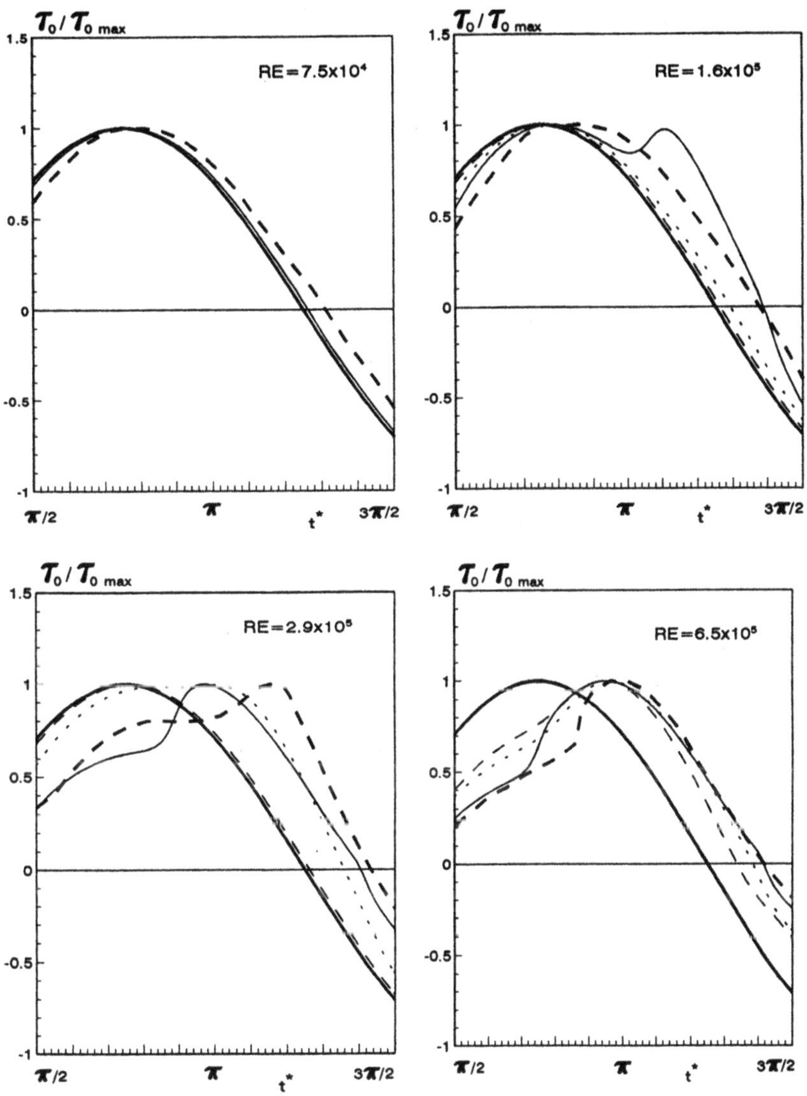

Fig.4 Temporal Variation of Bottom Shear Stress. Theory; Laminar (thick solid line), Present computations; Jones and Launder Model (thin solid line), Hoffman's Model (dotted line), Chien's Model (thin dashed line), Experimental Data; Jensen et al., [9] (thick dashed line).

A very important shortcoming of the models of Hoffman and Chien is that they do not produce a unique solution of the governing equations. Because, the computations were performed by providing various initial conditions, and, it was observed that for both of these models, if the initial condition was laminar, the predicted quantities, even for higher values of Reynolds Number, showed laminar properties, such as laminar shaped velocity profile and negligible turbulent kinetic energy etc. The present results were obtained by using turbulent initial conditions obtained from Jones and Launder Model.

4.3 Turbulent Flow

The friction factor values obtained by Jones and Launder model are closer to those by other two models, in turbulent zone. Similar behavior is depicted in phase difference diagram.

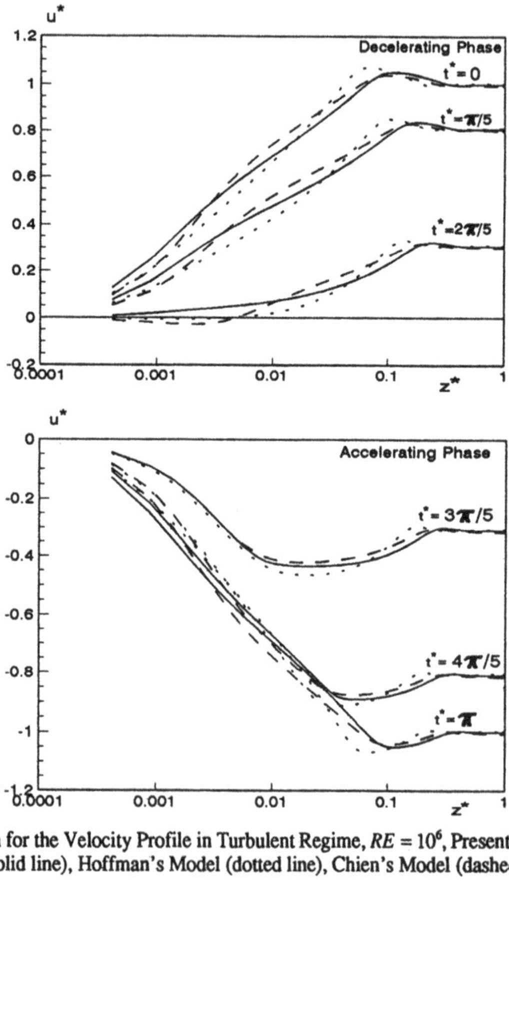

Fig. 5 Comparison for the Velocity Profile in Turbulent Regime, $RE = 10^6$, Present computations; Jones and Launder Model (solid line), Hoffman's Model (dotted line), Chien's Model (dashed line).

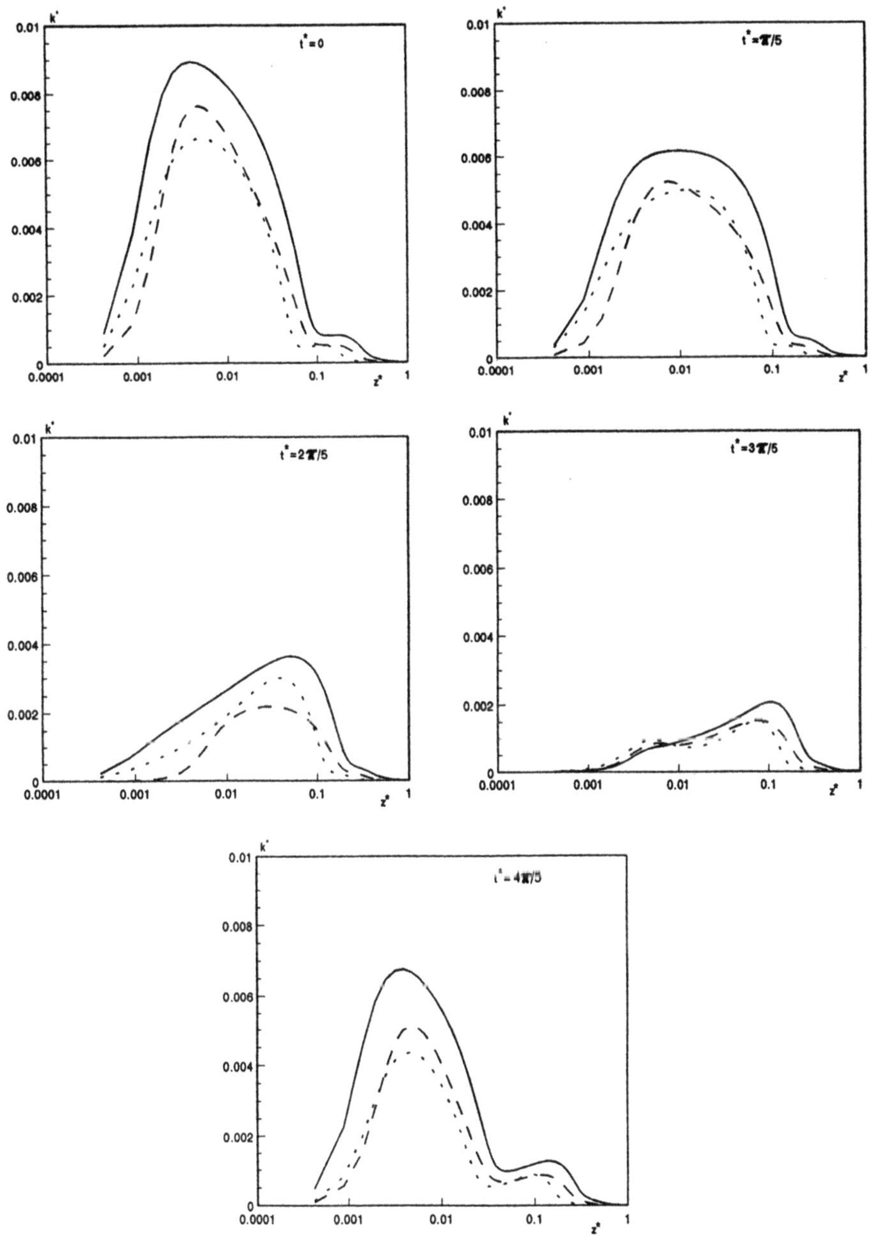

Fig.6 Comparison for the Turbulent Kinetic Energy in Turbulent Regime, $RE = 10^6$, Explanation of the lines and symbols is the same as in Fig.5.

The comparison of velocity profile for $RE = 1 \times 10^6$ (Fig. 5) reveals that Jones and Launder model shows an overlap layer clearly showing the logarithmic variation. The Chien's model also shows this behavior with a lesser thickness of overlap layer. But, with Hoffman's model this layer is even thinner. The thickness of wave boundary layer as depicted in the velocity profile (Fig.5) is the least obtained by Hoffman's model, during whole wave cycle. While, Jones and Launder Model and Chien's model show very close values.

The turbulent kinetic energy predicted by Jones and Launder model is the highest at all phases (Fig.6). Its concentration near the wall, during the deceleration, and spreading in cross-stream direction during acceleration, is qualitatively similar for the three models. In the initial part of deceleration phase, Chien's model depicts higher values of turbulent kinetic energy than those predicted by Hoffman's model.

5. Conclusion

After detailed comparison of important flow characteristics of wave boundary layer it is inferred that the low Reynolds number $k-\varepsilon$ model by Jones and Launder [10] is the most efficient as compared to the other two models under observation, i.e. those by Hoffman [8] and Chien [4]. Although, both of the later models, which are in fact the modified forms of the former one, are more economical computationally and have lesser degree of non-linearity. However, their predictions do not conform as well with the available experimental data as those by the former model do.

The uniqueness of the solution of Jones and Launder model is an important merit, which proves its reliability in comparison with the other two models.

A more elaborate and detailed comparison may be made on the basis of velocity profile and wave boundary layer thickness, which would depict the ability of these models to predict finer details of the wave boundary layer. Further extension of the Jones and Launder model to rough turbulent case and to wave-current combined flow is also essential from practical viewpoint.

6. References

1. ASANO, T., And IWAGAKI, Y., *19th Conference on Coastal Engineering, ASCE,* **3** (1984) 2397.
2. ASANO, T., GODO, H., And IWAGAKI, Y., *Coastal Engineering in Japan,* **30** No.2 (1988) 1.
3. AYDIN, I., And SHUTO, N., *Coastal Engineering in Japan,* **30** No.2 (1988) 11.
4. CHIEN, K.Y., *AIAA Journal,* **20** No.1 (1982) 33.
5. CHRISTOFFERSEN, J.B., And JONSSON, I.G., *Ocean Engineering,* **12** No.5 (1985) 387.
6. GRANT, W.D., And MADSEN, O.S., *Journal of Geophysical Research,* **84** No.C4 (1979) 1797.
7. HINO, M., KASHIWAYANAGI, M., NAKAYAMA, A., And HARA, T., *Journal of Fluid Mechanics,* **131** (1983) 363.

8. HOFFMAN, G.H., *The Physics of Fluids*, **18** No.3 (1975) 309.

9. JENSEN, B.L., SUMER, B.M., And FREDSOE, J., *Journal of Fluid Mechanics*, **206** (1989) 265.

10. JONES, W.P., And LAUNDER, B.E., *International Journal of Heat and Mass Transfer*, **15** (1972) 301.

11. JONES, W.P., And LAUNDER, B.E., *International Journal of Heat and Mass Transfer*, **16** (1973) 1119.

12. JONSSON, I.G., And CARLSEN, N.A., *Journal of Hydraulic Research*, **14** No.1 (1976) 45.

13. JONSSON, I.G., *A New Approach to Oscillatory Rough Turbulent Boundary Layers*, Series Paper No.**17**, (ISVA, Technical University of Denmark, Lyngby, 1978) p.79.

14. JUSTESEN, P., *Turbulent Wave Boundary Layers*, Series Paper No.**43**, (ISVA, Technical University of Denmark, Lyngby, 1988) p.226.

15. KAJIURA, K., *Bulletin, Earthquake Research Institute*, **46** (1968) 75.

16. KAMPHUIS, J.W., *Journal of The Waterways, Harbors and Coastal Engineering Division*, **101** No.WW2 (1975) 135.

17. MYRHAUG, D., *Ocean Engineering*, **11** No.3 (1984) 315.

18. SANA, A., *Application of a Turbulence Model to an Oscillatory Boundary Layer*, (Master of Engineering Thesis No.**WA-92-1**, Asian Institute of Technology, Bangkok, Thailand, 1992).

19. SATO, S., UEHARA, H., And WATANABE, A., *Coastal Engineering in Japan*, **29** (1986) 65.

20. SAWAMOTO, M., And SATO, E., *Coastal Engineering in Japan*, **34** No.1 (1991) 1.

21. SLEATH, J.F.A., *Journal of Fluid Mechanics*, **182** (1987) 369.

22. SLEATH, J.F.A., *The Sea*, **9B**, (1990) 693.

23. TANAKA, H., And SHUTO, N., *Coastal Engineering in Japan*, **24** (1981) 105.

24. TANAKA, H., CHIAN, C., And SHUTO, N., *Coastal Engineering in Japan*, **26** (1983) 19.

25. TANAKA, H., *Third International Symposium on River Sedimentation*, (The University of Mississippi, U.S.A., 1986) 334.

On the Mass Transport Near the Bed and the Turbulence Characteristics in Steep Waves

LIDONG ZHAN
and
WILHELM DURSTHOFF
Franzius-Institut, Universität Hannover, Nienburger Str. 4
30167 Hannover, Germany

ABSTRACT

The drift and turbulence components of the velocity field beneath progressive gravity waves are investigated in order to understand the basic mechanism of sediment transport due to waves in coastal environment. In this paper, two different stages of the mass transport process and the turbulence development are presented on the basis of results from experimental observations, and it is shown that the dynamical process in the entire wave field is determined by the interaction of these two components of motion. A qualitative description for the characteristic flow condition is given.

1 Introduction

In order to understand the basic mechanism of the sediment transport in off-shore region, the wave-induced mass transport and turbulence motion in transitional water depth evoked interests of many researchers in the past. A series of theoretical works and experimental studies can be found in previous publications.

As shown in earlier reports, the mass transport velocity induced by waves of small amplitude can be predicted well by the existing theory. In the classic paper by Longuet-Higgins [5] and many later works of other authors providing some modified versions of Longuet-Higgins's theory [3][7] supported by a few experimental observations [6][9], following results were treated as evident facts:

- There exists always a strong negative Eulerian mass transport with a parabolic distribution in the interior of the wave field, while the drift velocity near the bed is in the direction of the wave propagation.
- Due to the viscous effect there is a phase shift of the oscillatory motion in the bottom boundary layer against the interior field, and a positive Reynolds shear stress will

then be established in this flow region. This is the reason for the forward drift current near the bed.

- The turbulence generated in the boundary layer exerts only a very weak influence on the distribution of the drift velocity. The discrepances between the theory and the experimental data are caused through neglecting non-linear terms of higher order.

On the basis of results from the measurements using LDV and PIV measuring techniques in two closed wave flumes of different size, it should be shown in the present paper that the dynamical condition in a wave field may be much more complex than described in the existing theory, and phemonena can be observed under this condition which are entirely different as the theoretical predictions.

2 Experimental Observations

The experiments have been performed in two wave tanks of different size in the University of Hanover and the University of Edingburgh using LDV and PIV measuring techniques as well as flow visualization. Detailed information about the measuring systems and the experimental set-up is given in [11].

When the experimental data were analyzed, the drift component of the velocity field at each measuring level was not considered as a time-averaged quantity, but rather as the direct-current part (DC) of the wave motion according to a modified assumption for the wave data decomposition introduced by the authors. This is, as showed in Fig. 1, not necessarily a constant and can be estimated by means of a Kalman-Bucy processor [1].

A significant advantage of this processing method is the observation and estimation of system states using an algorithm for the recursive calculation in real time. Thus, the stochastic wave components need not to be treated as frequency responses and no statistical calculation is necesseray. Several biases can therefore be minimized.

In Fig. 2-II, the experimental drift profile from LDV measurements for a relatively long wave in a closed wave channel (dots) show good agreement to the theoretical calculation (solid line), even with quantitative deviations.

However, a more interesting phenomenon could be observed in short waves (Fig. 2-I): there is a strong drift current with a local maximum near the bed in the opposite direction to the wave propagation. This behaviour cannot be described either by the Stokes theory for irrotational flow, or by the Longuet-Higgins's "Conduction Solution" or any modified version of them.

The result of this experiment can also be fully reproduced using PIV in a wave tank of different size, and a similar distribution of the measured drift velocity is displayed in Fig. 3, and in Fig. 4 for the region near the bottom with a higher resolution.

The establishment of a negative drift current near the bottom was also reported by *C. Swan* in [8]. In his experimental work, the time effect on the mass transport was

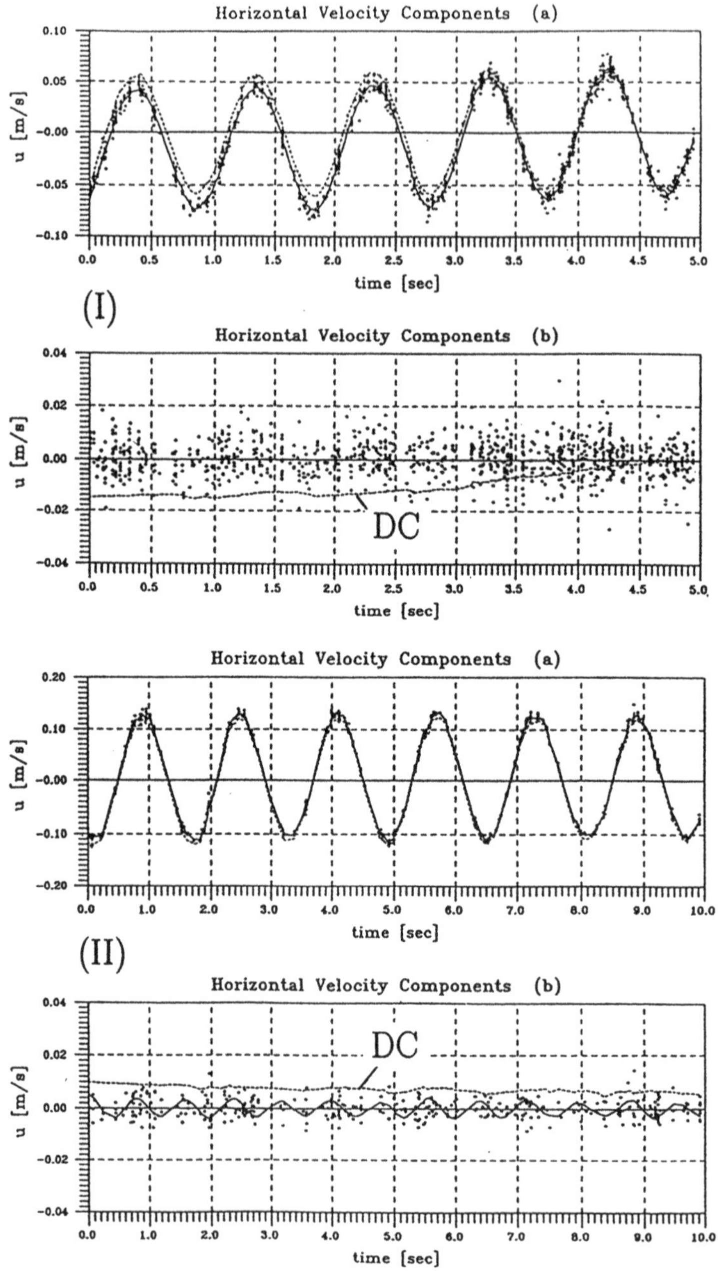

Figure 1: LDV time series of horizontal velocity components

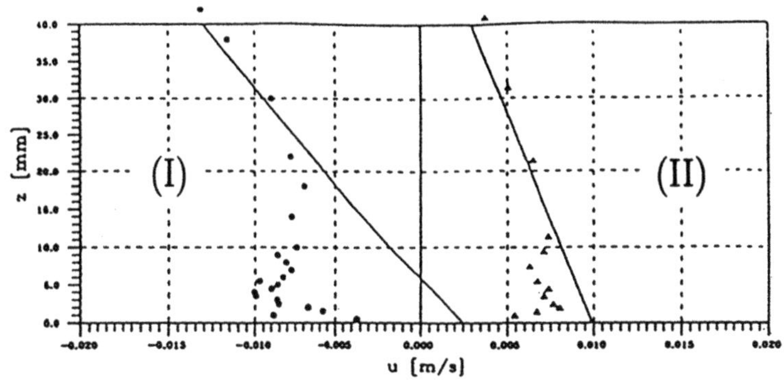

Figure 2: Profiles of the drift velocity near the bed (LDV)

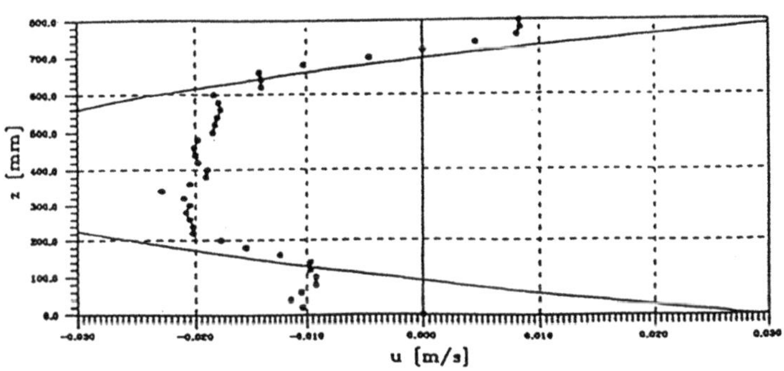

Figure 3: Profile of the drift velocity in a steep wave (PIV)

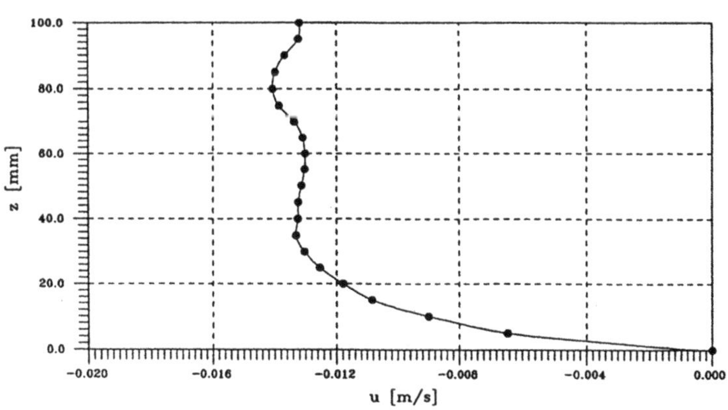

Figure 4: Drift profile near the bed in a steep wave (PIV)

examined and the importance of the horizontal convection of vorticity was pointed out. In the present paper, the authors try a different approach starting from the energy balance and the instability consideration for the turbulence development. This makes it possible to study the influence of different wave parameters.

It is important to consider a second fact indicated by LDV and PIV measurements in Fig. 1 and Fig. 5 implies the distributions of the vorticity for both waves, as well as by flow visualizations: the turbulence intensity in the short waves (I) is much higher than that in the longer waves (II), in spite of a smaller Reynolds number related to the boundary layer thickness!

According to additional information from the flow visualization it can be seen further that the dominant mode of the turbulence motion is transported from the beach by the backward drift current in the middle field, instead of a local production within the bottom boundary layer. This diffuses then into the bottom boundary layer, and will be amplified there due to the large local velocity gradient.

This suggests that the behaviour of the second order drift motion is decisively influenced by the turbulence developing process, which depends in turn on the convection of kinetic energy by the wave-induced drift current. Thus, the dynamical flow condition in the wave field is determined by the interaction of these two velocity components.

3 Theoretical Interpretation

The convection of the turbulence energy is similar to that in a free turbulent shear flow (Fig. 6), for which no assumption of homogeneity and isotropy of the chaotic turbulence motion can be made, and there is a close connection between the turbulence and the mean-velocity profile at any section, even those upstream (on the beach). In such a more complicated situation the local instability condition correponding to the eigenvalues of the *Orr-Sommerfeld* equation and the transport processes corresponding to the terms in the balance equation for the turbulence energy must be taken into account [2]:

$$(\bar{u}_m - c)\left(\frac{d^2\phi'}{dz_m^2} - \alpha^2\right) - \phi'\frac{d^2\bar{u}_m}{dz_m^2} = -\frac{i}{\alpha Re}\left(\frac{d^4\phi'}{dz_m^4} - 2\alpha^2\frac{d^2\phi'}{dz_m^2} + \alpha^4\phi'\right) \tag{1}$$

$$\begin{aligned}\underbrace{\frac{d}{dt}\frac{\overline{q^2}}{2}}_{total\ rate} &= \underbrace{\frac{\partial}{\partial t}\frac{\overline{q^2}}{2}}_{local\ rate} + \underbrace{\overline{U_j}\frac{\partial}{\partial x_j}\frac{\overline{q^2}}{2}}_{convection} = \underbrace{-\frac{\partial}{\partial x_i}\overline{u_i\left(\frac{p}{\varrho} + \frac{q^2}{2}\right)}}_{diffusion} - \underbrace{\overline{u_i u_j}\frac{\partial \overline{U_j}}{\partial x_i}}_{generation} \\ &+ \underbrace{\nu\frac{\partial}{\partial x_i}\overline{u_j\left(\frac{\partial u_i}{\partial x_j} + \frac{\partial u_j}{\partial x_i}\right)}}_{work\ by\ viscosity} - \underbrace{\nu\overline{\left(\frac{\partial u_i}{\partial x_j} + \frac{\partial u_j}{\partial x_i}\right)\frac{\partial u_i}{\partial x_j}}}_{dissipation}\end{aligned} \tag{2}$$

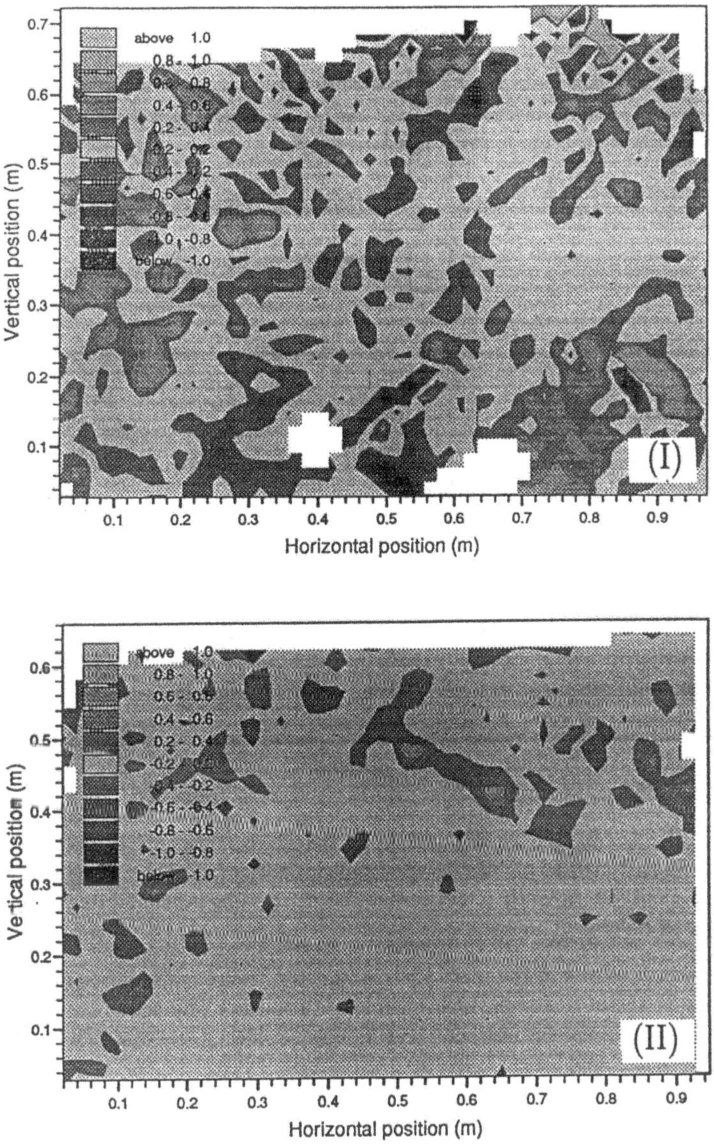

Figure 5: Distribution of the vorticity in the wave fields

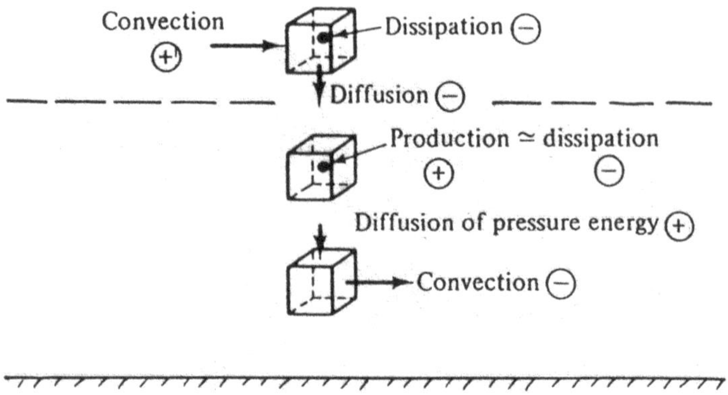

Figure 6: Turbulence developing processes in a convective field

From the shape of free turbulent shear flow and from the observed flow phenomena, some certain features can be summarized for physical understanding of dynamical processes in wave field:

- For the upper part of the flow field, the Reynolds number which implies the ratio of local turbulence production to the turbulence dissipation is larger in a steep wave than that in a long wave, but in the present case it is still far away from the value for equilibrium condition.

- With decreasing wave length the amplification of the periodic disturbances increases, so that the local linear instability of the drift motion may occur which is regarded as a necessary condition for the laminar/turbulent transition.

- On the beach, the kinetic energy of the wave motion is entirely fed into the nonlinearity-stochastic motion of turbulence before being transfered into heat. The turbulence energy is then transported downstream by horizintal convection with the second order drift velocity.

- The instabilities of the mean motion and the transported turbulence as a trigger may be the origins of the intensified local turbulence production even in a flow field with lower Reynolds numbers. The dissipative energy loss can then be balanced by this energy supplement.

- If the turbulence is maintaind in the wave field, the phase shift between the fluid motion in the bottom boundary layer and the free flow outside is compensated by the intensive exchange of mass, momentum and kinetic energy. The positve steady shear stress of a purely viscous flow will be reduced and the reversal of the drift velocity can then occur.

The critical condition for such a case could not be characterized solely by Reynolds number. Yet, both the instability amplification c_i in Eq. 1 and the convected turbulence energy according to Eq. 2 are functions of the wave steepness:

$$(c_i)_{max} \leftrightarrow a^2\kappa^m; \qquad \overline{q^2} = \overline{q_0^2}\, e^{-x/L_g} \quad \text{with} \quad L_g = 5\nu\, a^2\kappa\sigma\, \lambda_g^2 \tag{3}$$

The overall condition of a wave field seems to be sensitive to the chaotic motion from the beach, whose kinetic energy per unit of mass increases to second order with the wave steepness. According to the instability theory, the critical Reynolds number for the transition from a laminar flow to the turbulent one may be lower by strong disturbance. In this case the wave steepness is probably the more suitable parameter than the non-constant critical Reynolds number.

Due to the local instability in the bottom boundary layer triggered by the turbulent disturbance from the outer flow, the so-called "turbulence spots" are produced. As implied from flow visualization, these spots spread from the boundary layer into the interior of the wave field in the opposite direction of the wave propagation with a certain angle to the bed. This can be the reason for the existence of the negative local maximum of measured drift profiles under steep waves.

4 Further Considerations

According to the interpretation above, an interesting aspect for the suspension and transportation of sediments in the near shore region can be illustrated: subject to the increasing steepness of the incident waves, a strong seaward sediment transport could be expected because of the interaction between the wave-induced drift current close to the bed and the intensive turbulence motion coming from the beach, even at a large distance from the breaking zone.

These results provide a possible explanation of why, under the action of long waves, sediments tend to be carried on in the direction of the wave propagation and build up on the beach, while considerable erosion of the beach can be observed under storm conditions [10]. The remaining question of whether this is the actual mechanism of the real behaviour should be handled in further investigations of the flow field beneath steep waves above a movable bed.

In addition to the practical relevance already mentioned, such a study is also important for the theoretical description of the complete velocity field under the waves. It is demonstrated here that intensive turbulence fluctuations may not only be generated by very long waves with large Reynolds number (comparable to the steady flow condition), but can also be expected in a convective field beneath steep wave. And this may play a dominant role in forming other wave components, for example, the distribution of the Eulerian drift velocity near the bottom.

In order to estimate the mass transport velocity correctly and to quantify the turbulence developing process in relation to different wave parameters, a more detailed and more extensive investigation of individual wave components and their interaction is required. This contains theoretical analysis and especially accurate measurements of production, dissipation and transportation of turbulence energy as well as the spatial turbulence structure in steep waves.

References

1. W. Dursthoff and L. Zhan: *The Analysis of Irregular LDV Time Series of Waves Using the Kalman-Bucy Algorithm*, Optics and Lasers in Eng., **14** (1992)
2. J. O. Hinze: *Turbulence*, (Second Edition, McGraw-Hill Inc., New York, 1975)
3. B. Johns: *On the Mass Transport Induced by Oscillatory Flow in a Turbulent Boundary Layer*, J. Fluid Mech., **43** (1970), Part 1, pp. 177-185
4. H. T. Kim, S. J. Kline and W. C. Reynolds: *The Production of Turbulence near a Smooth Wall in a Turbulent Boundary Layer*, J. Fluid Mech., **50** (1971), Part 1, pp. 133-160
5. M. S. Longuet-Higgins: *Mass Transport in Water Waves*, Phil. Trans. of the Royal Soc. of London, No. 345 (1953), pp. 535-581
6. R. C. H. Russell and J. D. C. Osorio: *An Experimental Investigation of Drift Profiles in a Closed Channel*, Proc. 6th Conf. Coastal Engng. (1985), ASCE, pp. 171-183
7. C. Swan and J. F. A. Sleath: *A Second Approximation to the Time-Mean Lagrangian Drift Beneath a Series of Progressive Gravity Waves*, Ocean Engng., **17** (1990), No. 1/2, pp. 65-79
8. C. Swan: *Convection within an Experimental Wave Flume*, J. Hydr. Res., **28** (1990), No. 3, pp. 273-282
9. Ü. Ünlüata and C. C. Mei: *Mass Transport in Water Waves*, J. Geophys. Res., **75** (1970), pp. 7611-7618
10. G. E. Vincent: *Contribution of the Sediment Transport on a Horizontal Bed due to Wave Action*, Proc. 6th Conf. Coastal Engng. (Florida, 1958), pp. 326-335
11. L. Zhan, W. Dursthoff, T. Bruce and C. A. Greated: *An Experimental Study of the Drift Velocity in Progressive Waves*, Proc. 6th Symp. Appl. of Laser Tech. to Fluid Mech. (Lisbon, 1992), pp. 326-335

Numerical and Analytical Modelling

"SUSPENSION OF SAND IN OSCILLATORY FLOW ABOVE RIPPLES: DISCRETE-VORTEX MODEL AND LABORATORY EXPERIMENTS"

M.E. Block - A.G. Davies
School of Ocean Sciences, University of Wales (Bangor)
Marine Science Laboratories, Menai Bridge, Gwynedd, LL59 5EY, U.K.

and C.Villaret
EDF, Laboratoire National d'Hydraulique
6 quai Watier, 78400 Chatou, France

ABSTRACT

A discrete vortex model of the type proposed by Longuet-Higgins (1981) is used to simulate the formation and shedding of vortices above a rippled bed in oscillatory flow. This time-stepping model is coupled with an unsteady boundary layer model which, in association with a bedload formula, is used to estimate the rate of sediment entrainment into suspension at the ripple crest. The paths of suspended sediment particles are followed using a simple Lagrangian tracking scheme, and thus the model has been used to predict the time-varying concentration above both ripple crest and trough positions. Here these results are compared with some new, detailed, measurements made in a large wave flume. The phase of vortex release at the ripple crest, and the later arrival of a sediment-laden vortex above the trough, are quite well predicted. However the time-mean concentration tends to be underpredicted, partly due to the lack of vertical diffusion in the model. Despite this, the experimental evidence provides generally good support for the discrete vortex approach adopted.

1. Introduction

Vortex shedding above steep sand ripples in oscillatory flow is a potent mechanism for both the entrainment of sediment and the dissipation of wave energy. During each wave half-cycle, a vortex is formed above the lee slope of each ripple. This traps sediment passing over the ripple crest, and also sediment mobilized beneath the vortex. When the flow reverses, the sediment-laden vortex is ejected from the bed, giving up some of its load as it passes over the crest and moves away. Meanwhile, new vortices are formed above the lee slopes, which are later ejected from the bed, and so on. The motion of each ejected, decaying vortex is governed by both the oscillatory free-stream flow and the other vortices present. The suspended sediment grains tend to follow the trajectories of the fluid particles, while settling to the bed under gravity. Ultimately, the locations on the bed at which the grains settle out determine whether or not the ripples grow, or are destroyed, by the wave action. Stable equilibrium ripples have been generated by regular waves in many laboratory investigations (Nielsen, 1981), and the time-varying sediment concentrations in such flows have been measured by Nakato *et al* (1977) and others.

Time-stepping models of a field of discrete (potential) vortices have been used in studies of the oscillatory flow over steep, two-dimensional, ripples by Longuet-Higgins (1981) and Macpherson (1984). More recently, Hansen *et al* (1992) have used a "cloud in cell" discrete-vortex approach, accompanied by a boundary layer model (including near-bed sediment motion) and a Lagrangian description of the motion of grains in suspension. In addition, more fundamental studies of the oscillating flow over ripples have been carried out by, for example, Blondeaux and Vittori (1991).

Here a discrete-vortex model is used to simulate the evolving vortex and sediment concentration fields above a bed of round-crested, 2-D ripples. In Section 2, the formulation is discussed, and some brief comments are made about the associated near-bed oscillatory boundary layer model. In Section 3, the methods used to model both the bedload and suspended load are discussed briefly. In Section 4, some new laboratory experiments are discussed in which the time-varying sediment concentration has been measured above the crests and troughs of equilibrium sand ripples. In Section 5, the model results are presented and compared with the observations.

2. Hydrodynamical Model

Discrete vortex model

The aim of discrete vortex models is to simplify the equations of motion, thereby allowing a simulation of high Reynolds number flows. Here such a model is used to predict the evolution (in 2D-space and time) of the vortex field above a rippled bed in oscillatory flow. The model represents a development of the approach of Longuet-Higgins (1981) and Macpherson (1984), based on the methods of classical potential flow theory. Only an outline of the model and its assumptions is given here (see Block (1994) for a complete account).

The calculations are made in the complex 'transform' ζ-plane, and the flow is obtained in the 'physical' z-plane by a one-to-one conformal mapping of the upper half of the transform plane. No computational grid is involved in this model, and the mapping rule used is as follows:

$$z = \zeta + ib.\exp(il\zeta)$$

where $z=x+iy$ and $\zeta=\xi+i\chi$. This function is suitable for a deep flow since $\zeta=\infty$ and $z=\infty$ are corresponding points for $\chi>0$. The (impermeable) bed surface ($y=y_b$) is defined as $\chi=0$, such that

$$y_b = b \cos l\xi$$

where $b(=\eta/2)$ is the ripple amplitude and $l(=2\pi/\lambda)$ is the ripple wavenumber. For small b, $x\cong\xi$ and the bed is purely sinusoidal. However, for increasing values of b, the ripple crests become more peaked, and the troughs flatter and longer, in accordance with the profiles of natural sand ripples.

The flow above the rippled bed is given by the complex potential $\Omega = \phi + i\psi$, where ϕ is the velocity potential and ψ the streamfunction, and the velocity components (u,v) in the x and y directions are given at any point in the z-plane by $u - iv = -d\Omega/dz$. Nonseparating oscillatory flow above a rippled bed is represented in the upper ζ-half plane by the complex potential $\Omega = -U\zeta\sin(\sigma t)$ where U is the free-stream velocity amplitude and σ is the wave frequency. The presence of vortices is modelled by superimposing a row of potential vortices, parallel to the ξ-axis in the upper ζ-half plane, one vortex of strength κ, say, being placed at an equivalent point above each ripple. Since the ξ-axis (i.e. the bed) is required to be a streamline, a symmetrical vortex is placed at the complex conjugate point in the lower ζ-half plane for each vortex in the row. The complex potential for the oscillating flow, the vortex row (one vortex being at point ζ_0), and the image row is

$$\Omega = -U\zeta\sin(\sigma t) - i\kappa \log \sin\{\pi(\zeta-\zeta_0)/\lambda\} + i\kappa \log \sin\{\pi(\zeta-\overline{\zeta_0})/\lambda\}$$

Further rows of vortices may be added by extension of this argument.

The present time-stepping model predicts the movement of each vortex row as a result of both the underlying oscillatory motion and also the presence of other rows. Since the calculations are made in the transform plane, it is necessary to correct the vortex positions at each timestep using Routh's rule. The time-varying solution is the same above each ripple, since the (deep) outer flow is assumed to be horizontally uniform.

The potential flow model makes no prediction of the vorticity generated in the bottom boundary layer and so, following Longuet-Higgins (1981), Prandtl's rule is used here to determine the rate of generation of vorticity at the ripple crest position:

$$d\kappa/dt = Q^2/4\pi \qquad \text{where } Q = \text{velocity above ripple crest}$$

The vorticity generated in this way during each model timestep (normally 500 timesteps per cycle) is released at a height of $\eta/8$ directly above the crest (η= ripple height). This height (which is also the height at which Q is evaluated) has been chosen for consistency with the predictions of the boundary layer model discussed later.

Starting from the instant t_n of (the nth) flow reversal at the crest, the new vorticity generated during each timestep is consolidated into a single 'growing' vortex. [The 'growth phase' lasts until the next flow reversal at the crest.] As the flow speed increases, this vortex increases in strength and size, and starts to occupy the region above the ripple lee slope. From its inception, each growing vortex suffers viscous decay (associated with growth of the core radius ~ $(t-t_n)^{1/2}$). The vortex decay rate has been assumed to increase linearly from $t=t_n$ until the end of the growth phase, after which it has been set such that vortex strength κ is reduced every timestep by 0.9% of its instantaneous value (assuming 500 timesteps per cycle). The practical effect of the decay rule is that the strength of each vortex, one wave period after its inception, is about 7% of the maximum value achieved during its growth phase.

After maximum velocity in the half-cycle has occurred, the flow speed decreases and the growing vortex starts to move towards the ripple crest (mainly under the influence of its image). In general, it travels over the crest, and is ejected from the bed prior to flow reversal in the free-stream. Following flow reversal at the crest, a new vortex starts to form above the opposite ripple lee slope, and the ejected vortex travels above this, following a trajectory dependent upon the ripple steepness (η/λ) and the ratio (D/λ) where $D=2U/\sigma$ is the orbital diameter. As the ejected vortex moves away, it becomes a progressively weaker feature of the flow. Each vortex is retained in the calculation for one and a half wave periods from its inception, unless its core touches the bed before this, in which case it is removed from the computation. In addition, if the cores of two vortices of the same sign touch, the vortices are amalgamated.

For a given set of input parameter settings, comprising η, λ, U and σ, a converged model solution has been achieved typically after 5-10 wave cycles. The model solutions have been used to predict, for example, the time-varying drag on a rippled bed, and also wave drag coefficients. The question of the validity of the hydrodynamical model is not addressed directly here (see Block, 1994). However, it may be inferred from the later comparisons with observations of suspended sediment concentration that the main features of vortex motion above a rippled bed are modelled realistically.

Boundary Layer Model

The discrete vortex model makes no prediction about the structure of the near-bed oscillatory boundary layer. However it does predict the time-varying velocity at the edge of this layer. It is assumed here that the outer flow region (discrete vortex model) can be decoupled from the near-bed region (oscillatory boundary layer model), which requires the boundary layer thickness to be small compared with the ripple height. The physical picture is then of the two models interacting with one another. The discrete vortex model drives the boundary layer model, and Prandtl's rule provides vorticity at the ripple crest which feeds back into the discrete vortex solution.

For the purpose of making morphological computations, the ripple profile has been discretized into 40 horizontal shelves per ripple wavelength. Here we refer only to the oscillatory boundary layer calculations made for the shelf corresponding to the ripple crest, from which sediment entrainment into suspension is assumed to occur. On this shelf, the time-varying bed shear stress, and hence bedload transport rate (see Section 3), are determined. In view of the rather irregular time-varying nature of the bottom velocity, the rough-turbulent, oscillatory, boundary layer model of Fredsoe (1984) has been adopted for convenience. Since this model (which is not discussed here) has been derived for a flat bed, its use here is subject to the assumption that the boundary layer is sufficiently thin for the bed to be treated as locally flat. The model has been run with an equivalent bed roughness of k_s=2.5d, where d is the median grain diameter, and has been used to predict features of the flow such as the phase lead of bed shear stress over free-stream velocity (important in connection with bedload estimates) and the boundary layer thickness. The earlier choice of $\eta/8$ as the height of injection of the vorticity calculated by Prandtl's rule is consistent with the predictions of oscillatory boundary layer thickness from Fredsoe's model. More generally, for the parameter ranges corresponding to equilibrium ripples, the thickness may be expected to lie in the range 0.1η to 0.2η. To this extent, the decoupling of the discrete vortex model and the boundary layer model is justified.

3. Sediment Model

The time-varying, local, bed shear stress $\tau(t)$ given by the model of Fredsoe (1984) is used, instant by instant, to determine the bedload transport rate at the ripple crest. The formula of Meyer-Peter and Muller (1948) has been used for this purpose. At each timestep, the mobile sediment at the crest is consolidated into a notional sediment 'particle', which is released above the crest at the same height (i.e. $\eta/8$ above the crest) as used for the introduction of vorticity. Up to 500 particles are released in this way during each wave cycle. The motion of each particle in suspension is governed by a simple Lagrangian tracking scheme. In particular, particles are assumed to move at each timestep with the local fluid velocity $\underline{u}$ =(u,w) subject to a correction for the (still water) settling velocity of the sediment [$\underline{u}_p = \underline{u} - (0,W_s)$]. Particles which enter rotational vortex cores may become trapped in these moving cores. However, in general, the sediment particles having followed a complicated trajectory eventually settle to the bed. At this time they are removed from the computation, and their positions are recorded for the purposes of any later morphological calculations (not discussed here).

The model simulation commences with no sediment in suspension and, as in the case of the hydrodynamical solution, the model is run until the (space- and cycle-averaged) suspended concentration converges to a steady state, usually after 5-10 wave cycles. The model is strictly deterministic, allowing no diffusion of the sediment particles in suspension, and no break-up of the 'particles' into individual grains. The computational simplicity thus achieved is at the expense of a more realistic, uniformly distributed, suspended load. The validity of the approach adopted may be judged from the comparisons with the sediment concentration data in Section 5.

4. Laboratory experiments

A series of experiments on wave and current flows above rippled beds has been carried out in a large wave flume (79 x1.5 x1.5m) at the Laboratoire National d'Hydraulique

(E.D.F.). Details of the experimental set-up and of the results obtained have been presented by Villaret and Latteux (1992) and Villaret and Perrier (1992). The two experiments reported here were carried out in 1992-3 with the same set-up.

Both experiments were carried out with regular, progressive waves (reflection coefficient < 0.1) above a bed of fine sand (median grain diameter d= 90μm, settling velocity W_s = 0.7cm/s). After equilibrium ripples had formed, measurements of the time-varying velocity and sediment concentration fields were made above both the crest and trough positions, using ultrasonic velocimeters and optical turbidity probes (OPCON), respectively. In Test 1 the ripples were long-crested (i.e. 2-D), while in the more vigorous conditions of Test 2 the ripples displayed a less regular pattern (from 2- to 3-D). For this reason, the later comparisons are made predominantly with the results from Test 1. The experimental parameter settings and measured ripple dimensions were as follows:

	T (s)	h (cm)	H (cm)	U (cm/s)	η (cm)	λ (cm)
Test 1	2.0	45	9	16	1.0	6.3
Test 2	2.0	60	19	26	0.7	6.3

where T= wave period, h= water depth, H= wave height, U= velocity amplitude, η= ripple height and λ= ripple wavelength. These values correspond to the dimensionless parameters

	D/λ	η/λ	RE	M
Test 1	1.62	0.16	8150	17.6
Test 2	2.63	0.11	21500	46.4

where $RE=U^2/\sigma\upsilon$ and $M=U^2 / (\gamma gD)$ (here υ= kinematic viscosity; γ=s-1, s= relative sediment density; and g= acceleration due to gravity). The observed ripple heights and wavelengths are consistent with the findings of Nielsen (1981) for the mobility numbers (M) in question. In Test 1 the measured residual velocities were small, while in Test 2 there was a near-bed drift in the wave direction of 1-2 cm/s. In the later comparisons, the waves have been assumed to be purely sinusoidal and drift velocities have been ignored.

Examples of the measured time-varying velocity and sediment concentration data for Tests 1 and 2 are shown in Figure 1. For Test 1 the (horizontally-averaged) velocity at height above the crest level Y (= y-b) = 1.0cm is compared with the sediment concentration measured at the same height above both the ripple crest and trough (i.e. at heights above the *local* bed level of 1.0 and 2.0cm, respectively). For Test 2 similar results are shown for height Y=0.5cm (i.e. at heights above the local bed level of 0.5cm at the crest and 1.2cm at the trough). Each data point represents a phase-averaged concentration over 50 wave cycles. The results show distinct concentration peaks associated with the passage of sediment-laden vortices. The appearance of a peak in concentration above the ripple crest at 10°-20° after flow reversal for Test 1, and at about 20° before reversal for Test 2, is

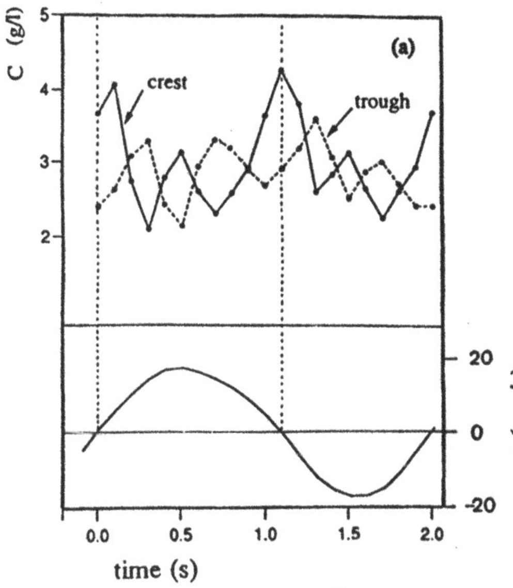

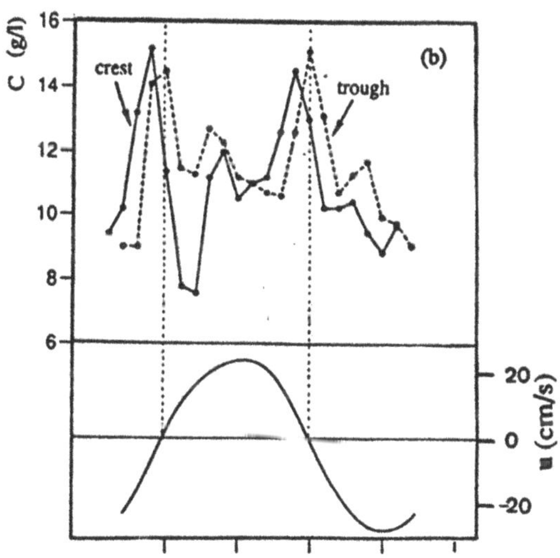

Figure 1. Phase-averaged measurements of velocity (u), and suspended sediment concentration (c) above both the ripple crest and trough, (a) at height Y= 1.0cm for Test 1 and (b) at height Y= 0.5cm for Test 2. The vertical dotted lines indicate the times of flow reversal.

followed, after a short interval in each case, by a peak above the neighbouring trough. Such observations have been shown to be repeatable from one ripple to the next during a particular experiment. The ejection of sediment-laden vortices at different phase angles in the wave cycle for Tests 1 and 2 is associated with the different values of D/λ (=1.62 and 2.63) in the two cases.

5. Model Results and Comparisons with the Observations

In Figure 2a, model results are presented showing the positions and strengths of vortices at successive instants through a wave half-cycle. The parameter settings correspond to those of Test 1 (with D/λ=1.62). As a new vortex (g) grows above each (right-hand) ripple lee slope, an old ejected vortex (e) from the previous half-cycle passes above it, and then passes over the next ripple crest. In Figure 2b, the associated suspended sediment distributions are shown. The entrained 'particles' circulate clockwise around the growing vortex core, while particles associated with the old vortex move over the trough, ultimately becoming absorbed into the vortex growing above the neighbouring ripple lee slope.

In Figure 3, comparisons are made for Test 1 between the model results and the measurements of instantaneous, phase-averaged, concentration at Y=0.5cm and 1.0cm above both the ripple crest and trough positions. Only the fluctuating components of concentration are plotted (the mean components being as shown in the table). The model concentrations have been calculated at each timestep by summing the sediment particles in a sampling 'box' drawn to mimic the OPCON sensor (see Figure 2b in which the centre of the box is 1cm above the ripple crest). Although this method for the estimation of concentration introduces some uncertainty, the model results show general agreement with the observations. The phase angles of the two concentration peaks in each half cycle (in relation to the flow reversals at t/T=0 and 0.5), and the delay in the arrival of the respective peaks above the troughs, are quite well predicted. Also, the amplitudes of the time-varying concentrations are of the correct magnitude, though the model predicts a greater decay rate in the fluctuations with increasing height than was observed. The double-peaked concentration maxima are explained by the passage through the averaging 'box' of empty vortex core areas (i.e. the measured concentrations were more uniform spatially than the predicted concentrations). Comparisons similar to those in Figure 3, for the more complicated case of Test 2, are presented and discussed by Block (1994). Here again the agreement with the time-varying concentration data is reasonable, and observed features such as the earlier release of the ejected vortex ahead of flow reversal (see Section 4) are reproduced qualitatively.

For Test 1, the *local*, measured, cycle-mean concentrations (see table in Figure 3) are larger than those predicted by factors of about 2 and 6 at the respective heights Y= 0.5cm and 1.0cm above the crest, and by more than an order of magnitude at these heights above the trough. For Test 2, the model also underpredicts the measured values, though rather

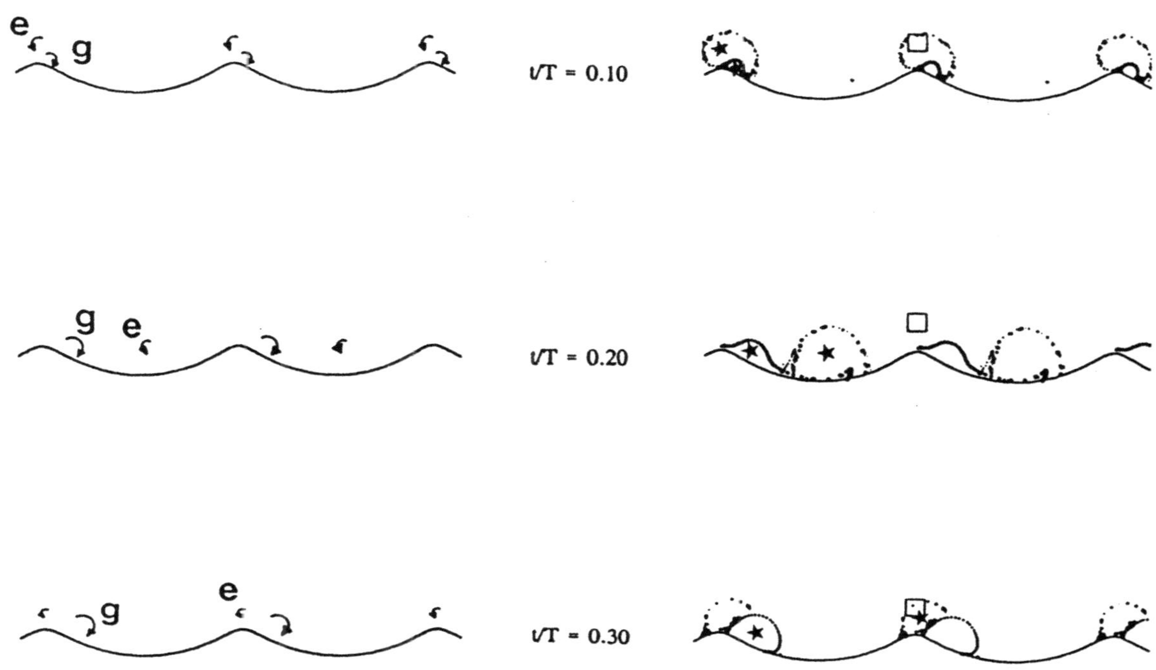

Figure 2. (a)

Figure 2. (b)

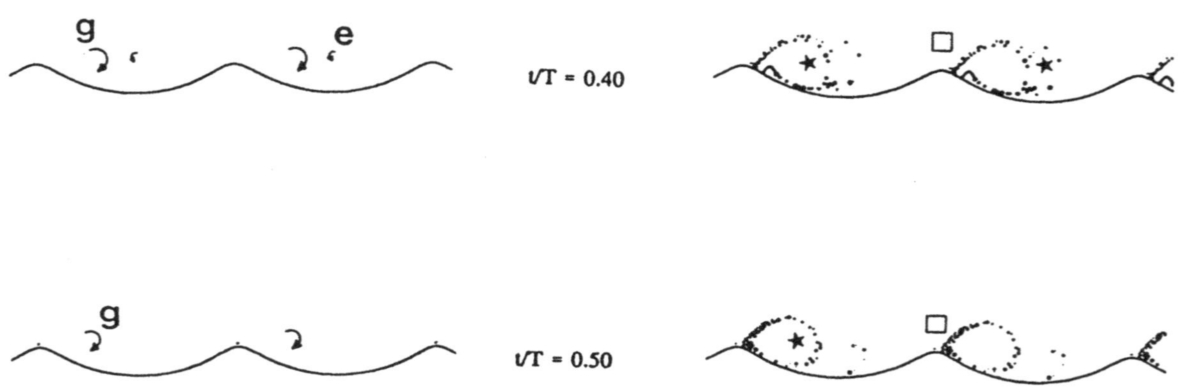

Figure 2. (a)

Figure 2. (b)

Figure 2. (a) Vortex positions and strengths during a wave half cycle for Test 1. The ripple profile is indicated by the full line, and the free-stream flow ($u=U\sin(2\pi t/T)$) is from left to right. Each arc represents a vortex: arc radius is proportional to the square-root of the vortex strength, and the sense of rotation is indicated by the arrow head.

Figure 2. (b) Sediment 'particle' positions for Test 1, corresponding to the vortex field in (a). Each dot has an area proportional to the volume of sediment which it represents. The positions of vortices 'g' and 'e' in (a) are indicated by stars. Parameter settings for Test 1: ripple height η=1.0cm, wavelength λ=6.3cm (steepness η/λ=0.16); median grain size d=90μm (settling velocity W_s=0.7cm/s); wave period T=2s, and D/λ=1.62, where D=orbital diameter.

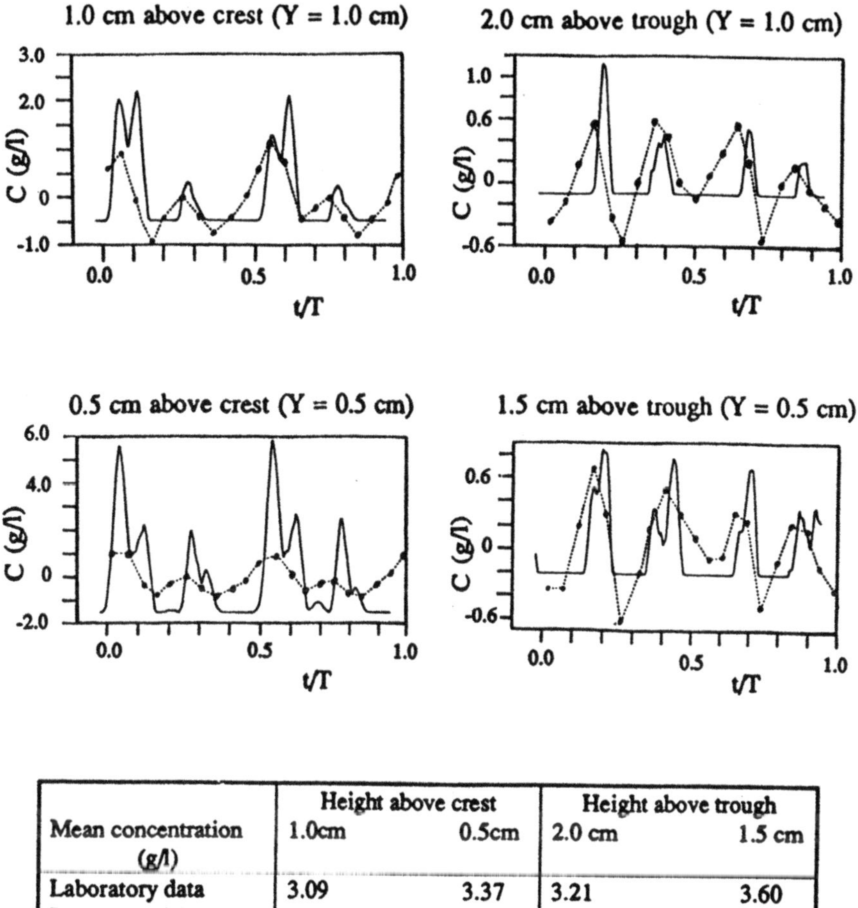

Mean concentration (g/l)	Height above crest 1.0cm	0.5cm	Height above trough 2.0 cm	1.5 cm
Laboratory data	3.09	3.37	3.21	3.60
Present model	0.48	1.52	0.10	0.22

Figure 3. Comparison of the fluctuating components of sediment concentration through a wave cycle, at two heights above both ripple crest and trough positions, for Test 1. Flow reversal in the free-stream occurs at t/T=0 and 0.5. The circles show the measured, phase-averaged, sediment concentrations. The full line represents the equivalent concentrations from the model, estimated at each timestep by summing the sediment particles in a box drawn to mimic the OPCON sensor. The 'local', cycle-mean concentrations at the respective positions are shown in the inset table; here the model estimates have been obtained using the sampling box.

better agreement is obtained above the trough in this case. The *local* predicted and measured values at height Y=0.5cm are 3.96 and 11.15 (g/l) at the crest, and 5.93 and 10.22 (g/l) at the trough, respectively.

In Figure 4, predicted, cycle-mean, concentration profiles corresponding to the parameter settings for Tests 1 and 2 are compared with the exponential profile of Nielsen (1992). Although this profile was derived from measurements made above ripple crests, it is compared here with cycle-mean profiles which have been obtained by averaging the predicted concentrations in horizontal strips extending over the full ripple wavelength. [This approach gives more representative profiles than those obtained from averages based on the sampling box.] Nielsen's profile is expected to be most applicable when the dominant mechanism of sediment entrainment is convection and, for the parameter range in question (U/W_s>18), it may be expressed:

$$\overline{C} = C_0 \exp\{-Y/1.4\eta\} \quad \text{where } C_0 = 0.005\vartheta_r^3 \text{ and } \vartheta_r = \vartheta/(1-\pi\eta/\lambda)^2$$

where Y (=y-b) =height above ripple crest level and ϑ= Shields parameter (skin friction). For Test 1 (D/λ=1.62), the predicted mean concentration displays a more rapid decay rate than the exponential profile, with concentration falling to zero at Y/η=1.25. For Test 2 (D/λ=2.63), the general shape of the model profile is in better agreement with the exponential profile, with concentration falling to zero at Y/η=2.5. [The 'spikes' in this profile may be explained by the absence of vertical diffusion in the model.]

Mean Sediment Concentration Above Crest (g/l)	Height Above Ripple Crest Y=0cm	Y=0.5cm	Y=1.0cm
Test 1			
Present model	2.28	0.56	0.32
Nielsen (1992)	2.18 (=C_0)	1.52	1.07
Laboratory data	-	3.37	3.09
Test 2			
Present model	8.80	3.50	2.53
Nielsen (1992)	5.12 (=C_0)	3.07	1.85
Laboratory data	-	11.15	-

In the table above, the (horizontally-averaged) model estimates and the present measurements of mean concentration above the crest are shown together with predictions based on Nielsen's exponential profile. For Test 1, the very good agreement between C_0 given by Nielsen's formula and the model prediction at Y=0 may be contrasted with the discrepancies (by factors of 6 and 10) between the modelled and measured mean concentrations at the respective heights Y=0.5cm and 1.0cm (i.e. Y/η=0.5 and 1.0). For Test 2, the predicted value of C_0 is 70% larger than Nielsen's value at Y=0, while the

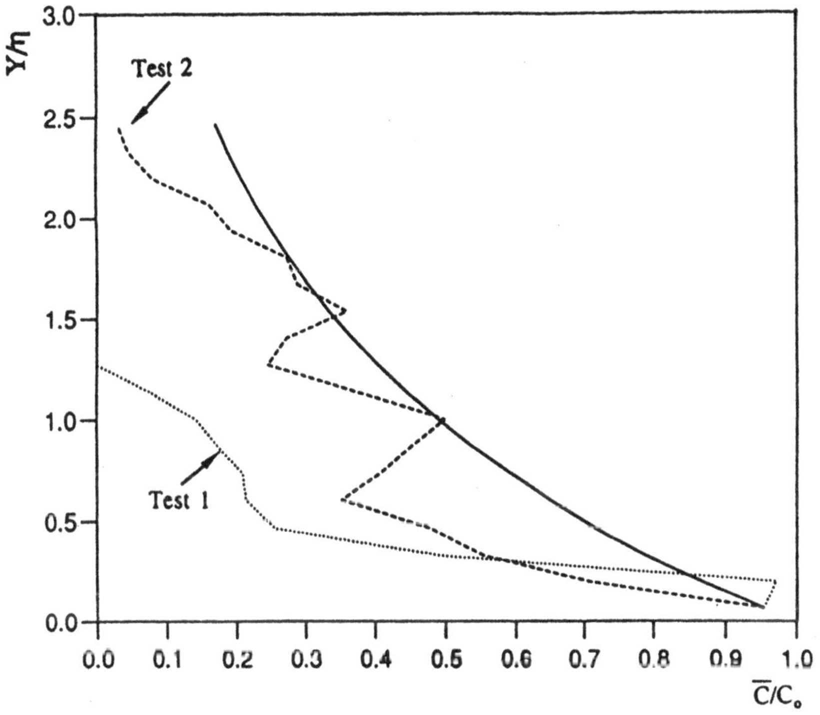

Figure 4. Horizontally-averaged, cycle-mean, profiles of concentration for Tests 1 and 2 (dashed and dotted lines, respectively), together with Nielsen's exponential profile (full line).

mean concentration is smaller (by a factor of 3) than the measured value above the crest at height Y=0.5cm (i.e. Y/η=0.71). Although the measured mean concentrations in both tests may appear somewhat large, they lie well within the expected range of experimental uncertainty (c.f. Nielsen, 1986, Fig.3). The underprediction of mean concentration by the present model, and the greater vertical uniformity in the measured concentrations compared with both the model and also Nielsen's profile, would appear to be associated with the absence of vertical diffusion in the formulation (i.e. upward convection of grains is insufficient on its own). The effects of diffusion are very likely to have been important for the fine-grained sediment (d=90μm) used in the present experiments (see Nielsen, 1992).

The use of the bedload formula of Meyer-Peter and Muller to determine the rate of sediment entrainment at the ripple crest (see section 3) is open to question. However, it has been found that the inclusion of a suspended load component in the *near-bed* transport rate (by the addition of a suspended sediment flux *within* the near-bed, oscillatory boundary layer given by Fredsoe's model) enhances the instantaneous sediment entrainment rate during the wave cycle by, at most, 30%. This would not accout for the differences in mean concentration between model and experiment referred to above.

In relation to the measured data, the contribution to the mean concentration of any residual turbidity background ('wash load') is not believed to have been significant. However, at least in the case of Test 2, there was a pronounced drift current which may have contributed to the upward dispersion of sediment. In addition, some uncertainty may have arisen in the comparisons in relation to the exact height of the OPCON probe, and its possible scouring effect on the bed.

6. Conclusions

The predictions of a discrete vortex model have been compared with some new experimental observations of the time-varying sediment concentration above a bed of equilibrium sand ripples in oscillatory flow. For two contrasting values of D/λ (D= orbital diameter, λ= ripple wavelength), the occurrence of ejected, sediment-laden, vortices above the ripple crests either just before, or just after, flow reversal in the free-stream is well predicted. Also, the delay in the arrival above the trough of the associated peak in sediment concentration is modelled convincingly. However, the time-mean concentrations above both the ripple crest and trough positions are underpredicted. This is due partly to the lack of vertical diffusion in the model, and possibly also to the (bedload) method used to model the rate of sediment entrainment into suspension. The agreement between the predicted and measured time-varying concentration provides generally good, albeit indirect, support for the discrete vortex approach adopted.

Acknowledgement

This work was partly-funded by the G8 Coastal Morphodynamics programme under EC contract number MAS2-CT92-0027.

References

Block, M.E., Wave generated flow over sand ripples and the resulting entrainment and suspension of sediment. *PhD Thesis, University of Wales*, (1994).

Blondeaux, P., and Vittori, G., 1991. Vorticity dynamics in an oscillatory flow over a rippled bed. *Journal of Fluid Mechanics*, **226**, (1991) 257-289.

Fredsoe, J., Turbulent boundary layers in wave-current motion. *Journal of Hydraulic Engineering*, **110**,(1984) 1103-1120

Hansen, E.A., Fredsoe, J., and Deigaard, R., Distribution of suspended sediment over wave generated ripples. *IAHR Int. Symp. on: The transport of suspended sediments and its mathematical modelling, Florence*, (1992) 111-128.

Longuet-Higgins, M.S., Oscillating flow over steep sand ripples. *Journal of Fluid Mechanics*, **107**, (1981) 1-35.

Macpherson, B., Flow and sediment transport over steep sand ripples. *PhD Thesis. Cambridge*, (1984).

Meyer-Peter, E., and Muller R., Formulae for bed load transport. *Proc. 2nd Congress I.A.H.R., Stockholm*, (June 1948).

Nakato, T., Locher, F.A., Glover, J.R., and Kennedy, J.F., Wave entrainment of sediment from rippled beds. *Journal of Waterway, Port, Coastal and Ocean Division*, **103**, WW1, (1977) 83-99.

Nielsen, P., Dynamics and geometry of wave generated ripples. *Journal of Geophysical Research*, **86**, (1981) 6467-6472.

Nielsen, P., Suspended sediment concentrations under waves. *Coastal Engineering*, **10**, (1986) 23-31.

Nielsen, P., Coastal bottom boundary layers and sediment transport. *World Scientific Publishing Co. Pte. Ltd., Singapore*, (1992) 324pp.

Villaret, C., and Latteux, B., Entrainment and transport of fine sand by combined waves and current: an experimental study. *Proc. 23rd Int. Conf. on Coastal Eng., Venice, In Press*, (1992).

Villaret, C., and Perrier, G., Transport of fine sand by combined waves and current: an experimental study. *Report HE-42/92.68, E.D.F., Laboratoire National d'Hydraulique, France*, (1992) 81pp.

SUBHARMONIC EDGE WAVE EXCITATION BY WAVES INCIDENT ON PARTIALLY ABSORBING BEACHES

PAOLO BLONDEAUX, GIOVANNA VITTORI
Hydraulic Institute, Genoa University
Via Montallegro 1, 16145 Genova - Italy

and

HUIB DE SWART
Institute for Marine & Atmospheric Research, Utrecht University
P.O. Box 80.005, 3508 TA Utrecht - The Netherlands

ABSTRACT

The excitation of edge waves by an incoming wave normally approaching a partially absorbing beach is considered. The incoming and partially reflected waves are found to transfer energy to edge waves through the resonance mechanism pointed out by Guza & Davis (1974) for totally reflected incident waves. However, the results obtained show that the energy transfer rate strongly depends on the beach geometry and on its reflection coefficient. These quantities determine also the most unstable edge wave mode. In accordance with Hasselmann's rule, only edge waves with angular frequencies smaller than the frequency of the incident wave are excited by this mechanism.

1. Introduction

Several types of beach features seem to have a rather regular, longshore pattern. This pattern may indeed be sufficiently uniform to be described in terms of a recognizable longshore wavelength. A likely explanation for such features lies in the motion of edge waves, which are surface waves trapped in the nearshore region. These waves, by themselves or through interaction with a normal, incoming surface wave can generate rhythmical steady currents which in turn can give rise to longshore morphological features having a characteristic wavelength related to the edge wave wavelength (Inman & Guza, 1982).

The dynamics of inviscid edge waves on a non erodable beach of constant slope extending into deep water has been discussed among others by Eckart (1951) and Ursell (1952). Ursell (1952) found a set of solutions to the exact linear equations representing finite motions confined to the nearshore region. Eckart (1951) attempted to describe trapped edge waves by using the approximate shallow water equations. He obtained a solution expressed in terms of Laguerre polynomials L_n (Abramowitz &

Stegun, 1965) and a dispersion relationship relating the frequency and wavenumber of the different modes. The solution based on the shallow water approximation is found to agree with Ursell's (1952) solution close to the shore if $(2n+1)\beta << 1$ (β being the slope of the beach and n the mode number of the edge wave), even though a discrepancy is present far offshore.

Several different mechanisms for generating edge waves on sloping beaches are possible. These trapped waves can be excited by external forces such as wind stress acting directly above the water surface or by uneven pressure distributions related to storms travelling parallel to the coast or by low-frequency waves which are initially seaward travelling and become trapped by refraction.

Edge waves may also be excited by a normally incident monochromatic wave through an instability mechanism. Guza & Davis (1974) showed, by using the shallow water approximation, that a wavetrain of angular frequency ω^*, which normally approaches a straight reflecting beach, can transfer energy to small disturbances in the form of edge waves characterized by an angular frequency equal to $\omega^*/2$. Later Minzoni & Whitham (1977) concluded that the non-uniformity of the shallow water approximation in the offshore region does not affect the main results obtained by Guza & Davis (1974) for small beach angles, the non-uniformity being mild.

The present contribution reconsiders the resonance mechanism highlighted by Guza & Davis (1974) by considering a different beach profile which allows the effects of energy absorbtion due to wave breaking to be included. The growth rates of the edge waves are found to be strongly influenced both by the beach profile and by the absorbing properties of the beach. While in the Guza & Davis' case (1974), the first subharmonic edge wave mode is always the most unstable, in the present paper it is shown that values of the parameters exist for which the second or the third mode become more unstable.

In the following section, the problem is formulated and in section 3 the wave field induced by the incoming wave partially reflected by the shore is determined. Then the structure of the free edge waves as well as the dispersion relationship relating the wavenumber and the frequency are determined. Section 5 describes the mechanism of excitation of the edge waves by the incoming wave and discusses some of the results.

2. Formulation of the problem

Let us consider a natural beach. Field observations show that the time averaged bottom slope is a decreasing function of the offshore coordinate x^*. Close to the coastline the slope is characterized by large values while far from it, it can be assumed almost constant (see figure 1a). Different approximating laws have been proposed in the past to describe the mean water depth as a function of x^* in the coastal region, the most widely used being perhaps the 2/3 power law by Dean (1991). We model natural beach profiles as a straight beach of constant small slope β delimited by a vertical wall (see figure 1b). The value of β and the position of the wall which model a natural beach in the best way can be determined by geometrical means, e.g. the least square method. Let us introduce a Cartesian coordinate system (x^*, y^*, z^*) with

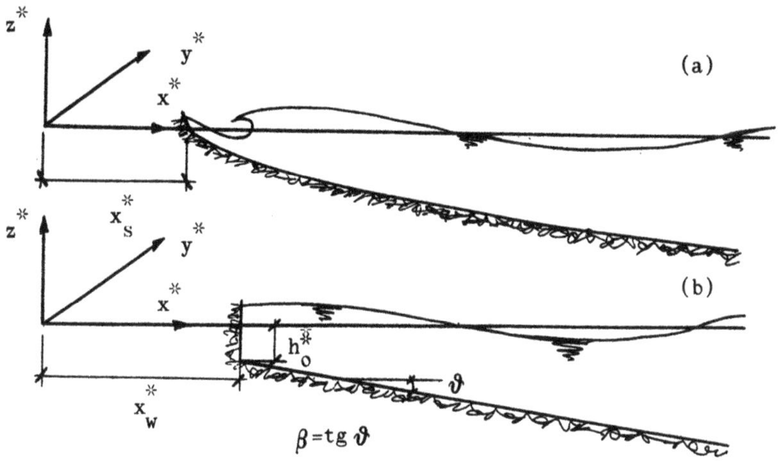

Figure 1: (a) Sketch of a real beach profile (b) beach geometry actually considered.

the x^*-axis lying on the still water level and pointing offshore, the y^*-axis parallel to the coastline and the z^*-axis pointing upwards (see figure 1b). The origin of the coordinate system is located at the intersection of the constant beach slope profile with the still water level. Let us denote with x^*_w the position of the vertical wall. For $x^* > x^*_w$ the local depth is described by

$$h^* = h^*_o + \beta(x^* - x^*_w) \tag{1}$$

where β is assumed to be constant and small.

Let us consider an incoming wave of small amplitude a^* and angular frequency ω^*. The following dimensionless variables are introduced

$$(x, y) = \frac{\omega^*}{\sqrt{gh^*_o}}(x^*, y^*) \qquad t = \omega^* t^*$$

$$h = h^*/h^*_o \qquad (u, v) = \frac{\sqrt{h^*_o}}{a^*\sqrt{g}}(u^*, v^*) \qquad \eta = \eta^*/a^* \tag{2}$$

where t^* is time, u^* and v^* denote velocity components in the cross-shore and long-shore directions respectively and η^* is the free surface displacement. Furthermore g denotes gravity. The incoming wave is assumed to propagate normally to the shoreline. If the wavelength of the incoming wave (L^*) is much larger than the local depth (h^*_o), the flow field can be described by means of the shallow water approximation. The dimensionless momentum and continuity equations read:

$$\frac{\partial \eta}{\partial t} + \frac{\partial}{\partial x}[u(h + a\eta)] + \frac{\partial}{\partial y}[v(h + a\eta)] = 0 \tag{3}$$

$$\frac{\partial u}{\partial t} + au\frac{\partial u}{\partial x} + av\frac{\partial u}{\partial y} = -\frac{\partial \eta}{\partial x} \tag{4}$$

$$\frac{\partial v}{\partial t} + au\frac{\partial v}{\partial x} + av\frac{\partial v}{\partial y} = -\frac{\partial \eta}{\partial y} \tag{5}$$

where $a = a^*/h_o^*$.

Appropriate boundary conditions should be forced both for $x = x_w$ and for $x \to \infty$. Since the boundary conditions are assumed to depend on the direction of wave propagation, they will be discussed separately for the incoming wave and the edge waves.

3. The incident wave

Let us assume $a << 1$ and determine the solution as a perturbation expansion in the small parameter a.

$$(u, v, \eta) = (u_{oo}, v_{oo}, \eta_{oo}) + O(a) \tag{6}$$

By substituting (6) into (3)-(5), at $O(a^o)$ the problem reads:

$$\frac{\partial \eta_{oo}}{\partial t} + \frac{\partial}{\partial x}\left(\frac{x}{x_w}u_{oo}\right) + \frac{\partial}{\partial y}\left(\frac{x}{x_w}v_{oo}\right) = 0 \tag{7}$$

$$\frac{\partial u_{oo}}{\partial t} = -\frac{\partial \eta_{oo}}{\partial x} \tag{8}$$

$$\frac{\partial v_{oo}}{\partial t} = -\frac{\partial \eta_{oo}}{\partial y} \tag{9}$$

Equations (7)-(9) can be combined in order to get the following equation which involves only η_{oo}

$$\frac{\partial^2 \eta_{oo}}{\partial t^2} - \frac{\partial}{\partial x}\left(\frac{x}{x_w}\frac{\partial \eta_{oo}}{\partial x}\right) - \frac{x}{x_w}\frac{\partial^2 \eta_{oo}}{\partial y^2} = 0 \tag{10}$$

The matching with the deep water region suggests assuming

$$\eta_{oo} = \hat{\eta}_{oo}(x)e^{it} + \text{c.c.} \tag{11}$$

By substituting (11) into (10), a Bessel equation is obtained and the solution can be expressed in terms of Hankel functions (Abramowitz & Stegun (1965)):

$$\hat{\eta}_{oo}(x) = c_1 H_o^{(1)}(2\sqrt{xx_w}) + c_2 H_o^{(2)}(2\sqrt{xx_w}) \tag{12}$$

The coefficients c_1 and c_2 can be obtained by matching (12) with the incoming wave field. In the deep water region the wave field can be represented as the superposition of an incoming wave of amplitude a_∞^* and a reflected wave the amplitude of which is ka_∞^*, where k is the complex reflection coefficient of the beach. It turns out

$$c_1 = 1 \qquad c_2 = k$$

if the amplitude a^* appearing in (2) is assumed to be equal to $a^*_\infty\sqrt{\pi/(2\beta)}$ (Mei (1989), pp. 527-530). Then the velocity components are obtained by using (8) and (9):

$$u_{oo} = i\sqrt{\frac{x}{x_w}}\left[H_1^1(2\sqrt{xx_w}) + kH_1^2(2\sqrt{xx_w})\right] \tag{13}$$

$$v_{oo} = 0 \tag{14}$$

The solution (12)-(14) models the behaviour of the incoming wave propagating over the real beach of figure 1a in the range (x_w, ∞), if the amplitude a^*_∞ far offshore is small enough to cause wave breaking for values of x^* smaller than x^*_w. Equation (12) along with a simple breaker criterion (e.g. wave height at the breaker line equal to 0.88 the local depth) allows us to evaluated the maximum amplitude of the incoming wave which can be considered in the model. For wind waves and natural sandy beaches, the maximum value of a^* turns out to be $O(1^{-1}$ m) and a^*_∞ is $O(10^{-1}$ m). Alternatively the constant c_1 and c_2 appearing in (12) can be determined by forcing a boundary condition at $x = x_w$ which should describe the energy dissipation associated with wave breaking along with all the other phenomena occurring in the surf-zone and give the same reflective properties as the natural beach.

4. The edge waves

A perturbation of small amplitude ϵ is then superimposed on the basic flow and its time development is determined. Because of the presence of the small parameter a, the perturbation is expanded in terms of a.

$$(u, v, \eta) = (u_{00}, v_{00}, \eta_{00}) + \epsilon[(u_{01}, v_{01}, \eta_{01}) + a(u_{11}, v_{11}, \eta_{11})] + 0(a, \epsilon a^2, \epsilon^2)$$

The perturbation is assumed to admit a Fourier representation and the sum on the index n accounts for the existence of different modes.

$$(u_{01}, v_{01}, \eta_{01}) = \int_{-\infty}^{+\infty}\left\{\sum_{n=0}^{\infty} A_n(\hat{u}_{01}^{(n)}, \hat{v}_{01}^{(n)}, \hat{\eta}_{01}^{(n)})e^{i(\ell y+\sigma_n t)} + c.c.\right\} d\ell \tag{15}$$

The functions $\hat{u}_{01}^{(n)}, \hat{v}_{01}^{(n)}, \hat{\eta}_{01}^{(n)}$ depend on the cross-shore coordinate x and the amplitude A_n on ℓ and the slow time scales $\tau_m = a^m t$. The form of $\hat{\eta}_{01}^{(n)}, \hat{u}_{01}^{(n)}, \hat{v}_{01}^{(n)}$ is obtained by solving the problem at $O(\epsilon)$ where no interaction between the disturbance and the incoming wave takes place:

$$\frac{d}{dx}\left(\frac{x}{x_w}\frac{d\hat{\eta}_{01}^{(n)}}{dx}\right) + \left(\sigma_n^2 - \frac{x}{x_w}\ell^2\right)\hat{\eta}_{01}^{(n)} = 0 \tag{16}$$

$$\hat{u}_{01}^{(n)} = \frac{i}{\sigma_n}\frac{d\hat{\eta}_{01}^{(n)}}{dx} \tag{17}$$

$$\hat{v}_{01}^{(n)} = -\frac{\ell}{\sigma_n}\hat{\eta}_{01}^{(n)} \tag{18}$$

After some algebra and taking into account that $\hat{\eta}_{01}^{(n)}$ should vanish as $x \to \infty$, since edge waves are modes confined to the nearshore region, it is possible to obtain

$$\hat{\eta}_{01}^{(n)} = e^{-\ell x} U(d_n, 1, 2\ell x) \tag{19}$$

with $d_n = \frac{1}{2} - \frac{\sigma_n^2 x_w}{2\ell}$ and U is the Kummer function (Abramowitz & Stegum (1965)). The value of d_n depends on the boundary conditions at x_w. For natural beaches, because close to the coastline the slope is characterized by large values, the dimensionless location x_s of the shoreline differs by a small amount from x_w (compare figure 1a with figure 1b). Hence the mass flux due to the edge wave at $x = x_w$ can be ignored. Indeed the order of magnitude of the volume of water which can be stored per unit time and unit length in the longshore direction between the coastline at x_s^* and the coordinate x_w^* is $a^*(x_w^* - x_s^*)$ and turns out to be negligible with respect to $a^*\sqrt{gh_o^*}$ which is the order of magnitude of the volume crossing a vertical section per unit length in the y direction and unit time. The vanishing of $\hat{u}_{01}$ at $x = x_w$ can be also inferred from the vanishing of the x-component of the edge wave velocity at the coast line and by shifting this condition at the breaker line by means of a Taylor expansion and by neglecting terms of order $(x_w^* - x_s^*)/L^*$. It is worth pointing out that the same argument cannot be applied to the incoming wave field, because shoaling effects cause large amplitudes, wave breaking and the failure of the Taylor expansion in the surf zone.

The following boundary condition is thus forced

$$U(d_n, 1, 2\ell x_w) + 2d_n U(d_n + 1, 2, 2\ell x_w) = f(\sigma_n, \ell) = 0 \tag{20}$$

Equation (20) turns out to provide the dispersion relation for the edge waves for this particular geometry. In figure 2, where f is plotted versus σ_n for a fixed value of ℓ, it can be seen that more than one value of σ_n satisfy (20). Therefore many modes with different angular frequencies are possible for a fixed longshore wavenumber. A more general boundary condition might be forced when $a^*(x_w - x_s^*)$ assumes values as large as $a^*\sqrt{gh_o^*}$ (i.e. when the edge velocity at the coastline does not vanish), where a linear combination of $\hat{\eta}_{01}$ and its derivative is set equal to zero. The constants appearing in it could be tuned to describe field observations. However, this attempt has not been pursued for the present.

5. The edge wave excitation

At order ϵa the incoming wavefield interacts with the edge waves, providing a mechanism for the excitation of edge waves. At this order of approximation it is easy to obtain:

$$\frac{\partial^2 \eta_{11}}{\partial t^2} - \frac{\partial}{\partial x}\left(\frac{x}{x_w}\frac{\partial \eta_{11}}{\partial x}\right) - \frac{\partial}{\partial y}\left(\frac{x}{x_w}\frac{\partial \eta_{11}}{\partial y}\right) = -\frac{\partial \eta_{01}}{\partial \tau_1} - \frac{\partial}{\partial x}(\eta_{00}u_{01} + \eta_{01}u_{00}) - \frac{\partial}{\partial y}(\eta_{00}v_{01}) \tag{21}$$

$$\frac{\partial u_{11}}{\partial t} + \frac{\partial \eta_{11}}{\partial x} = -u_{00}\frac{\partial u_{01}}{\partial x} - u_{01}\frac{\partial \eta_{00}}{\partial x} - \frac{\partial u_{01}}{\partial \tau_1} \tag{22}$$

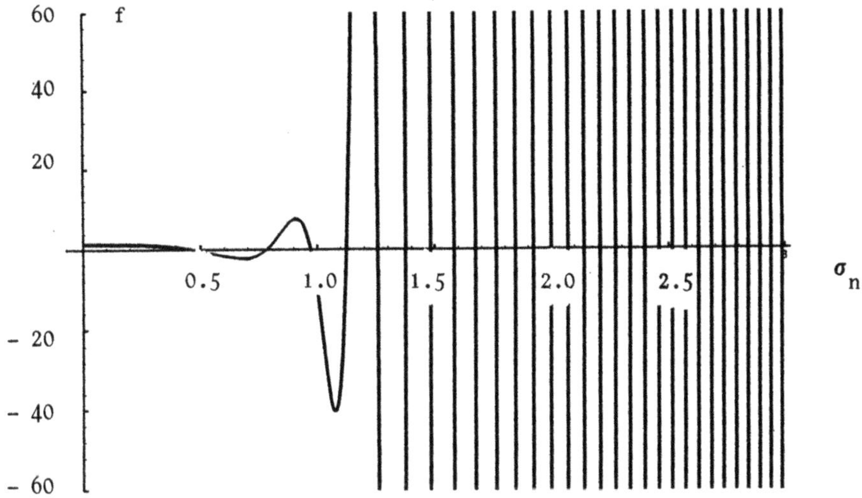

Figure 2: Plot of the function f (eq. 20) versus σ_n for $x_w = 2.0$ and $\ell = 0.27$.

$$\frac{\partial v_{11}}{\partial t} + \frac{\partial \eta_{11}}{\partial y} = -u_{00}\frac{\partial v_{01}}{\partial x} - \frac{\partial v_{01}}{\partial \tau_1} \tag{23}$$

where $\tau_1 = at$.

By substituting (15), (17), (18) and (19) into (21), the right-hand term can be recasted as:

$$\int_{-\infty}^{+\infty} \left\{ \sum_{n-0}^{\infty} A_n(\tau_1) \left[P_n(x) e^{i(\ell y + (\sigma_n+1)t)} + Q_n(x) e^{i(\ell y + (\sigma_n - 1)t)} \right] - 2i\sigma_n \frac{dA_n}{d\tau_1} e^{i(\ell y + \sigma_n t)} \right\} d\ell \tag{24}$$

where $P_n(x)$ and $Q_n(x)$ are complicated functions which are not given here for the sake of brevity.

When $\sigma_n \pm 1 \neq \sigma_m$ no problem arises in solving (21). However, if $\sigma_n \pm 1 = \sigma_m$, in order for (21) to have a solution, it is necessary to force a solvability condition that provides the time development of edge wave amplitudes

$$a_0 \frac{dA_n}{d\tau_1} = a_1 A_m \tag{25}$$

$$b_0 \frac{dA_m}{d\tau_1} = b_1 A_n \tag{26}$$

where

$$a_o = 2i\sigma_n \int_{x_w}^{\infty} \left[e^{-\ell x} U(d_n, 1, 2\ell x) \right]^2 dx \tag{27}$$

$$a_1 = \int_{x_w}^{\infty} Q_m(x) e^{-\ell x} U(d_m, 1, 2\ell x) dx \tag{28}$$

$$b_0 = 2i\sigma_m \int_{x_w}^{\infty} \left[e^{-\ell x} U(d_m, 1, 2\ell x) \right]^2 dx \tag{29}$$

$$b_1 = \int_{x_w}^{\infty} P_n(x) e^{-\ell x} U(d_n, 1, 2\ell x) dx \tag{30}$$

The solution of (25)-(26) is

$$A_n(\tau) = A_n(0)e^{\gamma\tau_1}$$

with $\gamma = \pm\left(\frac{a_1 b_1}{a_0 b_0}\right)^{1/2}$ and $A_n(0)$ the initial amplitude of the disturbance.

The resonance condition $\sigma_n \pm 1 = \sigma_m$ is satisfied, among other values, by $\sigma_n = \pm 1/2$ and $\sigma_m = \pm 1/2$. Therefore, the present mechanism can give rise to the growth of subharmonic edge waves. However, the excitation mechanism, for the beach geometry we have considered, leads to results which are quantitatively and qualitatively different from those obtained by Guza & Davis (1974). Indeed, because Guza & Davis (1974) considered only purely reflecting beaches, it can easily be appreciated that their theory can be applied only to waves characterized by periods much longer than those of wind waves when reasonable values of the amplitude of the incoming wave are considered. Moreover, the partial absorbtion of energy at $x = x_w$ which is considered in the present paper has considerable influence on the energy transfer from the incoming wave to the edge wave modes. In figure 3 the growth rate γ is plotted versus x_w, which is a free parameter of the model related to beach geometric characteristics and wave parameters. Indeed since $x_w^* = h_o^*/\beta$ and $x_w = x_w^* \omega^*/\sqrt{g h_o^*}$, it turns out that $x_w = \omega^* \sqrt{h_o^*}/(\beta\sqrt{g})$. Because typical values of ω^* for wind waves are O (1 sec^{-1}) and natural sandy beaches are charaterized by values of h_o^* and β which range about 1m and between 10^{-2} and 10^{-1} respectively, the parameter x_w varies between O(1) and O(10). Three different curves corresponding to the first three subharmonic modes are shown in figure 3. The mode number which is characterized by the maximum growth rate depends on the value of x_w. For example when $x_w = 1.5$, the first mode has an amplification rate which is larger than for the second and third modes. On the other hand, for $x_w = 2.4$ the second and third modes prevail over the first one and the second mode is characterized by the maximum growth rate. In figure 4 only the first subharmonic mode is considered and the different curves are drawn for different values of the reflection coefficient k. As previously pointed out, k equal to one describes a totally reflected incoming wave. On the other hand, $k = 0$ describes a pure progressive wave. A decrease of k produces a smoothing of the function $\gamma(x_w)$ and for $k = 0$ only one maximum is present for x_w about 2. For a fixed value of x_w the reflecting beach does not always give rise to the maximum growth rate. Figures 5a and 5b, which show γ as a function of k for $x_w = 5.$ and $x_w = 2.4$ respectively, show more clearly the influence of the reflecting properties of the beach on the growth rates for the first three subharmonic modes.

From the results presented here it is evident that both the geometric characteristics (x_w) and the reflective properties of the beach (k) contribute to determining the mode number which will be observed, i.e. the flow pattern and in particular the form of the steady currents which are generated by nonlinear effects and by the interaction of the edge waves with the incoming wave. Because it is known that the beach profile depends on sediment characteristics, it can be inferred that a strong correlation exists also among sediment properties, edge waves, steady currents and beach cusps. However in order to get quantitative results it would be necessary to remove the assumption of a fixed bottom and to consider explicitly its time development.

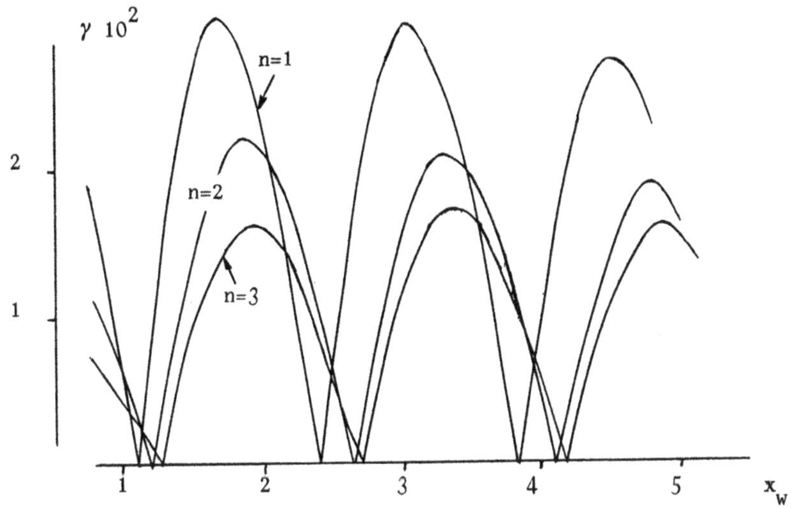

Figure 3: Plot of the amplification rate γ of subharmonic edge waves versus x_w for $k = 1$ and different mode number n.

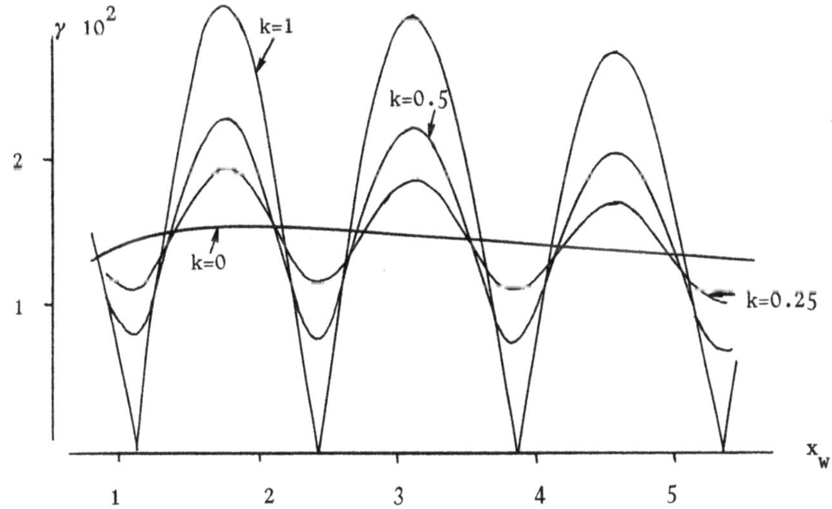

Figure 4: Plot of the amplification rate γ of subharmonic edge waves versus x_w for the first mode and different values of k.

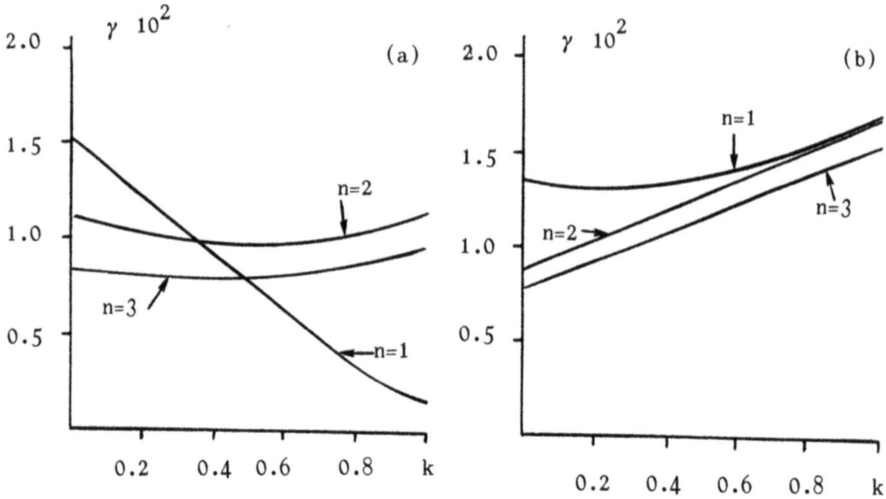

Figure 5: Plot of the amplification rate γ of subharmonic edge waves versus k for different mode numbers n (a) $x_w = 2.4$ (b) $x_w = 5.0$.

6. Conclusions

In the present paper the resonance mechanism described by Guza & Davis (1974) is extended to a geometry where the absorbing properties of the beach can be taken into account. According to Hasselman's (1967) rule, the incoming wave is found to be unstable only with respect to edge waves which are characterized by an angular frequency smaller than that of the primary wave. Therefore, the present theory is unable to explain the growth of edge waves synchronous with the incoming wave.

By carrying out the theory at the order ϵa^2, a similar mechanism can be found in which synchronous edge waves grow due to their interaction with the set down and with the $O(a^2)$ ultraharmonic part of the incoming wavefield. This mechanism of excitation considered by Rockliff (1978) for reflecting beaches is currently being studied for the partially absorbing case.

Synchronous edge waves are of particular interest in coastal morphodynamics due to their 'strong' interaction with the incoming wave which gives rise to steady horizontal recirculating cells at the first interaction. Such steady currents, periodic in the longshore direction, may strongly contribute to the erosion of the shoreline and to the appearance of beach cusps.

An alternative (nonlinear) mechanism by which synchronous edge waves can be excited is described in Blondeaux & Vittori (1994). The growth of synchronous modes is due to their resonant interaction with subharmonic modes which extract energy from the incoming wave. This nonlinear resonance mechanism is found to produce a rapid growth both of subharmonic and synchronous edge waves.

Acknowledgement

The Authors are indebted to Prof. H. de Vriend for useful comments on a first draft of the paper.

This work has been carried out as part of the MAST G8M Coastal Morphodynamics research program and has been funded jointly by the Commission of the European Communities, Directorate General for Science Research and Development, under MAST Contract No. MAS2-CT92-0027 and by the "Ministero dell'Università e della Ricerca Scientifica e Tecnologica".

References

[1] Abramowitz, M., & Stegun, I. (1965). "Handbook of mathematical functions". Dover, New York.

[2] Blondeaux, P., & Vittori, G. (1994). "Nonlinear excitation of synchronous and subharmonic edge waves". Submitted for publication.

[3] Dean, R.G. (1991). "Equilibrium beach profiles: characteristics and applications". J. Coastal Res., 7(1), 53-84.

[4] Eckart, C. (1951). "Surface waves on water of variable depth". Wave Rep. 100 Scripps Inst. of Oceanogr. Univ. of Calif., La Jolla.

[5] Guza, R.T., & Davis, R.E. (1974). "Excitation of edge waves by waves incident on a beach". J. Geoph. Research, 79 (9), 1285-1291.

[6] Hasselman, K. (1967). "A criterion for nonlinear wave stability". J. Fluid Mech., 30, 737-739.

[7] Inman, D.L. & Guza, R.T. (1982). "The origin of swash cusps on beaches". Marine Geology, 49, 133-148.

[8] Mei, C.C. (1989). "The applied dynamics of ocean surface waves". World Scientific.

[9] Minzoni, A.A. & Whitham, G.B. (1977). "On the excitation of edge waves on beaches". J. Fluid Mech., 79, 273-287.

[10] Rockliff, N. (1978). "Finite amplitude effects in free and forced edge wave". Math. Proc. Camb. Phil. Soc. 83, 463-479.

[11] Ursell, F. (1952). "Edge waves on a sloping beach". Proc. Roy. Soc. London, Ser. A, 214, 79-97.

THE DYNAMICS OF PARTICLE-LADEN FLUID ELEMENTS

ANDREW J. HOGG†, HERBERT E. HUPPERT†

†Institute of Theoretical Geophysics
Department of Applied Mathematics & Theoretical Physics
Silver Street, Cambridge CB3 9EW, United Kingdom

and

RICHARD L. SOULSBY‡

‡HR Wallingford Ltd., Howbery Park, Wallingford
Oxfordshire OX10 8BA, United Kingdom

ABSTRACT

This work presents a new framework for modelling the suspension of sediment within a turbulent boundary layer. It aims to provide a link between the 'bursting phenomenon', which consists of the abrupt ejection of fluid away from the boundary, and the process by which sediment is resuspended. We focus on the advective transport of sediment by coherent motions within the boundary layer and propose a phenomenological model to describe a discrete ejection of particle-laden fluid. As the fluid element rises, it expands due to the entrainment of ambient fluid. This reduces its vertical velocity and particle concentration. The particle concentration is also reduced by gravitational settling, because the heavy particles sediment out of the fluid element. The model accounts for the dynamics of these particle-laden fluid elements by considering the conservation of their mass, momentum and particle concentration. Results indicate that this framework may provide valuable insight to the suspension of sediment by the 'bursting' process.

1. Introduction

Many recent studies have identified three dimensional coherent motions within turbulent boundary layers and have indicated that such motions are responsible for the production of turbulent kinetic energy. Comprehensive reviews of the dynamics of these motions have been published by Grass *et al.* (1991), Robinson (1991) and Smith *et al.* (1991). The existence of these coherent motions was first identified by Kline *et al.* (1967) in an experimental study of turbulent flow over a smooth boundary. They identified streaks within the viscous sub-layer, with alternating narrow elongated zones of high and low velocity fluid. They linked this structure in the viscous layer to a bursting event, which consisted of a sudden ejection of fluid away from the boundary, followed by a compensating inrush of fluid towards the boundary. These ejection and inrush events form the quasi-periodic patterns of the coherent motions within the flow and will henceforth be termed ejection and sweep events. The sweep event transports high momentum fluid towards the boundary, whereas the ejection transports low momentum fluid into the body of the flow. Experimental studies have demonstrated that these coherent motions are responsible for the major contributions towards turbulent energy production and Reynolds stress momentum transport. Similar conclusions are deduced from the field studies of Gordon (1974) and Heathershaw (1974). They noted the intermittent nature of the transport of momentum through a boundary layer of geophysical dimensions, which is consistent with the laboratory observations made at much smaller scales.

The role of turbulence in the process of suspending and entraining particles from an erodible boundary has also received considerable study. Studies have sought to form a link between coherent motions within the turbulent boundary layer and the suspension process (Grass 1974, Jackson 1976, Sumer & Oguz 1978, Deigaard & Sumer 1981). Jackson (1976) reasoned that the bursting process is a plausible mechanism for the suspension of sediment because it provides a means for the maintenance of the vertical anisotropy of turbulence. The bursts impart an upward momentum flux on the particles which exceeds the downward flux from the return flow. Grass (1974) filmed the suspension process by towing a flat bed of sand through an otherwise quiescent fluid. He identified coherent flow structures within the boundary layer and calculated the velocities of the particles advected by these motions. This observation links sediment suspension with the ejection of fluid away from the boundary. Further experimental studies have been performed by Sumer & Oguz (1978) and Sumer & Deigaard (1981), who observed the motion of heavy particles near to the bottom boundary of a turbulent channel flow. These observations confirm the role of the bursting process as a mechanism for sediment suspension.

Soulsby *et al.* (1987) have demonstrated an important link between the bursting process and the suspension of sediment within boundary layers of geophysical scale. They made simultaneous measurements of the high frequency fluctuations of the concentration of sand, suspended by a tidal current, and the horizontal and vertical components of the water velocity, above a sandy bed of an estuary. They showed that large upward fluxes of sediment are associated with ejection events.

We present in figure 1 some data collected by Grass (1974) which gives typical vertical and horizontal velocities for sand particles, which are suspended within an ejection event. We note that these profiles have the correct velocity signature for an ejection event, namely the particles are moving upward and their horizontal velocity lags the mean flow. Further we note that the horizontal velocity of the ejection lags the mean flow by an approximately constant value and that the vertical velocity reaches a peak at a height of approximately 300 viscous length units above the lower boundary. Grass (1983) attributes this peak to a complex vortical interaction between the coherent structures. The dynamics of this interaction are incompletely understood.

The present work is concerned with the fundamentals of sediment transport and focuses on the suspension of sediment by coherent structures within the boundary layer. We develop a phenomenological model of the advection of sediment by an erupting turbulent burst as it is ejected away from the the boundary region. This framework is similar to the approach of recent studies by Nielsen (1991) and Deigaard (1991) who also consider the advection of sediment by ejection events. However, here we account for the vertical transport of momentum and particles by the ejection.

The classical approach to the modelling of suspended sediment is to assume it follows gradient-diffusion behaviour. The non-isotropic characteristics of turbulence

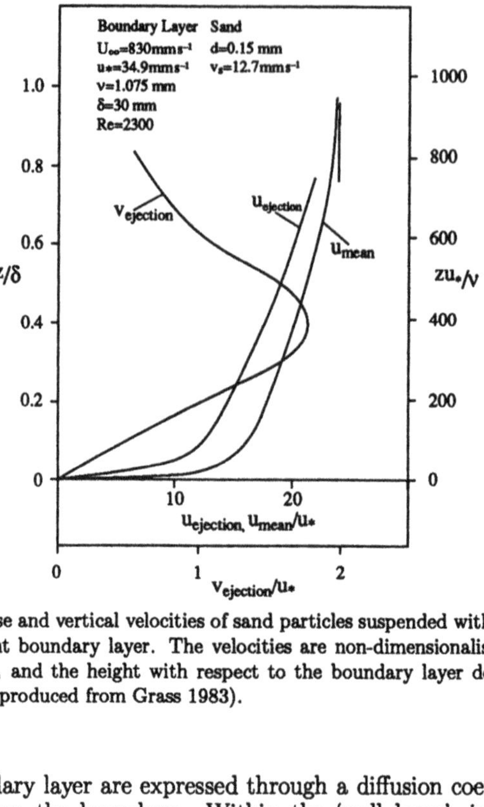

FIGURE 1 Streamwise and vertical velocities of sand particles suspended within ejection events from a flat-plate turbulent boundary layer. The velocities are non-dimensionalised with respect to the 'friction' velocity u_* and the height with respect to the boundary layer depth δ and the viscous length unit ν/u_* (reproduced from Grass 1983).

within the boundary layer are expressed through a diffusion coefficient which varies with distance from the boundary. Within the 'wall layer', in which the flow is turbulent but directly influenced by viscosity (Hinze 1975), the diffusion coefficient is often prescribed to vary linearly with distance from the boundary, reflecting the typical turbulent eddy dimension. Central to gradient-diffusion modelling is the principle of averaging over a number of random excursions which occur over scales significantly smaller than those over which the characteristics of the domain vary. However, within the turbulent boundary layer, we find these two scales are of a comparable magnitude. This is bourne out by the observations of motions which are coherent throughout the entire boundary layer. Hence transport by these large-scale advective events can not be neglected. Such a conclusion led Hinze (1975) to conclude that transport could not be described in a satisfactory way by means of gradient-diffusion alone.

The approach used here is to model the dynamics of one advective event and account for the resulting transport of particles and momentum. We develop a simple phenomenological model which aims to capture some of the fundamental physics of the suspension process. We demonstrate that within certain asymptotic regimes, this framework reproduces some 'well-known' results which are usually derived from a different set of assumptions. In the following section, we define the problem and introduce variables to describe a particle-laden fluid element. In §3, we present equations to describe the conservation of mass and momentum of the suspending phase and we introduce the entrainment assumption, following the work of Morton *et al.* (1956). This assumption is vital to quantify how the fluid element mixes with the ambient fluid. These equations are integrated and the results presented in §4. We consider the particulate phase in §5 and study the implication for the average concentration profile. Finally, we present some conclusions in §6, noting that this analysis reproduces some of the experimental results of Grass (1974) for the vertical and horizontal velocities of an ejection event.

2. Particle-laden fluid elements

We consider the suspension of sediment under the simplest turbulent flow conditions, namely a steady mean flow $[u(z), 0, 0]$, driven by a steady pressure gradient, $\partial p/\partial x$. We have noted above that even within this simplest flow, coherent motions are observed within the boundary layer, the dynamics of which are incompletely understood. Introducing the Reynolds decomposition of the Navier-Stokes equation by writing the turbulent velocity components as (u', v', w') and neglecting viscous terms, we can write the time-averaged conservation of fluid momentum as

$$0 = -\frac{1}{\rho}\frac{\partial p}{\partial x} - \frac{\partial}{\partial z}\left(\overline{u'w'}\right). \tag{1}$$

In this expression the Reynolds shear stress $\tau/\rho = -\overline{u'w'}$. Since the average pressure gradient driving the mean flow is constant and the shear stress vanishes at the free-surface $z = D$, we may integrate this equation to give

$$\tau = \rho u_*^2 (1 - z/D), \tag{2}$$

where D is the flow depth, u_* is the friction velocity, defined by the Reynolds shear stress at the boundary ($\tau_0 = \rho u_*^2$) and the z-axis is taken away from the solid boundary. We note that for ejections $u' < 0$ and $w' > 0$ while for sweeps $u' > 0$ and $w' < 0$. Both make positive contributions to the shear stress.

We develop a model of a particle-laden fluid element and use it to describe the evolution of a turbulent ejection of fluid away from the boundary. The model describes the size and velocity of the event. As the fluid element is ejected away from the boundary, it entrains some of the ambient fluid through which it rises and it is advected downstream. As a result of the entrainment, the element not only mixes with ambient fluid, but also grows in size. Particles that are advected with the fluid element undergo gravitational settling and some fall out of it, whilst others, suspended in the ambient, are entrained into it. The framework developed below provides a simple way of accounting for this process and enables the calculation of the particle concentration within an ascending fluid element.

We describe the particle-laden fluid element using the variables shown in figure 2. We denote the radius of the spherical fluid element by b and its vertical velocity by w. The horizontal velocity of the element is denoted by v, while the mean horizontal velocity of the ambient is denoted by u. The entrainment velocity is denoted by u_e and represents the velocity of the ambient fluid across the surface of the expanding fluid element. We denote the volume concentration of particles within the fluid element and the ambient by C and C_a, respectively, both of which will be assumed to be small. The particles have a settling velocity given by V_s.

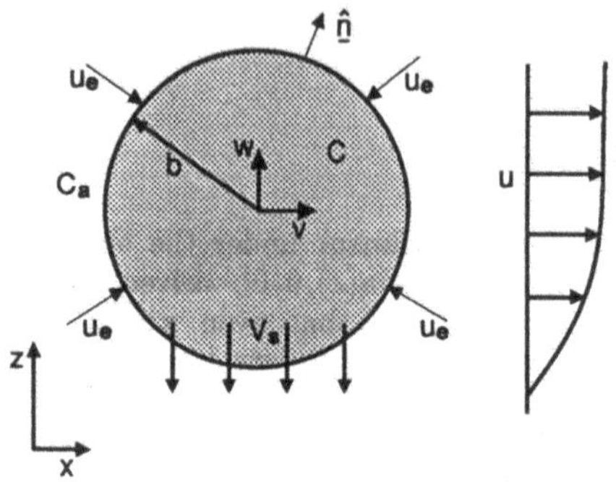

FIGURE 2 Variables used to describe the particle-laden fluid element (see §2).

The variables defined above are averaged over the timescale of eddies causing the entrainment into the fluid element. We treat the fluid element as an evolving sphere. Although at any instant the interface between the fluid element and the ambient is highly contorted by the presence of small-scale eddies causing mixing with the ambient, in an averaged sense the fluid element has a more regular shape. Three-dimensional shapes other than a sphere could be modelled by the inclusion of appropriate geometrical parameters, although this does not alter the inherent structure of the equations. We assume that the small-scale turbulence within the fluid element is sufficient to keep the particles well-mixed, so that we may use an average particle concentration. This framework for modelling the fluid elements follows the ideas first introduced by Morton *et al.* (1956).

3. Hydrodynamic model

We present equations to model the conservation of mass, horizontal and vertical momentum of the fluid element. We assume that the particle concentration is small ($C \ll 1$). This implies that the presence of the particles does not significantly affect the momentum of the fluid element. The particles are accounted for only in an equation expressing their conservation (see §5).

Central to the framework developed here is the entrainment assumption. This models the inflow to the fluid element as it ascends through and entrains ambient fluid. The model specifies the magnitude of the inflow velocity to be proportional to the vertical velocity of the element. The constant of proportionality is denoted by α and is termed the entrainment coefficient. Hence

$$\mathbf{u}_e = \alpha w \hat{\mathbf{n}}. \tag{3}$$

This assumption was first developed by Morton *et al.* (1956) and used to model the inflow to plumes and thermals. The use of the entrainment assumption postulates that the mixing with the ambient is driven by the vertical velocity.

3.1 Conservation of mass

As the fluid element ascends it entrains ambient fluid and so its volume increases. The principle of conservation of mass equates the rate of change of the volume with the rate of inflow over the boundary. Hence, in accordance with Morton *et al.* (1956), we find that

$$\frac{\mathrm{d}}{\mathrm{d}t}\left(\tfrac{4}{3}\pi b^3\right) = 4\pi\alpha b^2 w. \tag{4}$$

By noting that $\mathrm{d}z/\mathrm{d}t = w$, we conclude that

$$\frac{\mathrm{d}b}{\mathrm{d}z} = \alpha. \tag{5}$$

Hence the dimension of the ejections grows linearly with distance from the lower boundary. This result follows directly from the entrainment assumption and is independent of the actual velocity of the element. It is in accord with the observation of the typical eddy dimension within turbulent boundary layers which also is observed to increase linearly with distance from the boundary.

3.2 Conservation of vertical momentum

The fluid element is ascending through ambient which has no mean vertical velocity. Hence it does not entrain any net vertical momentum. Therefore the rate of change of the vertical momentum of the element is equal to a vertical forcing which we denote by F_v,

$$\rho\frac{\mathrm{d}}{\mathrm{d}t}\left(\tfrac{4}{3}\pi b^3 w\right) = F_v, \tag{6}$$

If the fluid element were impulsively hurled from the lower boundary, then there would be no vertical forcing. However observations of turbulent ejections suggest that they accelerate during some initial phase. Dynamically this is due to some complex vortical interaction and hence we require a vertical forcing, the magnitude of which may be expected to decrease as the element ascends. We discuss a possible form of this forcing in §3.6. Observations indicate that ejections may impinge upon the free-surface of the flow (Jackson 1976). However, interaction with the free-surface of the channel flow is not included in this model and we find that the element has some small vertical velocity as it reaches the surface. Observations of a vortex pair impinging on a free-surface indicate that the surface is deformed (Ohring & Lugt 1991). However we do not include this effect as little sediment is advected to this height above the boundary.

3.3 Conservation of horizontal momentum

The ambient fluid has net horizontal momentum, which the fluid element entrains as it ascends. The element is also acted upon by a drag force, which we denote by F_h, and which is the horizontal component of the force explicitly discussed in §3.6. We expect that sufficiently distant from the boundary, the rate of change of horizontal momentum is balanced by the entrainment of ambient momentum, rather than the drag force F_h. The equation expressing this conservation relation is given by

$$\rho \frac{\mathrm{d}}{\mathrm{d}t}\left(\tfrac{4}{3}\pi b^3 v\right) = 4\pi\alpha\rho b^2 w u + F_h. \tag{7}$$

3.4 Global conservation of mass

In order to preserve the global conservation of mass within the channel, the ascending ejection drives a vertical return flow. If we assume that the average areal separation on the lower boundary of ejection events, which are initiated at the same time, is A, then the return flow w_r is given by,

$$\left(A - \pi b^2\right) w_r = \pi b^2 w. \tag{8}$$

The area A is a statistical quantity and may be a function of the outer flow variables. This continuity expression assumes that the return flow velocity is constant over the entire area A. Also tacitly assumed is that the spatial frequency of the ejections is such that they do not overlap (i.e. $A > \pi b^2$ throughout the channel depth).

3.5 Shear stress

The Reynolds shear stress is the rate of vertical transfer of horizontal momentum by turbulent motion. Using the framework developed here, we calculate the shear stress contributed by an ejection. The ascending fluid element carries fluid moving with horizontal velocity $v(z)$. Conversely the return flow driven by the element carries fluid moving with velocity $u(z)$. Hence the shear stress due to one of these elements passing is given by,

$$\tau_{ejection}/\rho = -\frac{1}{A}\left(\pi b^2 w v - \left(A - \pi b^2\right) w_r u\right),$$

$$= \frac{\pi b^2 w}{A}(u-v)\,. \tag{9}$$

We wish to link the shear stress caused by ejections to the mean shear stress within the channel, which varies linearly with distance from the boundary. This necessarily involves some averaging over the magnitude of ejection events. Observations suggest that the burst and sweep events contribute most of the turbulent shear stress and that they occur for a fairly constant fraction of time. Soulsby (1983) finds that ejection events occur for 10-13% of the time at heights of 30cm and 140cm above the bed and that they contribute 55% of the shear stress. We assume that ejections contribute a fixed proportion of the shear stress during a fixed proportion of time and denote these fractions by β and f respectively. Then the average turbulent shear stress is given by

$$\tau_{average}/\rho = \frac{f}{\beta A} b^2 w\,(u-v)\,. \tag{10}$$

3.6 Drag Force

As part of the discussion of coherent motions (§1), we noted that it is postulated that ejections are initiated by vortical motions associated with earlier ejection events. Such vortical motions interact with fluid near the boundary, causing the eruption of the boundary layer. This behaviour is found in the phenomenological studies of the interaction of an inviscid vortex with a laminar boundary layer (Dogalski & Walker 1984). Their study demonstrates that the boundary layer erupts within a finite time.

We assume that the force exerted on the erupting fluid element, the vertical and horizontal components of which we have denoted by F_v, F_h, may be modelled by a non-linear drag law involving the velocity field associated with an overpassing ejection, initiated upstream at an earlier time ($t = -T$). The force is linked to the velocity difference between the overpassing eruption and the ambient fluid. Thus we propose,

$$\mathbf{F} = 4\pi\rho C_D b^2 |\mathbf{U}_I| \mathbf{U}_I, \tag{11}$$

where C_D is a constant, representing a drag coefficient, b is the radius of the erupting ejection and $\mathbf{U}_I$ is the velocity field associated with the overpassing ejection. The velocity field $\mathbf{U}_I$ is treated as a vortex of radius $b(t+T)$, with a constant radius of the inner vortical core a, moving with the velocity of an ejection initiated at $t = -T$, relative to the ambient flow. Hence we find that

$$\mathbf{U_I} = \frac{a}{b(t+T)}\mathbf{u}'(t+T), \tag{12}$$

where $\mathbf{u}' = (u-v, 0, w)$. We emphasise that this expression for the force on an erupting fluid element represents the vortical interaction with other events and not the interaction with the ambient, which is included via the process of entrainment. As noted by Grass (1983), the actual form of this interaction is complex and not fully understood. However we demonstrate that this model permits the derivation of well-known asymptotic results within the constant stress layer.

4. Hydrodynamic results

The equations (2)–(12) form a system of equations which may be integrated to find the evolution of the dimension and velocity of the fluid element. The equations involve a number of physical constants for which we prescribe values from a preliminary study of experimental measurements. If the system of equations is non-dimensionalised with respect to the initial velocity and radius of the fluid element, we find there are seven non-dimensional entities in the equations (1)–(12). It is emphasised, though, that the inherent structure of the equations remains unchanged by the chosen values of the constants. Graphs displaying the results of a numerical integration of the equations are presented in figures 3–5. In these figures, the velocities are non-dimensionalised with respect to the initial vertical velocity and height is non-dimensionalised with respect to the flow depth. The values of the constants used for this integration are given in table 1. Many of these values have some statistical distribution and a preliminary study of field data suggests that many are functions of the friction velocity u_*. This observation is reasonable, since the magnitude of many of these parameters will be determined by the magnitude of the turbulent fluctuations, which in turn is reflected in the magnitude of u_*.

TABLE 1 Parameter values used to produce figures 3–5

Entrainment coefficient	α	0.2
Friction velocity	u_*/w_0	1.03
Drag coefficient	$C_D a^2/b_0^2$	0.33
Flow depth	D/b_0	50
Proportion of stress due to ejections	β	0.5
Proportion of time with ejections occurring	f	0.1
Areal frequency	A/b_0^2	1000

We observe that the vertical velocity of the ejection increases to some peak value at around $z = b_0$ before decaying. In physical terms, the vertical forcing initially increases the momentum of the element by increasing its vertical velocity. However the ascending ejection is also entraining fluid which has to be accelerated to the upward velocity. Far from the boundary, there is a balance between the acceleration of entrained fluid and the forcing and so the overall vertical velocity of the element decays. We find that the decay with height is proportional to z^{-2}, which is consistent with the experimental data presented by Grass (1983).

The horizontal velocity deficit of the ejection is approximately constant within a region close to the bottom boundary. Again this mirrors the experimental observations of Grass (1983). Furthermore we find there is a region in which the

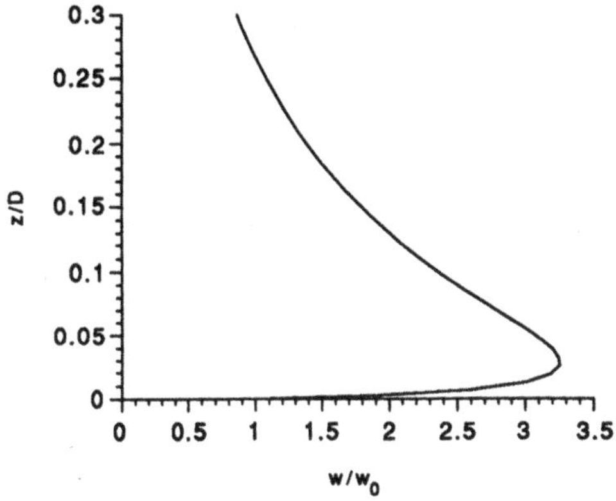

FIGURE 3 Vertical velocity profile of an ejection event.

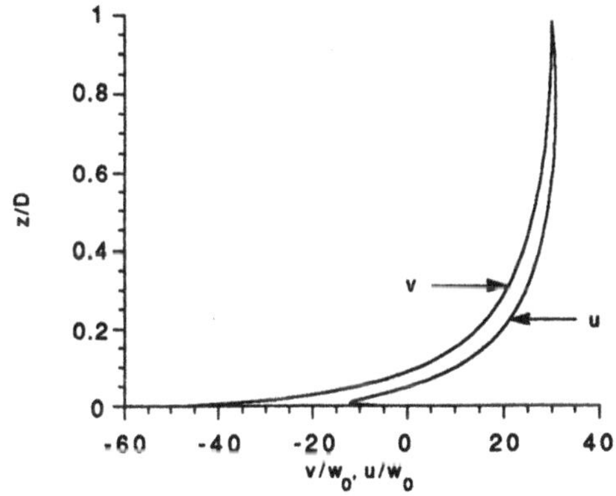

FIGURE 4 Horizontal velocity profile of an ejection event (v) and of the ambient (u).

ambient velocity profile exhibits a logarithmic dependence on distance from the boundary. This reproduces the classical 'wall layer' result.

We find the following asymptotic results in the region, corresponding to the constant stress layer, $D \gg Z \gg b_0$

$$b \sim \alpha z, \tag{13}$$

$$w \sim u_* \sqrt{\frac{3C_D a^2 A\beta}{f\alpha^5}} \frac{1}{z^2}, \tag{14}$$

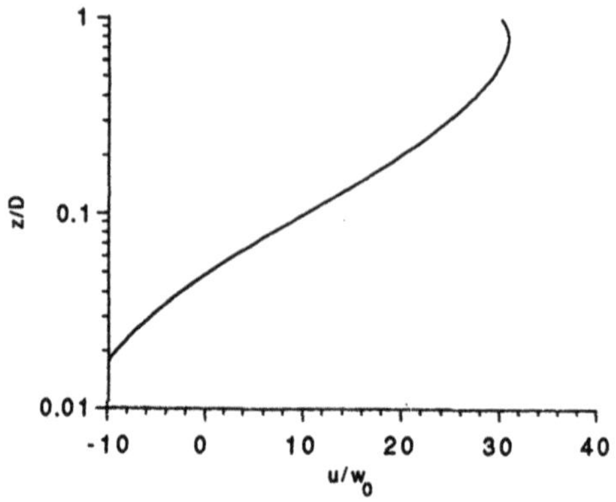

FIGURE 5 Horizontal velocity profile of the ambient shown on a logarithmic scale for distance from the boundary.

$$u - v \sim u_* \sqrt{\frac{A\beta\alpha}{3C_D a^2 f}}, \tag{15}$$

$$u \sim 4u_* \sqrt{\frac{A\beta\alpha}{3C_D a^2 f}} \log z. \tag{16}$$

These asymptotic results confirm the graphical observations made above. Furthermore within the framework of this analysis, the eddy viscosity is given by $b^3w/A \sim zu_*\sqrt{3\alpha C_D a^2/fA}$; it varies linearly with distance from the boundary which reproduces the traditional diffusion result.

5. Particle concentration

We derive the following particle conservation equation for the fluid element

$$\frac{\mathrm{d}}{\mathrm{d}t}\left(\tfrac{4}{3}\pi b^3 C\right) = 4\pi\alpha b^2 w C_a - \pi V_s b^2 C. \tag{17}$$

This equates the rate of change of the number of particles within the fluid element to the number entrained from the ambient minus the number which settle out. We assume that the turbulence within the fluid element is sufficient to keep the particles well-mixed and that they sediment with their settling velocity once they have left the element. Bonnecaze *et al.* (1993) present a similar approach to modelling the particle concentration within a turbulent flow, although they applied the technique to a different problem.

The fluid element is rising through an ambient in which particles are suspended. Therefore we include the entrainment of these particles into the element. We apply a steady-state principle of the global conservation of particles (cf. §3.4). We introduce an area A over which just one ejection event occurs and invoke the steady-state balance of the sedimentation of particles within the ambient with those brought up by the sporadic ejections, with a typical period T. Hence we find that

$$\int_{z}^{z+V_sT} AC_a\,\mathrm{d}z = \tfrac{4}{3}\pi b^3 C. \tag{18}$$

This closes the system of equations for the concentrations of particles in both the ambient and fluid element. Since we assume that ejections only occur for a constant fraction of any interval, we may calculate the average concentration profile. This is a temporal average of the concentration within the ambient and the fluid element as it ascends. Also we take a spatial average, because there is only one ejection event per area A, at any time. Denoting the average ambient concentration of particles by $\overline{C}_a$, we find that

$$\overline{C}_{average} = (1-f)\overline{C}_a + f\left(\left(1-\pi b^2/A\right)\overline{C}_a + \pi b^2 C/A\right), \tag{19}$$

where $\overline{C}_{average}$ is the average particle concentration in the flow. We present some typical concentration profiles in figure 6, for varying values of the settling velocity of the particles. We note that the value of the parameter V_s/w_0 determines the shape of the concentration profile. This is equivalent to the role of the parameter V_s/u_*, which is used in the classical diffusion model of suspended sediment. It expresses the ratio of the settling velocity of the particles to the magnitude of turbulent velocity fluctuations. We note the concentration profiles are convex upwards.

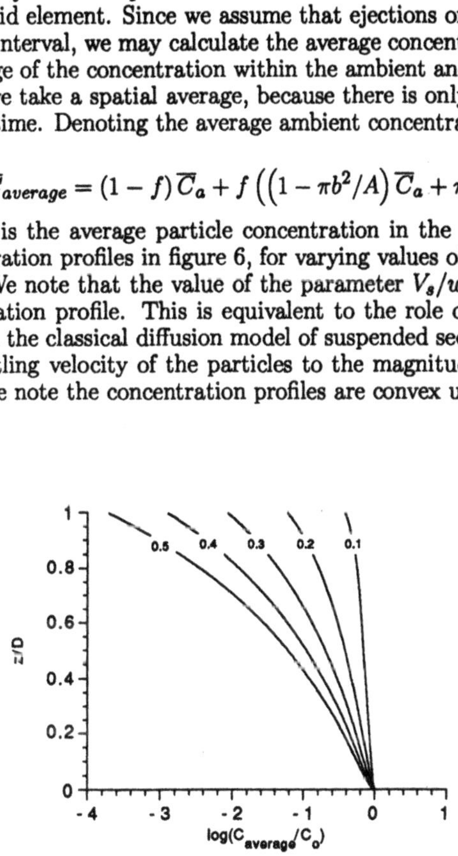

FIGURE 6 Average concentration profiles, non dimensionalised with respect to the concentration at the bed, for various ratios of the settling velocity to the initial vertical velocity of the ejection (V_s/w_0).

We may also perform asymptotic analysis to find an expression for the concentrations within the fluid element and within the ambient. In the regime

$z \gg V_s T$, we find that

$$C_a = \frac{\frac{4}{3}\pi b^3}{AV_s T} C. \tag{20}$$

Furthermore, within the constant stress layer $D \gg z \gg b_0$ and denoting the initial concentration within the ejection by C_0, we find that

$$C \sim \frac{b_0^3 C_0}{\alpha^3 z^3} \exp\left(-\frac{3V_s}{8u_*} \sqrt{\frac{f\alpha^3}{3C_D a^2 A\beta}} z^2 + \frac{4\pi\alpha^3}{3AV_s T} z^3 \right). \tag{21}$$

6. Conclusions

We have developed a model of sediment suspension that is based on the advective transport of particles by large-scale coherent events within turbulent boundary layers. This model accounts for the vertical transport of both particles and momentum by studying the dynamics of particle-laden fluid elements. We have demonstrated that by using this analysis we may reproduce some of the experimental observations made by Grass (1983). Furthermore, within the constant stress layer, we recover the logarithmic profile for the ambient velocity.

Whilst the model develops a framework for the analysis, there are many additional features of the flow which are not included at present. We have considered only the ejection events and have included various parameters, the values of which we have determined by only a preliminary study of data. Also we have included a forcing expression (§3.6) which requires further justification.

This framework could be applied to the modelling of the suspension of sediment for non-planar beds with waves and currents. For these scenarios, the initiation and scale of the coherent events which advect the particles away from the boundary are different, but it may be possible to apply the principles developed here to model sediment transport.

Acknowledgements

The authors acknowledge the assistance of Dr. W.B. Dade in the development of this model and thank Dr. A.J. Grass for permission to reproduce figure 1. AJH acknowledges the financial support of a CASE studentship from SERC and HR Wallingford. RLS acknowledges funding from UK Department of the Environment, Construction Sponsorship Directorate and the Commission of the European Communities, Directorate General for Science, as part of the MAST G8 Coastal Morphodynamics research programme, Contract No. MAS2-CT92-0027.

References

BONNECAZE R.T., HUPPERT H.E. & LISTER J.R. 1993 Particle-driven gravity currents *J. Fluid Mech.* **250**, 339–369

DEIGAARD R. 1992 On the turbulent diffusion coefficient for suspended sediment *MAST Abstracts (Extended) 1992*, 3–19

DOLIGALSKI T.L. & WALKER J.D.A. 1984 The boundary layer induced by a convected two dimensional vortex *J. Fluid Mech.* **139**, 1–28

GORDON C.M. 1974 Intermittent momentum transport in a geophysical boundary layer *Nature* **248**, 292-294

GRASS A.J. 1971 Structural features of turbulent flow over smooth and rough boundaries *J. Fluid Mech.* **50**, 233–255

GRASS A.J. 1974 Transport of fine sand on a flat bed: turbulence & suspension mechanics *EUROMECH 48: Transport, erosion & deposition of sediment in turbulent streams.*

GRASS A.J. 1983 The influence of boundary layer turbulence on the mechanics of sediment transport *EUROMECH 156: Mechanics of sediment transport* ed. Sumer & Muller

GRASS A.J., STUART R.J. & MANSOUR-TEHRANI M. 1991 Vortical structures and coherent motions in turbulent flow over smooth and rough boundaries *Phil. Trans. R. Soc. Lond.* **A336**, 36–65

HEATHERSHAW A.D. 1974 Bursting phenomena in the sea *Nature* **248**, 294–295

HINZE J.O. 1975 *Turbulence*, McGraw-Hill, pp586

JACKSON R.G. 1976 Sedimentological and fluid dynamical implications of the turbulent bursting phenomenon in geophysical flows *J. Fluid Mech.* **77**, 531–560

KLINE S.J., REYNOLDS W.C., SCHRAUB F.A. & RUNSTADLER P.W. 1967 The structure of turbulent boundary layers *J. Fluid Mech.* **30**, 741–773

MORTON B.R., TAYLOR G.I. & TURNER J.S. 1956 Turbulent gravitational convection from maintained and instantaneous sources *Proc. Roy. Soc.* **A234**, 1–23

NIELSON P. 1991 Combined convection and diffusion: a new framework for suspended sediment modelling *Coastal Sediments '91*, 418–431

OHRING S. & LUGT H.J. 1991 Interaction of a viscous vortex pair with a free surface *J. Fluid Mech.* **227**, 47–70

ROBINSON S.K. 1991 Coherent motions in the turbulent boundary layer Ann. Rev. Fluid Mech. **23**, 601–639

SMITH C.R., WALKER J.D.A., HAIDARI A.H. & SOBRUN U. 1991 On the dynamics of near-wall turbulence *Phil. Trans. R. Soc. Lond.* **A336**, 131–175

SOULSBY R.L. 1983 The bottom boundary layer of shelf seas *Physical oceanography of coastal and shelf seas* ed. B. Johns, Elsevier

SOULSBY R.L., ATKINS R. & SALKIELD A.P. 1987 Observations of the turbulent structure if a suspension of sand in a tidal current *EUROMECH 215: Mechanics of sediment transport in fluvial and marine environments*

SUMER B.M. & OĞUZ B. 1978 Particle motions near the bottom in turbulent flow in an open channel *J. Fluid Mech.* **86**, 109–127

SUMER B.M. & DEIGAARD R. 1981 Particle motions near the bottom in turbulent flow in an open channel - part 2 *J. Fluid Mech.* **109**, 311–337

SEDIMENT SUSPENSION MODELLING IN TERMS OF COMBINED CONVECTION DIFFUSION

PETER NIELSEN

Department of Civil Engineering, University of Queensland, St Lucia, Australia 4072.

Experimental evidence shows that pure gradient diffusion is not an appropriate model for suspended sediment distributions in steady flows and under waves. Some problems like the need to apply different diffusivities for different grain sizes have been overcome traditionally by the use of the ad hoc β-function, but this is shown to be unsatisfactory on the basis of theory and other types of diffusion experiments. Other phenomena like differences in distribution shapes for different grain sizes in the same flow do not lend themselves to simple ad hoc solutions within the gradient diffusion framework. To explain such phenomena a combined convection diffusion model is suggested, where the term convection refers to movement of sediment in one step through a considerable fraction of the considered vertical range. This occurs rather obviously over sand ripples under waves but it also seems to be the dominant mechanism in oscillatory sheet flow. The combined convection diffusion approach is generally recommende because it represents the actual physical processes rather well. For oscillatory sheet flow with grain sizes greater than *0.2mm* and periods longer than

1 The problems with pure diffusion

Since the Nineteen Thirties sediment suspensions in both rivers and coastal environments have been modelled using the concept of gradient diffusion. That is, the upward sediment flux is assumed to be given by

$$q_D = -\varepsilon_s \frac{\partial c}{\partial z} \tag{1}$$

and the sediment diffusivity ε_s is expected to be closely related to the diffusivity of momentum i e the eddy viscosity ν_t of the flow. The question of whether one should expect the sediment diffusivity to be smaller, greater or equal to the eddy viscosity was raised and subsequently addressed theoretically by Yudine, 1959 and Csanady, 1963. They noted two effects which would both reduce the diffusivity of dense particles: the *crossing trajectories effect* and the *inertia effect*. The idea behind the crossing trajectories effect is that the Integral Lagrangian time scale T_L for dense particles is less than for fluid elements because dense particles cross fluid trajectories and hence loose correlation more quickly. The effect of particle inertia, on the other hand, is to reduce the velocity fluctuation magnitude u' relative to that of the fluid. However, since the diffusivity can be written as

$$\varepsilon_s = u'^2 T_L \tag{2}$$

(Taylor, 1921) both effects are seen to reduce the diffusivity of dense particles relative to that of fluid elements and momentum.

These theoretical results were subsequently confirmed by dispersion experiments by Snyder & Lumley, 1971 who found that heavy particles dispersed more slowly than almost-neutral particles in the spanwise direction in grid turbulence.

Csanady, 1963 had argued that the crossing trajectories effect is more important than the inertia effect for particles in the atmosphere, and this was supported by the experiments of Wells & Stock, 1983. Wells and Stock studied the dispersion of electrically charged particles and were thus able to vary the the crossing trajectories effect by changing the strength of the electrical field while keeping the particle inertia (mass) constant.

Hence, theory and cloud spreading in grid turbulence both indicate that the diffusivity of dense suspended particles should be less than that of neutrally bouyant particles and momentum.

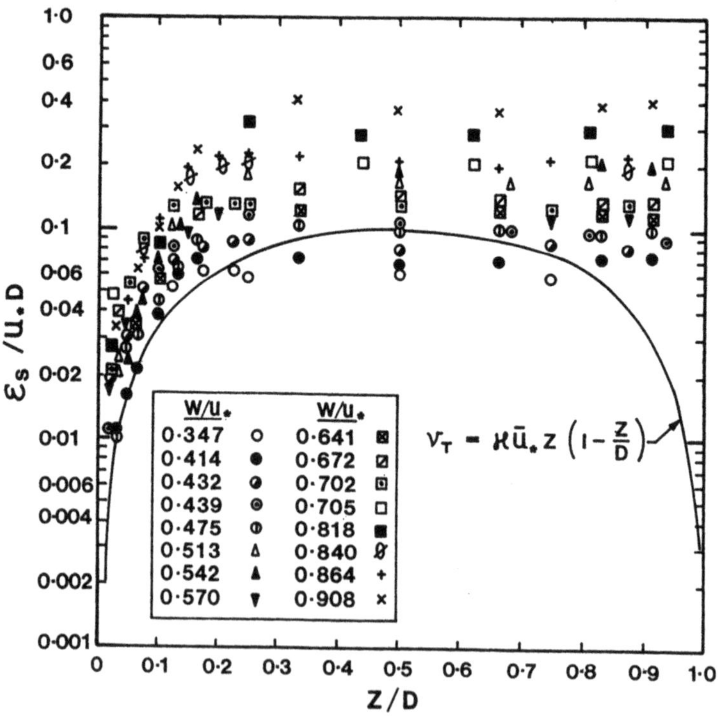

Figure 1: Sediment diffusivities derived through Equation (3) from sediment concentration profiles measured in steady open channel flow. Data from Coleman, 1970.

However, diffusivities derived from measured sediment concentration profiles via the one dimensional, time averaged diffusion equation

$$\varepsilon_s \frac{d\bar{c}}{dz} + w_o \bar{c} = 0 \tag{3}$$

do not verify this expectation. In fact, early measurement by Coleman, 1970 contradict it rather strongly. See Figure 1.

Coleman's data shows first of all that the sediment diffusivity $\varepsilon_{(3)}$ derived through Equation (3) is generally larger than the eddy viscosity ν_t. Secondly, it shows that the ratio $\varepsilon_{(3)}/\nu_t$ increases very rapidly with the relative settling velocity w_o/u_*. In hindsight, the message is quite clear: Equation (3) is not an appropriate model for the vertical distribution of suspended sediment in steady open channel flow.

However, no alternative was at hand so, practical modellers kept on using Equation (3) with the ad hoc function

$$\beta(\frac{w_o}{u_*}) = \frac{\varepsilon_s}{\nu_t} \tag{4}$$

linking the required diffusivity to the eddy viscosity.

Although the value of β would generally have to be greater than unity (van Rijn, 1984, van de Graff, 1988) in conflict with the theories of Yudine, 1959 and Csanady, 1963 and with the dispersion experiments of Snyder & Lumley, 1971 and Wells & Stock, 1983, this praxis is still prevailing.

Changes in effective settling velocity due to the presence of turbulence as discussed by Nielsen, 1993 are not likely to save the diffusion model either. The changes that have been observed are small reductions *(20-40%)* for large particles $(u'/w_o \lesssim 3)$, and increases by up to a factor *4* for small particles $(u'/w_o \gtrsim 10)$. These changes would through Equation (3) correspond to $\varepsilon_{(3)}$-values much smaller than ν_t for fine sand and marginally greater than ν_t and for large particles but that is not what Coleman's data indicates.

More trouble for Equation (3) arrived however, with the field data of Nielsen, 1983 and the laboratory data of McFetridge & Nielsen, 1985 both measured over rippled sand beds under non breaking waves.

These data show that the required diffusivity to model the observed $\bar{c}(z)$ profiles through Equation (3) would not only have to be of different magnitude for different sand sizes, as mimiced by Equation (4), but different sand sizes in the same flow require different distributions of $\varepsilon_s(z)$. The fine fractions show upward convex $\bar{c}(z)$-profiles which correspond to $\varepsilon_s(z)$ decreasing with z while the coarse fractions show upward concave $\bar{c}(z)$-profiles corresponding to an increasing $\varepsilon_s(z)$.

It was thus evident that equation (3) where the upward sediment flux is represented by pure gradient diffusion is inadequate for the modelling of $\bar{c}(z)$-profiles over rippled sand beds under waves and an alternative was suggested by Nielsen, 1991.

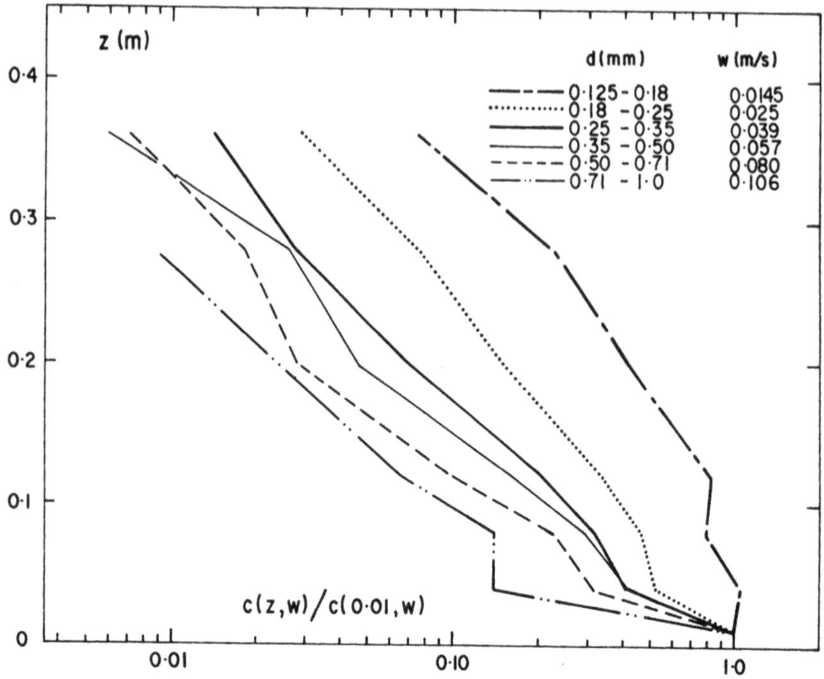

Figure 2: Normalised time averaged concentration profiles for different sand sizes in the same flow over a rippled bed under non-breaking waves, after Nielsen, 1983.

2 Combined convection diffusion.

The assumption (1) of pure gradient diffusion is justified if the particles move as in brownian motion where the individual steps are very small compared to the scale over which the concentration profile is measured. However, for rippled beds under waves the sand moves in a quasi coherent fashion from the lee of the ripples all the way to the top of the boundary layer (Bijker et al, 1976). This sand is trapped in a travelling vortex (Nielsen 1979, 1992) and the corresponding upward flux of sand is not necessarily related to the local concentration gradient as assumed by Equations (1) and (3). The essential requirement for a suitable modelling frameweork is therefore that it can represent processes like the upward transport of sand by travelling vortices, see the illustrations of Bijker et al, 1976. At the same time diffusion caused by small scale velocity fluctuations may have to be accounted for. Nielsen, 1991 therefore suggested a combined convection diffusion model where the total upward sediment flux consists of a convective flux q_C and the usual diffusive flux q_D.

2.2 Quantification of convective entrainment flux

A quantitative description of the convective, upward sediment flux $q_C(z,t)$ in accordance with Figure 3, was derived by Nielsen, 1991.

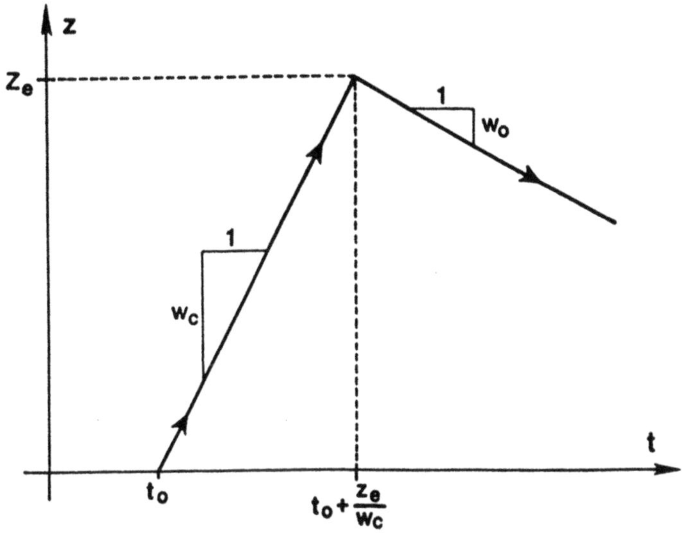

Figure 3: The upward flux of sand from the bed at time t is $p(t)$. This sand travels upwards with constant speed w_c until it reaches its entrainment level z_e. The amount of sand entrained at time t which travels beyond a certain level z is $F(z)\, p(t)$.

For the process described in Figure 3 the upward convective flux through the level z at time t is

$$q_C(z,t) \;=\; p(t - \frac{z}{w_c})\, F(z) \tag{5}$$

where $p(t)$ is the amount of sand which is picked up from the bed at time t, w_c is the velocity with which the sediment is convected upwards and the function $F(z)$ describes the extent to which sand is retained in the travelling vortex: $F(0) = 1$, $F(z) \rightarrow 0$ for $z \rightarrow \infty$.

2.3 The new continuity equation

The continuity equation which expresses conservation of sand in the combined convection diffusion process is

$$\frac{\partial c}{\partial t} \;=\; w_o \frac{\partial c}{\partial z} - \frac{\partial q_C}{\partial z} - \frac{\partial q_D}{\partial z} \tag{6}$$

which, with the expression (5) inserted for the convective upward flux, and with the upward diffusive flux $q_D = -\varepsilon_s \frac{\partial c}{\partial z}$ becomes

$$\frac{\partial c}{\partial t} - w_o \frac{\partial c}{\partial z} - \frac{\partial}{\partial z}(\varepsilon_s \frac{\partial c}{\partial z})$$

$$= \frac{1}{w_c} p'(t - \frac{z}{w_c})\, F(z) - p(t - \frac{z}{w_c})\, F'(z) \tag{7}$$

from which we see that the homogeneous equation is the usual diffusion equation, while the convective flux enters only through the forcing terms on the right hand side.

3 $\overline{C}(Z)$-profiles in combined processes

The solution to the time-averaged version

$$w_o \bar{c} + \varepsilon_s \frac{d\bar{c}}{dz} = \bar{p}\, F(z) \tag{8}$$

of the continuity equation (7) has been discussed in detail by Nielsen 1991 and 1992. It has been shown that the essential qualitative feature of the rippled-bed-data in Figure 2 can be modelled. That is, for appropriate $\varepsilon_s(z)$ and $F(z)$, the fine sand fractions tend to show upward convex profiles while coarser fractions show progressively more upward concave profiles.

More work is needed to establish the form and magnitude of the diffusivity (and the eddy viscosity) in oscillatory boundary layers over natural sand beds. Fortunately however, it is not always needed for practical $\bar{c}(z)$-modelling and the convective parameters, which seem more important, are easier to evaluate experimentally.

For sufficiently coarse sand the $\bar{c}(z)$-profile is indistinguishable (Nielsen, 1992 p 261) from the pure convection solution

$$\bar{c}(z) = \frac{\bar{p}}{w_o} F(z) \tag{9}$$

which emerges from (8) when $\varepsilon_s = 0$. This means that, in praxis, the shape of the distribution function $F(z)$ can be inferrred from the distributions of the coarser sand fractions. Nielsen found that for oscillatory flow over both rippled and flat beds $F(z)$ can be reasonably fitted by

$$F(z) = \frac{1}{(1 + z/z_1)^2} \tag{10}$$

where z_1 is related to the bed roughness r and the water particle semi-excursion A by $z_1 = 0.09\sqrt{rA}$ for $r/A > 0.06$.

The practical range of "coarse" sand for which the pure convection solution (9) is indistinguishable from the complete solution to (8) was suggested to be

$$\frac{w_o z_1}{\varepsilon_s} \gtrsim 2 \tag{11}$$

Nielsen pointed out that this range included all the available oscillatory sheet flow data with wave periods greater than *6s* so that $\bar{c}(z)$-profiles from oscillatory sheet flow can in praxis be reasonably predicted by the pure convection solution (9) + (10). This agrees with the fact, revealed in Figure 4, that also the time dependent components of $c(z,t)$ in such sheet flows can be satisfactorily modelled by the pure convection model.

The pure convection nature of the entrainment over flat beds under waves also explains why all sand size fractions showed similar distributions in the flat bed experiments of Ribberink & Al Salem, 1992. In the pure convection solution (9) all size fractions have the same distribution shape given by $F(z)$.

The dimensionless parameter $w_o z_1/\varepsilon_s$ of Equation (11) is (with $\varepsilon_s = 0.5 z_1^2 \omega$) proportional to the profile shape parameter

$$S = \frac{w_o T}{\sqrt{r A}} \tag{12}$$

which was used by Nielsen, 1992 and Nielsen, 1992b to explain why the shape of $\bar{c}(z)$-profiles vary systematically with the wave period as well as with grain size for oscillatory flows over both rippled and flat beds.

3.2 Starved $\bar{c}(z)$-profiles

If the sand supply at the bed is limited, the convective term in equation (8) seems to be less important than for natural beds of loose sand. Thus, both McFetridge & Nielsen, 1985 and van de Graaf, 1988 who measured $\bar{c}(z)$ over artificial wooden ripples with just a small amount of sand available found that under such "starved" conditions, the distributions of different sand sizes behaved in much closer agreement with the pure diffusion model (3) than for natural sand beds. The obvious explanation for this is that when the sand supply is limited, there is less sand to be picked up. Hence $\bar{p}$ decreases and the term on the right hand side of Equation (8) becomes less significant.

4 Time dependent concentrations

Similarly to the analysis above of time averaged concentration profiles it turns out that also time dependent concentration profiles over natural beds are better represented by a pure convection model than by a pure gradient diffusion model.

Please note that the following discussion compares only with measurements over flat beds because the horizontally averaged $c(z,t)$-models, which are considered here, do not model the time variation of measured point concentrations $c(x,y,z,t)$ over rippled beds. Over rippled beds, the time-variation of concentrations measured at a point results mainly from the "spotted carpet effect"; that is, from the horizontal movement of sediment clouds, see e g Nakato et , 1977.

5 $C(z,t)$ in pure gradient diffusion

The time dependent diffusion equation equation

$$\frac{\partial c}{\partial t} - w_o \frac{\partial c}{\partial z} - \frac{\partial}{\partial z}(\varepsilon_s \frac{\partial c}{\partial z}) = 0 \tag{13}$$

which emerges from Equation (7) when the convective entrainment flux is zero was solved for constant diffusivity and with the boundary conditions

$$-\varepsilon_s \frac{\partial c}{\partial z} = p(t) \quad \text{for } z=0$$
$$c(z,t) \to 0 \quad \text{for } z \to \infty \tag{14}$$

by Nielsen et al, 1978. The solution is in the form of a Fourier series and based on separation of the variables. Each concentration component $c_n(z,t)$ which is generated by the corresponding component $p_n(t) = P_n e^{in\omega t}$ of the pickup function was found to have the form

$$c_n(z,t) = \frac{P_n}{w_o \alpha_n} e^{-\alpha_n \frac{w_o z}{\varepsilon_s}} e^{in\omega t} \tag{15}$$

where

$$\alpha_n = \frac{1}{2} + \sqrt{\frac{1}{4} + i\frac{n\omega\varepsilon_s}{w_o^2}} \tag{16}$$

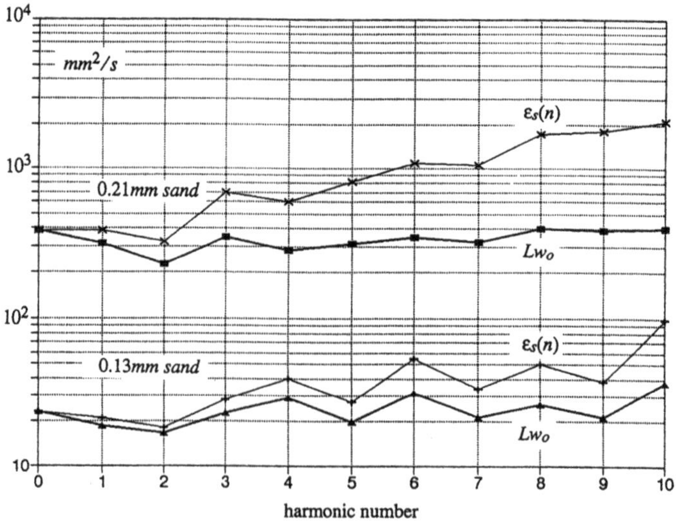

Figure 4: Comparison of entrainment parameters $\varepsilon_s(n)$ for a pure diffusion model with constant diffusivity and w_oL for a pure convection model with $F(z) = \exp(-z/L)$. The parameters are derived from the lower, almost exponential, parts of the $c(z,t)$ profiles measured in oscillatory sheet flow by Ribberink & Al Salem (1993).

Based on this solution, it is possible to derive a value of the pure-diffusion-diffusivity $\varepsilon_s(n)$ for each harmonic component $c_n(z,t)$ provided exponential curves can be fitted with reasonable confidence to $|c_n(z,t)|$.

If those diffusivity values are found to be equal, Equation (13) may be said to be a reasonable model. Differences between the values would indicate that Equation (13) is not a good model.

Figure 4 shows such $\varepsilon_s(n)$-values derived from concentrations measured over flat beds by Ribberink & Al Salem, 1993. It is seen that different harmonics correspond to different diffusivities indicating that the pure diffusion model is not an internally consistent description of sediment suspension over flat sand bed under waves.

6 $C(z,t)$ in pure convection

For a process with purely convective entrainment, a Fourier series solution has been obtained by Nielsen, 1991, 1992 in analogy with the pure diffusion solution above. The pure convection process does however, also lend itself to easy analytical solution without the periodicity assumption. Analytical solutions can be obtained for direct use in the time domain with irregular waves. This will be shown in the following subsection. The periodic solution is subsequently described in subsection 6.3.

6.2 General solutions for the pure convection case

Because of the simplistic nature of the convective entrainment process outlined in Figure 3, the local, time dependent concentration $c(z_o,t)$ can be expressed as a fairly simple convolution integral.

In the purely convective process the sand concentrations at the level z_o are made up by material which has been entrained to levels above z_o and is settling through. Since all the material is assumed to settle with settling velocity w_o, the relation between settling flux and concentration is $q_s = -w_o\, c$.

Sand which is settling through z_o at time t after being entrained to the level z must have reached z at time $t' = t - (z-z_o)/w_o$ and, travelling upwards at speed w_c, it must have been picked up from the bed at time $t'' = t' - z/w_c = t - (z-z_o)/w_o - z/w_c$. Its settling flux contribution dq_s therefore amounts to the fraction $-(-F'(z)\,dz)$ of what was picked up at time t''

$$dq_s \;=\; p\left(t-\frac{z-z_o}{w_o}-\frac{z}{w_c}\right)\left(-\frac{dF}{dz}\right) dz \tag{17}$$

and hence the total settling flux through z_o from all levels above is

$$q_s \;=\; \int_{z_o}^{\infty} p\left(t-\frac{z-z_o}{w_o}-\frac{z}{w_c}\right)\frac{dF}{dz}\, dz \tag{18}$$

and the concentration is

$$c(z_o,t) = \frac{q_s}{-w_o} = \frac{1}{-w_o}\int_{z_o}^{\infty} p(t - \frac{z-z_o}{w_o} - \frac{z}{w_c})\frac{dF}{dz}\,dz \qquad (19)$$

For most functions $p(t)$ and $F(z)$ this integral is easy to calculate numerically compared to solving a time dependent diffusion equation. This pure convection solution, which (given $p(t)$) is easily applicable to irregular waves, therefore provides a very attractive alternative to the traditional diffusion approach. Particularly for waves over flat beds where the pure convection model is often a good representation of the physics.

For the inverse problem of finding $p(t)$, from known $c(z_o,t)$ and $F(z)$ we must solve (19) as an integral equation in $p(t)$, and this may sometimes be difficult. The mathematical difficulties are however vastly reduced if $p(t)$ and $c(z_o,t)$ can be considered in terms of Fourier series.

Further simplification is achieved if the entrained sand is assumed to move instantaneously from the bed to z_e, cf Nielsen, 1988, and this may in fact often be acceptable for practical modelling.

6.3 Fourier series solutions for pure convection

The application of equation (19) is made somewhat simpler when the process can be assumed time periodic so that both $p(t)$ and $c(z_o,t)$ can be written in terms of Fourier series:

$$c(z_o,t) = \sum_{n=o}^{\infty} C_n(z_o)\, e^{in\omega t} \quad \text{and} \quad p(t) = \sum_{n=o}^{\infty} P_n\, e^{in\omega t} \qquad (20)$$

In that case, Equation (19) boils down to the following equation between the Fourier coefficients

$$C_n(z_o) = -\frac{P_n}{w_o}\int_{z_o}^{\infty} e^{-in\omega(\frac{z-z_o}{w_o} + \frac{z}{w_c})}\, F'(z)\,dz \qquad (21)$$

which may be solved directly for either $C_n(z_o)$ or P_n, which ever is the unknown.

For the particular case of $F(z) = \exp(-z/L)$ and correspondingly $F'(z) = \frac{-1}{L}\exp(-z/L)$ we find

$$C_n(z_o) = C_n\, e^{-\beta_n z_o/L} \qquad (22)$$

where

$$\beta_n = 1 + i\frac{n\omega L}{w_c} \qquad (23)$$

and

$$C_n \;=\; C_n(0) \;=\; \frac{P_n}{w_o}\frac{\beta_n}{\beta_n + i\frac{n\omega L}{w_o}} \tag{24}$$

This solution was also obtained by Nielsen, 1991, 1992 by solving Equation (7) with $\varepsilon_s \equiv 0$ and $F(z) = \exp(-z/L)$.

The solution (22) can be written as

$$c_n(z,t) \;=\; |C_n|\, e^{-z/L} \cos[n\omega(t-z/w_c) + Arg\{C_n\}] \tag{18}$$

and this expression illustrates that all the harmonics of the pure convection solution decay at the same rate (as $F(z)$ in general, and as $exp(-z/L)$ in this particular case), and that all time lags grow at the same rate with z. Compared with the diffusion solution (15)+(16), this corresponds to a greater shape similarity between concentration time series at different elevations for the convective case. See also Nielsen 1992b.

Based on the pure convection solution (22) it is possible to extract the vertical scale L from measured concentrations provided an exponential curve can be fitted with reasonable confidence to the shape of $|c_n(z,t)|$. This has been done for the lower, almost exponential part of the concentrations measured by Ribberink & Al Salem, 1993 and the results are plotted in Figure 4. This figure shows that the estimates of L are consistent for all harmonics. Hence, as far as the comparison in Figure 4 goes, it may be concluded that the pure convection model is a better model of $c(z,t)$ over flat beds under waves than the pure diffusion model. This agrees of course with the observation of Nielsen, 1992, quoted in relation with (9), that $\bar{c}(z)$ is well described by a pure convection solution under such conditions.

Admittedly, the exponential shape of $|c_n(z,t)|$ which is assumed for the comparison in Figure 4 is not a perfect fit to the complete profiles, Ribberink & Al Salem, 1993 advocate the shape of z^{-2} for $\bar{c}(z)$ and Nielsen, 1992 suggests Equation (10). However, the fit is acceptable over the lowest few centimetres, and the assumption of different forms of $\varepsilon_s(z)$ and $F(z)$ wouldy not change the qualitative conclusion of the comparison in Figure 4. It would however make the algebra unreasonably complicated for illustrating the present point.

7 Conclusion

The problems with applying a pure gradient diffusion approach to sediment suspensions have been reviewed in Section 1. It was shown that sediment diffusivities derived from steady flow data through Equation (3) differ from the eddy viscosity of the flow in ways that are not explicable on the basis of theory and dispersion experiments. Those apparent diffusivities from steady flows are generally found to be greater than ν_t while the *crossing trajectories effect* and the *inertia effect* should both make them smaller. The amount of data is still limited however. The data of Coleman, 1970 on which the above is based should therefore be supplemented with experiments where suction samples from different elevations are sieved so that concentration profiles for different sand sizes in exactly the same flow can be plotted.

This has been done for waves over both rippled beds and flat beds. For the rippled beds, the results show not only the greater diffusivities for bigger particles which were shown by Coleman's data. For natural sand beds with unlimited supply, it is also clear that the diffusivities required for different sand sizes would have to have different shapes. In order to model the correspondingly different $\bar{c}(z)$ profiles for different size fractions a combined convection diffusion approach is necessary.

For oscillatory flows over flat beds, the combined approach is often not necessary. If the shape parameter $w_o T/\sqrt{r A}$ is sufficiently large, a pure convection approach is adequate. This is very convenient because the pure convection solutions for both $\bar{c}(z)$ and for $c(z,t)$ are much simpler to calculate than either pure diffusion solutions or combined convection diffusion solutions. Particularly for irregular waves.

Most of the existing sediment suspension models are still pure gradient diffusion models which apply the ad hoc β-function given by (4) to mimic the differences in diffusivity magnitude for different grain sizes. However, such models cannot possibly model the differences in profile shape which have been observed over natural ripples under waves. Whether these differences are significant in rivers remains to be seen. The most likely places to find them would be where large scale mixing or convection in the stile of Figure 3 coexists with significant small scale mixing. We note however, that similar distributions for all grain sizes does not necessarily mean that a pure gradient diffusion approach is the most appropriate. As in the case of oscillatory sheet flow it may well be that a pure convection model is preferable. If it is physically more realistic, it will also generally be much simpler.

Pure diffusion models with time dependent diffusivity have not been discussed above because they are all necessarily numerical. Such models have been used to give fairly realistic time histories of concentrations over flat beds under waves. However, as the measurements of Ribberink & Al Salem show the different grain sizes to have identical slopes $(\frac{d \ln \bar{c}}{dz})$ in such flows, a β-function roughly proportional to w_o must be applied. Such a β-function corresponds, of course, to a pure convection model in the stile of Figure 3 and Equation (9) where all sand sizes show the same vertical scale and hence the same value of ε_s/w_o. Choosing the more complicated diffusion approach thus seems totally illogical in this case. Admittedly, some of the features which can be modelled using time dependent diffusivities cannot be modelled with the pure convection model described in Figure 3. Thus, in order to model the fact that sand which is entrained during the deceleration phase travels further upwards than sand which is entrained during acceleration (see the data of Ribberink & Al Salem 1992) requires the distribution function to depend on time i e, $F = F(z,t)$. However, even with this complication added, pure convection models are vastly simpler than diffusion models.

Horikawa et al, 1982 performed sheet flow experiments with shorter wave periods and hence smaller values of $w_o T/\sqrt{r A}$ than those of Ribberink & Al Salem and the $\bar{c}(z)$-distribution shapes vary with $w_o T/\sqrt{r A}$ in a fashion that indicates a transition from diffusion dominated for small *T* and convection dominated for lager *T* (3.6s<T<6s), see Nielsen, 1992 p 217. This complete picture can, like the situation over ripples, only be understood in terms of a combined convection diffusion model.

8 References

Bijker, E W, E van Hijum & P Vellinga (1976): Sand transport by waves. Proc 15th Int Conf Coastal Engineering, Hawaii, pp 1149-1167.

Csanady, G T (1963): Turbulent diffusion of heavy particles in the atmosphere. *J Atmospheric Sciences, Vol 20*, pp 201- 208.

Coleman, N L (1970): Flume studies of the sediment transfer coefficient. *Water Resources Research, Vol 6, No 3*, pp 801-809.

Horikawa K, A Watanabe & S Katori (1982): Sediment transport under sheet flow condition. *Proc 18th Int Conf Coastal Eng*, Capetown, pp 1335-1352.

McFetridge, W F & P Nielsen (1985): Sediment suspension by non-breaking waves over rippled beds. *Tech Rep COEL- 85/005*, Coastal and Oceanographical Eng Dept, Univ of Florida, 132 pp.

Nakato, T, F A Locher, J R Glover and J F Kennedy (1977): Wave entrainment of sediment from rippled beds. *Proc ASCE, Vol 103 (WW1)*, pp 83-100.

Nielsen, P (1979): *Some basic concepts of wave sediment transport.* Series paper 20, Institute of Hydrodynamics and Hydraulic Engineering (ISVA), Technical Univ Denmark.

Nielsen, P (1983): Entrainment and distribution of different sand sizes in the same flow. *J Sedimentary Petrology, Vol 53, No 2*, pp 423-428.

Nielsen, P (1988): Three simple models of wave sediment transport. *Coastal Engineering, Vol 12*, pp 43-62.

Nielsen, P (1991): Combined convection and diffusion: A new framework for suspended sediment modelling, *Proc "Coastal Sediments '91"*, Seattle, pp 418-431.

Nielsen, P (1992): *Coastal bottom boundary layers and sediment transport.* World Scientific, Singapore, 324 pp.

Nielsen, P (1992b): Combined convection diffusion modelling of sediment entrainment. *Proc 23rd Int Conf Coastal Engineering, Venice*, pp 3202-3215.

Nielsen, P (1993): Turbulence effects on the settling of suspended particles. *J Sedimentary Petrology, Vol 63, No 5*, pp 835-838.

Nielsen, P, I A Svendsen & C Staub (1978): Onshore-offshore sediment transport on a beach. *Proc 16th Int Conf Coastal Eng*, Hamburg, pp 1475-1492.

van de Graaff, J (1988): *Sediment concentration due to wave action.* Dr Eng Thesis, Dept of Civil Engineering, Delft University of Technology.

Ribberink, J S & A Al-Salem (1992): Time dependent sediment transport phenomena in oscillatory boundary layer flow. *Delft Hydraulics Tech Rep no 840.20 Part VI.*

Ribberink, J S & A Al-Salem (1993): Sediment transport of fine sand under asymmetric oscillatory flow. *Delft Hydraulics Tech Rep no 840.20 Part VII.*

Snyder, W H & J L Lumley (1971): Some measurements of particle velocity autocorrelation functions in a turbulent flow. *J Fluid Mech Vol 48, Part 1,* pp 41-71.

Wells, M R and D E Stock (1983): The effects of crossing trajectories on the dispersion of particles in a turbulent flow. *J Fluid Mech Vol 136,* pp 31-62.

Yudine, M I (1959): Physical considerations on heavy particle diffusion. *Atmospheric diffusion and air pollution, Advances in Geophysics, Vol 6,* New York, Academic press, pp 185- 191.

BED LOAD TRANSPORT IN OSCILLATORY FLOW

J.F.A. SLEATH
Cambridge University Engineering Department,
Trumpington Street, Cambridge, England

ABSTRACT

A new model is proposed for bed load transport in oscillatory flow. The model is based on revised versions of the constitutive equations suggested by Bagnold (1954) and others. Expressions are derived for the thickness of the moving layer of sediment over a plane bed and for the sediment transport rate.

1. Introduction

Several formulae have been proposed for sediment transport by waves (e.g. Madsen and Grant, 1976, Sleath, 1978, Horikawa et al, 1982, Sawamoto and Yamashita, 1986, etc.) but all are basically empirical and none has achieved widespread acceptance. The aim of this paper is to attempt to derive a new expression for sediment transport from first principles.

2. Basic Assumptions

Bagnold (1954) carried out an extensive investigation into the shearing of fluid/ sediment mixtures. He observed two different flow regimes: "viscous" flow at low rates of shear and "inertial" flow at higher shear. Subsequent investigations have been carried out by Savage and McKeown (1983) and Hanes and Inman (1985). These suggest that the detailed formulae for particle-particle stress proposed by Bagnold should be treated with caution. However, there is agreement on the existence of the two flow regimes and on the general form of the expressions for particle-particle stress in each case. We thus adopt the following generalised expressions for particle-particle shear stress T. For "viscous" flow:

$$T = \mu f_1\left(\frac{C}{C*}\right)\frac{du}{dy}, \tag{1}$$

and for "inertial" flow:

$$T = \rho_s f_2\left(\frac{C}{C*}\right)D^2\left(\frac{du}{dy}\right)^2, \tag{2}$$

where C is concentration, C_* is the limiting concentration of the sediment, ρ_s is the dry density of the sediment, D is grain size, μ is the dynamic viscosity of the fluid and y is measured vertically upwards from the bottom of the mobile layer.

Our second assumption concerns the velocity and concentration profiles. The parameter $\rho U_0\omega/(\rho_s-\rho)g$ may be thought of as a measure of the relative importance of pressure gradient and gravitational forces acting on the sediment. At low values of this parameter the mobile layer of sediment is relatively thin, the effect of pressure gradient is negligible and the flow is driven by the shear stress exerted by the fluid above. We will refer to this situation as the "quasi-steady" regime. Under these conditions the velocity in the mobile layer varies from zero at the bottom of the layer to some reference velocity u_m at the top, while the concentration of sediment falls from C_* down to zero. Consequently, it is reasonable to assume velocity and concentration profiles of the form

$$\frac{u}{u_m} = f_3\left(\frac{y}{\delta}\right), \tag{3}$$

$$\frac{C}{C_*} = f_4\left(\frac{y}{\delta}\right), \tag{4}$$

where

$$\delta = \frac{1}{C_*}\int_{y=0}^{\infty} C dy. \tag{5}$$

On the other hand, when $\rho U_0\omega/(\rho_s - \rho)g$ is large the effect of pressure gradient on the flow in the mobile layer cannot be ignored. We will refer to this as the "pressure-gradient" regime. Although the mobile layer is no longer thin Dick and Sleath (1991) found that it is still possible to make use of similarity profiles for the velocity amplitude and phase. The same is probably true of the concentration profile. Consequently, we can again make use of Eqns. 3 and 4 although the functions f_3 and f_4 may be different from those for the quasi-steady regime.

Finally, following Bagnold, Savage and McKeown, Hanes and Inman and others we assume

$$T = KP, \tag{6}$$

where T and P are, respectively, the tangential and normal stresses, due to particle-particle interactions and K is the dynamic analogue of the static friction coefficient.

In the following analysis we will only make use of Eqns. 1,2 and 6 at $y = 0$ (the bottom of the moving layer of sediment). Thus K in Eq. 6 is not assumed to be constant throughout the flow and the dependence of T on du/dy, etc may differ from Eqns. 1 and 2 at other positions.

3. Thickness Of Moving Layer In Sheet Flow

3.1 Equilibrium Flow

We start by considering the situation where the distribution of sediment in the mobile layer is in equilibrium with the flow. Under these circumstances the normal stress due to particle-particle interactions will be equal to the weight of sediment above. Thus, at the bottom of the mobile layer

$$P = \int_{y=0}^{\infty} (\rho_s - \rho) g C dy, \tag{7}$$

where ρ is fluid density. Hence (from Eq. 5)

$$P = (\rho_s - \rho) g C_* \delta, \tag{8}$$

From Eq. 6

$$T = K(\rho_s - \rho) g C_* \delta, \tag{9}$$

Making use of the momentum equation, we can write the shear stress τ at any height y above the bottom of the mobile layer as

$$\tau = \tau_0 - \int_y^{\delta} (\rho_m \frac{\partial u}{\partial t} - \rho \frac{\partial u_\infty}{\partial t}) dy \tag{10}$$

Here ρ_m is the density of the fluid/sediment mixture, τ_o is the shear stress at the initial bed level ($y = \delta$), u_∞ is the value of u outside the bed boundary layer and t is time. We assume that

$$u_\infty = U_o \sin(\omega t), \tag{11}$$

where ω is angular frequency and U_o is a constant. Since C/C_* scales with y/δ (Eq. 4), ρ_m/ρ also scales with y/δ. Consequently, if u is simple harmonic throughout the mobile layer

$$u_m = R\{U_m \exp(i\omega t)\}, \tag{12}$$

where $R\{\}$ indicates the real part. Also, since ρ_m/ρ is a function of C and ρ_s/ρ and since u is simple harmonic it follows From Eqns. 3 and 4 that for given ρ_s/ρ

$$\int_0^{\delta} \rho_m \frac{\partial u}{\partial t} dy = const\ \rho U_m \omega \delta \cos(\omega t + \phi_1), \tag{13}$$

where ϕ_1 is a constant. When sediment concentration is low a significant proportion of the shear stress τ is transmitted through the fluid rather than by particle-particle interactions. However, at the bottom of the mobile layer the concentration of sediment is high. Consequently, for the moment, we assume that at the bottom of the mobile layer the particle-particle stress is much greater than the stress transmitted through the fluid. In other words

$$T = \tau. \tag{14}$$

From Eqns. 10, 11, 13, 14

$$T = \tau_0 + B \rho U_0 \omega \delta \cos(\omega t + \phi_2), \tag{15}$$

where ϕ_2 is a constant and

$$B = 1 - const\ (U_m/U_0). \tag{16}$$

The amplitude of the shear stress τ_o may be written in terms of the friction factor f_w as follows

$$\tau_{0m} = f_w \frac{1}{2} \rho U_0^2. \tag{17}$$

Assuming that τ_o varies with simple harmonic motion and that the phase shift between it and the second term on the right of Eq. 15 is ϕ, we obtain from Eqns. 9, 15 and 17

$$K(\rho_s - \rho) g C_* \delta_s = ((f_w \frac{1}{2} \rho U_0^2)^2 + (B \rho U_0 \omega \delta_s)^2 + 2(f_w \frac{1}{2} \rho U_0^2)(B \rho U_0 \omega \delta_s))^{1/2} \tag{18}$$

where δ_s is the maximum thickness of the mobile layer (measured from the still bed up to the initial bed level). This equation may be re-written as

$$\alpha^2 \left(B^2 (\frac{\delta_s}{a f_w})^2 + 0.25 + B(\frac{\delta_s}{a f_w}) \cos(\phi) \right) = (K C_*)^2 (\frac{\delta_s}{a f_w})^2 \tag{19}$$

where a is the semi-orbital excursion of the fluid particles outside the bed boundary layer and

$$\alpha = \frac{\rho U_0 \omega}{(\rho_s - \rho) g} \tag{20}$$

3.2 The Reference Velocity u_m In "Viscous" Flow

Differentiating Eq. 3 and substituting into Eq. 1 we obtain

$$T = \mu f_1 \left(\frac{C}{C_*} \right) \frac{u_m}{\delta} f_3'(\frac{y}{\delta}). \tag{21}$$

Thus, making use of Eq. 4,

$$T = \frac{\mu u_m}{\delta} f_5(\frac{y}{\delta}). \tag{22}$$

Substituting into Eq. 9 for $y = 0$ and considering only the condition where mobile layer thickness is maximum

$$\frac{\mu U_m}{\delta_s} = const\ K(\rho_s - \rho) g C_* \delta_s. \tag{23}$$

3.3 The Reference Velocity u_m In "Inertial" Flow

Substituting Eqns. 3 and 4 into Eq. 2 we obtain

$$T = \rho_s D^2 (\frac{u_m}{\delta})^2 f_6(\frac{y}{\delta}). \tag{24}$$

Substituting Eq. 24 into Eq. 9 for $y = 0$, making use of Eq. 12, and considering maximum δ, we obtain

$$\rho_s D^2 (\frac{U_m}{\delta_s})^2 = const\, K(\rho_s - \rho) g C_* \delta_s. \tag{25}$$

Substituting for U_m in Eq. 16 we find

$$B = 1 - A\, (\frac{\delta_s}{a f_w})^{3/2} (\alpha)^{-1/2}. \tag{26}$$

where

$$A = const\, \left(KC_* \frac{\rho}{\rho_s} (\frac{a}{D})^2 f_w^3 \right)^{1/2}. \tag{27}$$

In view of the restrictions on Eq. 13 it should be borne in mind that the *const* in Eq. 27 is only constant for given ρ_s/ρ.

3.4 Measurements Of Mobile Layer Thickness

Fig 1 shows the curves obtained from Eqns. 19 and 26 for various values of the parameter A. Computed curves are only shown for "inertial" flow because all test results are for this condition. The curves for "viscous" flow are very similar. The friction factor f_w was calculated from Jonsson's (1963) formula with Nikuradse roughness $k_s = 2D$.

The value of KC_* on the right of Eq. 19 has been taken as 0.34 and the value of ϕ as 68.5°. The reason for choosing this value of ϕ is that Dick and Sleath (1991) found that ϕ_2 was approximately 90° at the bottom of the mobile layer and the results of Sleath (1992) suggest that for fully turbulent flow the shear stress τ_o leads the free stream velocity by about 22.5°. In fact, the curves are very little different for values of ϕ down to zero. The value of KC_* was chosen to fit the results of Sawamoto and Yamashita in Fig 1. However, Hanes and Inman suggest that typical values of K range from about 0.5 to 0.6 and a typical value of C_* might be about 0.64 so the chosen value of KC_* is in reasonable agreement with what might have been anticipated.

We note that because of the shape of the curve of f_w versus relative roughness a/k_s the parameter A is almost independent of flow conditions. In addition, ρ_s/ρ does not usually vary widely. Consequently, we expect the parameter A to be nearly constant, at any rate for a given sediment. The results in Fig 1 suggest a value for A of about 0.02. There is clearly a need for more experimental data. When comparing data it is important to use the same definition for mobile layer thickness. The present analysis is based on the shearing of grains over each other. Consequently, the bottom of the mobile layer for present purposes is the lowest level at which grains shear over each other. This is probably also the criterion adopted by Sawamoto and Yamashita. In oscillatory flow some slight rocking of grains about a mean position may be observed below this level. If the bottom of the mobile layer were taken as the level at which even slight rocking disappeared the values obtained for δ_s would be

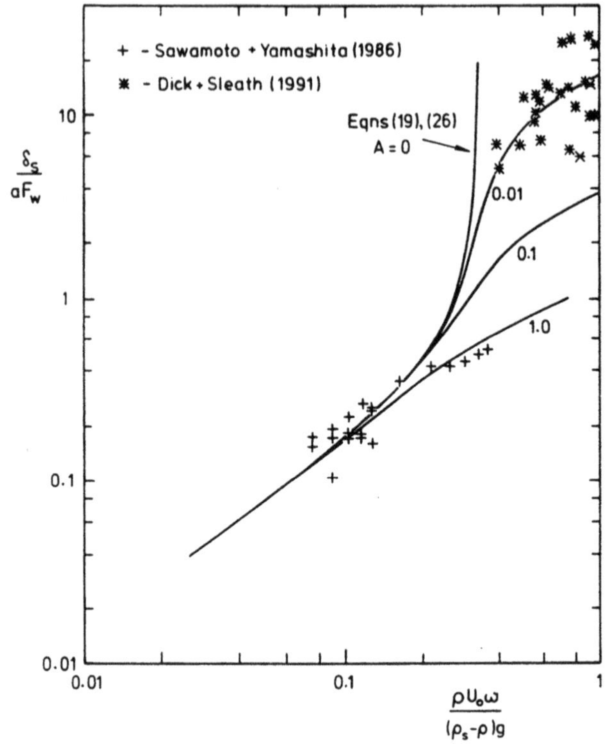

Fig 1. Variation of mobile layer thickness with $\rho U_0\omega/(\rho_s - \rho)g$

larger.

Fig 1 shows that at values of α less than about 0.3 the experimental results are quite well approximated by the straight line

$$\frac{\delta_s}{af_w} = 1.47\,\alpha \tag{28}$$

This may be re-written as

$$\frac{\delta_s}{D} = 2.94\frac{\tau_{0m}}{(\rho_s - \rho)gD}. \tag{29}$$

This result is based on the curves for "inertial" flow but exactly the same equation is obtained for small values of α for "viscous" flow. We could have obtained a result of this form directly if we had assumed that T was equal to the shear stress τ_0 acting on the upper surface of the mobile layer and that the influence of the pressure gradient and inertial terms was negligible. The dominance of τ_0 is due to the fact that for these low values of α the moving bed layer is relatively thin.

Finally, it was assumed in the derivation of Eq. 7 that the distribution of sediment was in equilibrium with the flow. Although this appears to be a reasonable approximation in the quasi-steady regime it is not necessarily true at each instant throughout the cycle in the pressure-gradient regime: normal stress may be greater than immersed weight at some moments and less at others. However, even in the pressure-gradient regime, it is still the case that the mean value of normal stress, averaged over a half cycle, must equal the mean value of immersed weight. So the above analysis still applies in terms of cycle-mean quantities.

4. Sediment Transport In The Quasi-Steady Regime

4.1 "Viscous" Flow

The sediment transport rate at any instant is

$$q_s = \int_0^\infty uC\,dy. \tag{30}$$

Substituting from Eqns. 3, 4 and 23 into Eq. 30

$$(q_s)_{max} = const\ \frac{1}{\mu}\delta_s^3(\rho_s-\rho)gC_*K\int_0^\infty f_7(\frac{y}{\delta_s})d(\frac{y}{\delta_s}), \tag{31}$$

$$(q_s)_{max} = const\ \left(\frac{\delta_s}{D}\right)^3\frac{(\rho_s-\rho)}{\rho\nu}gD^3, \tag{32}$$

where ν is kinematic viscosity and we have taken C_* and K to be constant. Making use of Eq. 29, the mean sediment transport rate (averaged over a half cycle) is given by

$$\frac{\rho\nu\overline{q_s}}{(\rho_s-\rho)gD^3} = const\left(\frac{\tau_{0m}}{(\rho_s-\rho)gD}\right)^3, \tag{33}$$

4.2 "Inertial" Flow

Making use of Eqns. 3, 4 and 25 in Eq. 30

$$(q_s)_{max} = const\ \delta_s^{5/2}\left(\frac{(\rho_s-\rho)gC_*}{\rho_sD^2}\right)^{1/2}\int_0^\infty f_8(\frac{y}{\delta_s})d(\frac{y}{\delta_s}), \tag{34}$$

$$(q_s)_{max} = const\left(\frac{\delta_s}{D}\right)^{5/2}\left(\frac{(\rho_s-\rho)gD^3}{\rho_s}\right)^{1/2}, \tag{35}$$

where, once again, we have taken C_* and K to be constant. Making use of Eq. 29, the mean sediment transport rate is

$$\overline{q_s}\left(\frac{\rho_s}{(\rho_s-\rho)gD^3}\right)^{1/2} = const\left(\frac{\tau_{0m}}{(\rho_s-\rho)gD}\right)^{5/2} \tag{36}$$

4.3 The Transition Between "Viscous" And "Inertial" Flow

Eqns. 33 and 36 may both be written in the form

$$\frac{\overline{q_s}}{\nu}\left((\frac{\rho_s}{\rho})^{3/2}\frac{(\rho_s-\rho)gD^3}{\rho\nu^2}\right)^2 = f\left((\frac{\rho_s}{\rho})^{1/2}\frac{u_*D}{\nu}\right) \tag{37}$$

where $u_* = (\tau_{0m}/\rho)^{1/2}$. For "inertial" flow

$$f\left((\frac{\rho_s}{\rho})^{1/2}\frac{u_*D}{\nu}\right) = const\left((\frac{\rho_s}{\rho})^{1/2}\frac{u_*D}{\nu}\right)^5 \tag{38}$$

and for "viscous" flow

$$f\left((\frac{\rho_s}{\rho})^{1/2}\frac{u_*D}{\nu}\right) = const\left((\frac{\rho_s}{\rho})^{1/2}\frac{u_*D}{\nu}\right)^6 \tag{39}$$

We see that the transition between the two regimes is a function of $(\rho_s/\rho)^{1/2}u_*D/\nu$. This is consistent with Bagnold's (1954) finding that the transition was a function of $G^2 = (\rho_s D^2 P/\mu^2\lambda)$ where λ is linear concentration. From Eq. 8 the value of Bagnold's parameter at the bottom of the mobile layer is

$$\left(\frac{\rho_s D^2 P}{\mu^2\lambda}\right) = \left((\frac{\rho_s}{\rho})\frac{(\rho_s-\rho)gD^2C_*\delta}{\rho\nu^2\lambda_{max}}\right) \tag{40}$$

Assuming that C_* and λ_{max} have fixed values at the bottom of the mobile layer and making use of Eq. 29 we obtain

$$\left(\frac{\rho_s D^2 P}{\mu^2\lambda}\right) = const\left((\frac{\rho_s}{\rho})^{1/2}\frac{u_*D}{\nu}\right)^2 \tag{41}$$

4.4 Measurements Of Sediment Transport

Fig 2 shows how the theory compares with existing experimental results for $\alpha < 0.3$.

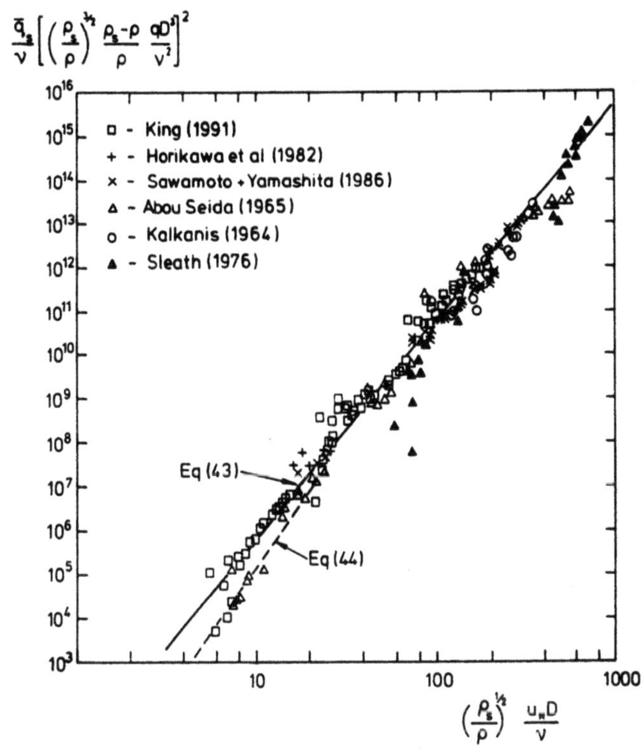

Fig 2. Variation of sediment transport rate with $(\rho_s/\rho)^{1/2}u_*D/\nu$

me of the measurements were made in oscillating trays and some in water tunnels. here does not seem to be any significant difference between the two sets of results. his is because the additional inertia effects associated with oscillating trays only come significant at larger values of α. Almost all of the experimental results in g 2 appear to correspond to "inertial" flow (Eqns. 37 and 38) but at low values $(\rho_s/\rho)^{1/2}u_*D/\nu$ there is a trend towards the curve (Eqns. 37 and 39) predicted : viscous flow. The solid line (for inertial flow) corresponds to

$$\frac{\overline{q_s}}{\nu}\left(\frac{(\rho_s-\rho)gD^3}{\rho\nu^2}\right)^2 = 5.67\left(\frac{\rho}{\rho_s}\right)^{1/2}\left(\frac{u_*D}{\nu}\right)^5 \tag{42}$$

and the broken line (viscous flow) to

$$\frac{\overline{q_s}}{\nu}\left(\frac{(\rho_s-\rho)gD^3}{\rho\nu^2}\right)^2 = 0.14\left(\frac{u_*D}{\nu}\right)^6 \tag{43}$$

It is suggested that Eq. 42 might be used for $(\rho_s/\rho)^{1/2}u_*D/\nu > 10$ and Eq. 43 for smaller values of $(\rho_s/\rho)^{1/2}u_*D/\nu$.

The fact that the transition between "viscous" and "inertial" flow depends on $(\rho_s/\rho)^{1/2}u_*D/\nu$ invites comparison with Shields curve. Taylor and Vanoni (1972) showed that the curves obtained by different investigators for initial motion depended on what level of sediment transport they were adopting as their criterion. Eqns. 33 and 36 may be re-written in the form

$$\frac{\tau_{0m}/(\rho_s-\rho)gD}{(\rho_s\overline{q_s}^2/(\rho_s-\rho)gD^3)^{1/5}} = f\left((\frac{\rho_s}{\rho})^{1/2}\frac{u_*D}{\nu}\right) \tag{44}$$

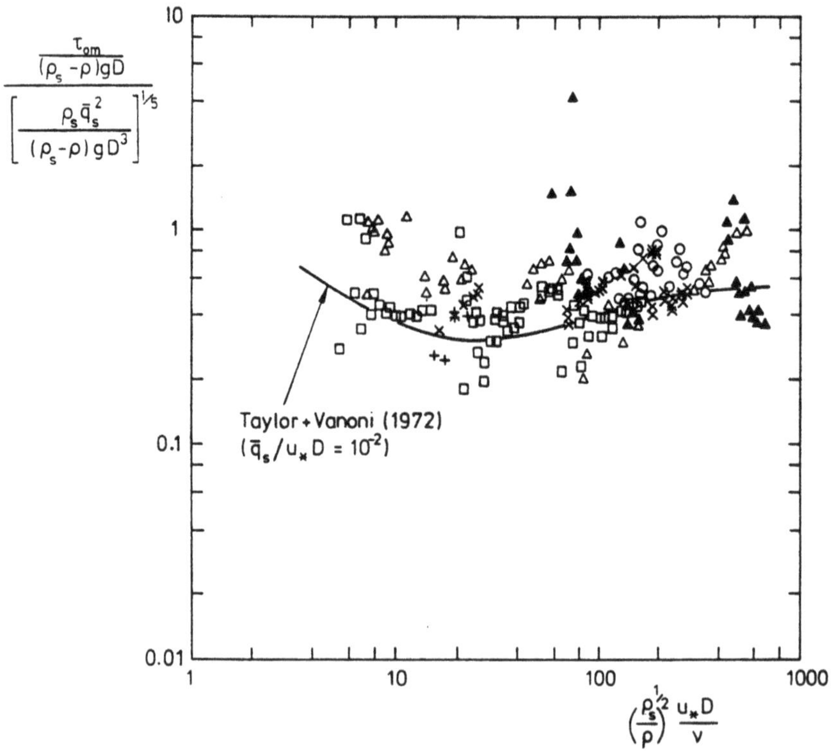

Fig 3. Variation of modified Shields parameter with $(\rho_s/\rho)^{1/2}u_*D/\nu$

The left hand side of Eq. 44 may be thought of as a form of Shields parameter modified to allow for sediment transport. Since the variation in $(\rho_s/\rho)^{1/2}$ is small, the right hand side is almost the same as a function of u_*D/ν. Fig 3 shows how the oscillatory flow measurements compare with Taylor and Vanoni's steady flow results for $q_s/u_*D = 0.01$. Despite the scatter in the oscillatory flow results the trends are similar. In other words, Bagnold's "inertial" flow is just the high u_*D/ν regime for which Shield's parameter is constant, while his "viscous" flow corresponds to small values of u_*D/ν for which Shield's parameter rises with decreasing u_*D/ν.

4.5 Low Sediment Transport Rates.

We have assumed that the stress transmitted through the fluid is negligible in comparison with the particle-particle stress. Although that assumption is justified at high sediment transport rates, the particle-particle stress clearly ceases to be dominant as sediment transport rate tends to zero. This effect is clearly shown by the measurements of Sleath (1976) in Fig 2: at high values of $(\rho_s/\rho)^{1/2}u_*D/\nu$ the points follow the theoretical curve quite well but at lower values (as τ_0 approaches τ_c) the measured sediment transport falls significantly below the predicted value. Bagnold (1956) suggested that we could allow for the stress transmitted through the fluid by writing

$$T = \tau - \tau_c \tag{45}$$

instead of Eq. 14. However, it seems improbable that at very high sediment transport rates the fluid stress at the bottom of the mobile layer remains equal to τ_c. There is clearly a need for more experimental data. In the meantime it is suggested that in the formulae derived for mean sediment transport we replace τ_{0m} by $\tau_{0m} - \tau_c$ when $\tau_{0m}/(\rho_s - \rho)gD$ is less than about 0.1, say.

4.6 The Calculation Of Friction Factor f_w For Rippled Beds

Some of the scatter in Figs 2 and 3 is due to the neglect of the shear stress transmitted through the fluid, as discussed in the previous section. Another reason for the scatter is uncertainty over the value of f_w. As mentioned earlier, this has been calculated from Jonsson's (1963) formula for flat beds. However, some of the results in Figs 2 and 3 were probably for rippled beds. If the bed is rippled the horizontal force on the bed is increased. Most of this increase is due to flow separation. (The free-stream pressure gradient produces a positive force on the ripple crests but a negative force on the troughs). This increase mainly affects the normal stress on the bed. When the thickness of the mobile layer is small compared with ripple height this normal stress is carried directly by the stationary sediment below the mobile layer. Consequently, under these circumstances it is probably correct to use the flat bed shear stress in the various formulae. This is, of course, the usual practice in calculations of sediment transport in steady flow. On the other hand, when the moving layer thickness is relatively large the increased stress is carried by the moving sediment. Under these circumstances ripple formation would increase sediment transport rate. However, this situation corresponds to the transition to sheet flow regime where ripples are relatively unimportant. Consequently, use of the flat bed formula for f_w may still be valid.

5. The "Pressure-Gradient" Regime

So far, the analysis has been based on the assumptions of the quasi-steady regime (low α). As mentioned earlier, similar assumptions apply to the pressure-gradient

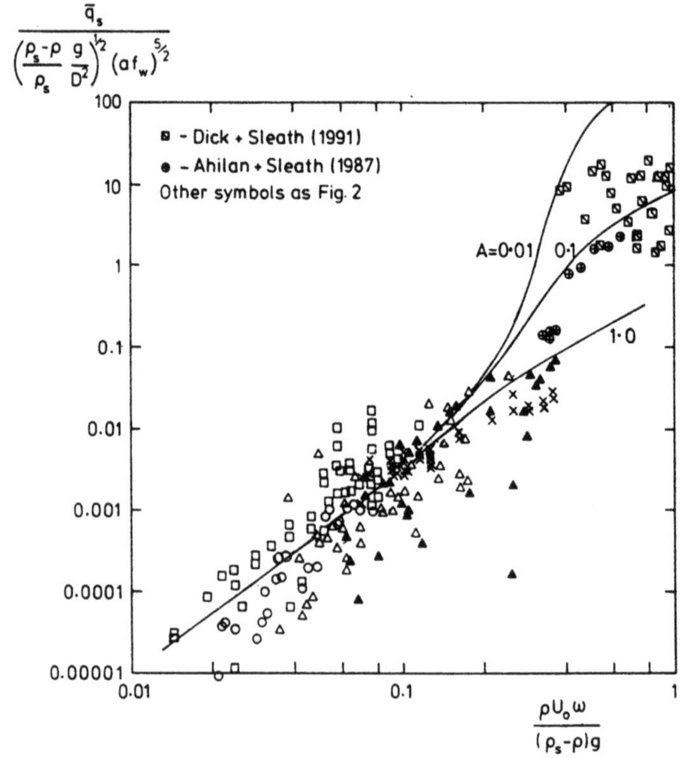

Fig 4. Variation of sediment transport rate with $\rho U_0 \omega/(\rho_s - \rho)g$

regime (high α) if we work in terms of cycle-mean rather than instantaneous parameters. Consequently, the general form of the solution is the same for each regime. However, since the various functions of y/δ are not the same the numerical constants obtained from the integration or differentiation of these functions are different in the two regimes. This means that if we just apply the quasi-steady curve to the whole range of α the best-fit value of A will not necessarily be the same in different applications. This is illustrated by the results shown in Fig 4. The theoretical curves were obtained by substituting the value of δ_s given by Eqns. 19 and 26 into Eq. 35. The best fit value of A is closer to 0.1 than the value of 0.02 which gave the best fit for mobile layer thickness in Fig 1.

Ideally, we ought to determine the best-fit curve for high α separately and then match it smoothly to the low α curve. However, there is too much scatter in the experimental data to allow us to do this. For the time being it is probably best to use the quasi-steady solution throughout (with the appropriate value of A). Since the shape of the curves is similar in the two regimes the error is probably not large.

6. Summary And Conclusions

(1) Generalised forms of Bagnold's constitutive equations have been used to obtain expressions for the mobile layer thickness and sediment transport rate. It should be emphasized that these constitutive equations (and Eq. 6) are only applied at the bottom of the mobile layer ($y = 0$); it is not necessary for them to hold throughout the layer.

(2) We identify two regimes of flow. At values of α ($= \rho U_0 \omega/(\rho_s - \rho)g$) less than about 0.3 the mobile layer of sediment is relatively thin and the flow is dominated by the shear stress exerted on the mobile layer by the fluid above. We call this the "quasi-steady" regime. At higher values of α the thickness of the mobile layer is much greater and pressure gradient and inertia forces are no longer negligible. We call this the "pressure-gradient" regime.

(3) For the quasi-steady regime ($\alpha < 0.3$) mobile layer thickness is given by Eq. 28 or Eq. 29, and mean sediment transport rate is given by Eq. 42 for $(\rho_s/\rho)^{1/2}u_*D/\nu >$ 10 and by Eq. 43 for $(\rho_s/\rho)^{1/2}u_*D/\nu < 10$.

(4) For the pressure-gradient regime ($\alpha > 0.3$) mobile layer thickness is given by Eqns. 19 and 26 with $A = 0.02$ and mean sediment transport rate is given by Eq. 35 and Eqns. 19 and 26 with $A = 0.1$. These value of A are only approximate. There is a need for further experimental data in this regime.

(5) At low sediment transport rates it is recommended that τ_{0m} be replaced by $\tau_{0m} - \tau_c$ in the various formulae.

(6) Bagnold (1954) observed different flow behaviour at high and low values of his parameter $G^2 = (\rho_s D^2 P/\mu^2 \lambda)$. It is suggested that, at the bottom of the mobile layer, Bagnold's parameter G is equal to a constant times $(\rho_s/\rho)^{1/2}u_*D/\nu$. In other words, the different flow behaviour is just that normally associated with high and low values of u_*D/ν.

7. References

1. M.M. Abou-Seida, Bed load function due to wave action, Rep. HEL 2-11, 78 pp., Hydraul. Eng. Lab., Univ. of Calif., Berkeley, 1965.
2. R.V., Ahilan, and J.F.A. Sleath, Sediment transport in oscillatory flow over flat beds, *J. Hydraul. Eng.*, Am. Soc. Civ. Eng., **113(3)**, 308-322, 1987.
3. R.A. Bagnold, Experiments on a gravity-free dispersion of large solid spheres in a Newtonian fluid under shear. *Proc. Roy. Soc. A.*, **225**, pp 49-63, 1954.
4. R.A. Bagnold, The flow of cohesionless grains in fluids. *Phil. Trans. Roy. Soc. A.*, **249**, pp 235-297, 1956.
5. J.E. Dick and J.F.A. Sleath, Velocities and concentrations in oscillatory flow over beds of sediment, *J. Fluid Mech.*, **233**, 165-196, 1991.
6. D.M. Hanes and D.L. Inman, Observations of rapidly flowing granular-fluid materials. *J.Fluid Mech.*, **150**, pp 357-380, 1985.
7. K. Horikawa, A. Watanabe, and S. Katori, Sediment transport under sheet flow conditions, *Proc. 18th Conf. Coastal Eng.*, 1335-1352, 1982.
8. I.G. Jonsson, Measurements in the turbulent wave boundary layer. *Proc. 10th Congress IAHR*, London, pp 85-92, 1963.
9. G. Kalkanis, Transportation of bed material due to wave action, *Tech. Memo. 2, Coastal Eng. Res. Cent.*, U.S. Army Corps of Eng., Washington, D.C., 1964.
10. D.B. King, Studies in oscillatory flow bedload sediment transport. Ph.D thesis, Univ. California, San Diego, 183 pp, 1991.
11. O.S. Madsen, and W.D. Grant, Sediment transport in the coastal environment, Rep. 209, Ralph M. Parsons Lab., Mass. Inst. of Technol., Cambridge, Mass., 1976.
12. S.B. Savage and S. McKeown, Shear stresses developed during rapid shear of concentrated suspensions of large spherical particles between concentric cylinders. *J. Fluid Mech.*, **127**, pp 453-472, 1983.
13. M. Sawamoto and T. Yamashita, Sediment transport rate due to wave action, *J. Hydrosci. Hydraul. Eng.*, **4(1)**, 1-15, 1986.
14. J.F.A. Sleath, Measurements of bed load in oscillatory flow. *Proc. ASCE. J. Waterway Port Coastal and Ocean Div.*, **104**, No. WW3, pp 291-307, 1978.
15. J.F.A. Sleath, The effect of jet formation on the velocity distribution in oscillatory flow over flat beds of sand or gravel. *Coastal Engineering*, **6**, pp 151-177, 1982.
16. B.D. Taylor and V.A. Vanoni, Temperature effects in low-transport flat-bed flows. *Proc. ASCE. J. Hyd. Div.*, **98**, No. HY8, pp 1427-1445, 1972.

NUMERICAL MODELLING OF THREE-DIMENSIONAL COHESIVE SEDIMENT TRANSPORT IN A ESTUARINE ENVIRONMENT

LEONOR CANCINO
Dept. of Mech. Engineering, Instituto Superior Técnico,
Av. Rovisco Pais, 1096 Lisboa, Portugal

and

RAMIRO NEVES
Dept. of Mech. Engineering, Instituto Superior Técnico,
Av. Rovisco Pais, 1096 Lisboa, Portugal

ABSTRACT

This paper describes the development, sensitivity analysis and application of a full three-dimensional finite- differences cohesive sediment transport model for coastal and estuarine environments. Circulation is described by the barotropic version of a three- dimensional hydrodynamic model and the transport of fine suspended matter is calculated solving the advection-diffusion equation. To quantify the exchange of cohesive material between the bed and the water column the algorithm incorporates the processes of flocculation, erosion, deposition and bed formation. An application of the model to the Western Scheldt estuary is made and provides considerable good agreement with field data.

1. INTRODUCTION

The study of cohesive sediment transport assumes great importance for the management and preservation of estuarine environments both in a direct and an indirect way. Directly because of its implication in morphological evolution, and indirectly because of the implications of light penetration and transport of chemicals on biological processes. The complexity of the processes and time scales involved in sediment transport turn the mathematical modelling into an important research tool to individualize processes and to perform sensitivity analysis. Mathematical modelling of sediment transport is an essential tool to predict morphological modifications due to *e.g.* coastal engineering works, allowing the study of different scenarios, and the prediction of consequent ecological impacts.

Following the works of Einstein and his collaborators (Einstein,[1]) describing the sediment dynamics at the bed-water interface, a large number of mathematical models have been developed. Presently depth-integrated (2D) and three-dimensional (3D) models are currently used for research and coastal engineering purposes.

The basic differences between the various numerical models are due to the number of dimensions considered, the numerical technique adopted and the complexity of the bed boundary conditions. The 2D models do not allow the description of vertical concentration profiles and its interaction with the vertical velocity profiles, but allow a finer horizontal grid. The three-dimensional models allow a more realistic simulation of the transport phenomena but require more computational time and larger amounts of input data.

In terms of numerical modelling the major distinction between suspended cohesive sediments and non-cohesive sediments lies in the calculation of the fall velocity of sediment particles (W_S), the exchanges across the sediment-water interface and the consideration of a bed compaction (Ariathurai and Krone,[2]; Odd,[3]; Nicholson,[4]; Hayter and Mehta,[5];Sheng,[6];Mehta et al,[7],[8]; Dyer and Evans,[9]; Van Rijn,[10]; Cancino and Neves,[11]). In the case of cohesive sediments (particles smaller than 62 μm) the effect of flocculation makes settling velocity a function of sediment concentration. For coarse particles W_S is closely related to the diameter of the particles, considered in general as spherical bodies to allow the use of Stokes' theory to compute the fall velocities.

Most of the existing cohesive sediment models are based on the assumption of a limiting critical value controlling the occurrence of erosion and deposition fluxes. To simulate the erosion fluxes the consolidation of deposited material must be known. In general this is done by solving an evolution equation for the bed density profile (Owen,[12]; Hayter and Mehta,[5]; Dyer,[13]; O'Connor and Nicholson,[14]; Le Hirt,[15]). Most of field work in sediment transport studies has been devoted to the characterization of sediment distributions, to flocculation and to critical shear stresses for erosion/ deposition. Nevertheless the bioturbation can be in certain areas an important factor on sediment dynamics (Mackin and Swider,[16]). A lacuna in cohesive models is due to the insufficient understanding of the role of biochemical processes on particles interaction and its ecological implications on the sediment dynamics. The need to quantify these processes allowing the incorporation of numerical models, should be a significant step towards a better understanding of the complex estuarine environment.

The aim of this work is to describe a numerical model developed to simulate the three-dimensional transport of cohesive sediments due to barotropic currents in a mesotidal estuary. The advective and diffusive fluxes are computed by the sediment transport model (SETMOD) using the 3D velocity field computed by the hydrodynamic model (TRID). The mass exchange balance between the water column and the bed due to erosion and deposition is performed, conducting to the evaluation of the newly deposited bed thickness.

The model is being used to simulate the sediment transport in the Western Scheldt estuary as part of a large inter-disciplinary research project. The final goal of this model is to simulate the sediment dynamics in the maximum turbidity zone, including baroclinic effects (Cancino and Neves,[11]).

2. MODELS DESCRIPTION

Hydrodynamics is the most important process involved in sediment transport, controlling the exchanges between the bottom and the water column and the horizontal transport. 3D models, computing the vertical velocity profile, allow the simulation of secondary flows originated by differential inertia and/or vertical density gradients and thus a better representation of the processes near the bottom. The hydrodynamic model used is a fully 3D-baroclinic model. It considers the hydrostatic and Boussinesq approximations, uses the vertical sigma coordinate with a staggered grid and a semi-implicit two-time level scheme (Santos and Neves,[17]). All the scalar values are computed in the center of the cell (Fig. 1), and the velocities are computed at their faces.

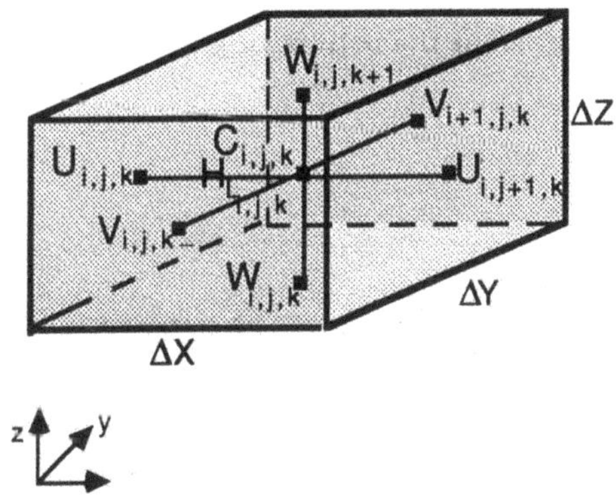

Fig. 1. Computational cell. U, V and W are the velocity components in x, y, z directions; C is the sediment concentration and H_L is the depth of the layer.

The nonlinear advective terms, horizontal diffusion and Coriolis term are solved explicitly. The model uses an implicit algorithm for the pressure terms and for the vertical diffusion. Horizontal viscosity is calculated using the Kolmogorov law and vertical viscosity is computed using a mixing length approach. At the artificial boundaries, the model uses a combination of open and specified boundaries allowing known external forcing to be imposed and the radiation of waves produced inside the modelling area. At the natural air-sea boundary the conditions are prescribed using known-atmospheric values. In addition to the momentum and continuity equations, the model can optionally solve two transport equations for salt and temperature and an equation of state to include the baroclinic effects.

The momentum and continuity equations in cartesian coordinates are:

$$\frac{\partial u}{\partial t}+\frac{\partial(uu)}{\partial x}+\frac{\partial(uv)}{\partial y}+\frac{\partial(uw)}{\partial z}-fv+\frac{1}{\rho}\frac{\partial p}{\partial x}-\frac{1}{\rho}\left(\frac{\partial \tau_{xx}}{\partial x}+\frac{\partial \tau_{yx}}{\partial y}+\frac{\partial \tau_{zx}}{\partial z}\right)=0 \quad (1)$$

$$\frac{\partial v}{\partial t}+\frac{\partial(vu)}{\partial x}+\frac{\partial(vv)}{\partial y}+\frac{\partial(vw)}{\partial z}+fu+\frac{1}{\rho}\frac{\partial p}{\partial y}-\frac{1}{\rho}\left(\frac{\partial \tau_{xy}}{\partial x}+\frac{\partial \tau_{yy}}{\partial y}+\frac{\partial \tau_{zy}}{\partial z}\right)=0 \quad (2)$$

$$\frac{\partial p}{\partial z}+\rho g=0 \quad (3)$$

$$\frac{\partial \xi}{\partial t}=-\int_{-h}^{\xi}\frac{\partial u}{\partial x}dz-\int_{-h}^{\xi}\frac{\partial v}{\partial y}dz \quad (4)$$

where: t - time; u, x, z - velocity components in u, x, z directions; f - Coriolis parameter; p- pressure; ρ - water density; τ- shear stress; g- acceleration of gravity; ξ - elevation of water surface.

As boundary condition at the bottom the friction shear stress is imposed:

$$\tau=\xi c_d\left|\bar{u}_+\right|\bar{u}_+ \quad (5)$$

$$c_d=k^2\left(\ln\frac{z_+}{z_\phi}\right)^{-2} \quad (6)$$

where: u_+ - horizontal velocity vector in x,y direction at some distance z_+ above the bottom; c_d - bottom drag coefficient; k - Von Karman constant and z_+ - physical roughness height.

The three-dimensional sediment transport model is a semi-implicit finite-difference model. The advection-diffusion equation is transformed to the new coordinate system in similarity with the sigma coordinate grid used by the hydrodynamic model. The advective and horizontal diffusive terms are solved explicitly, while the vertical diffusion and settling terms are solved implicitly for numerical stability reasons.At each time-step the inversion of a tridiagonal matrix is performed using the Thomas algorithm (Thomas,[18]).

The 3D - advection- diffusion equation can be written as:

$$\frac{\partial C}{\partial t}+\frac{\partial uC}{\partial x}+\frac{\partial vC}{\partial y}+\frac{\partial\left(w+W_S\right)C}{\partial z}=\frac{\partial}{\partial x}\left(\varepsilon_x\frac{\partial C}{\partial x}\right)+\frac{\partial}{\partial y}\left(\varepsilon_y\frac{\partial C}{\partial y}\right)+\frac{\partial}{\partial z}\left(\varepsilon_z\frac{\partial C}{\partial z}\right) \quad (7)$$

where C is the suspended sediment concentration, t is time, x, y, z are the longitudinal, lateral and vertical coordinates, $\varepsilon_x, \varepsilon_y, \varepsilon_z$ are the sediment mass diffusion coefficients, W_S the sediment fall velocity and u, v, w the flow velocity components in x, y, z directions.

The quality of the results of a 3D model solving equations (1) - (4) and (7) depends on its ability to describe the vertical processes. The sigma coordinate introduced in meteorology by Phillips [19], and used in marine systems by a number of authors (Beckers,[20]; Lardner and Song,[21]; Santos and Neves,[17]) allows the use of the same number of layers independently of the local depth and is used in this model.

In Figure 2 it is described how the water column in an irregular real topography (Fig. 2a) is described in a cartesian coordinate model (Fig. 2b), and in a sigma model (Fig. 2c). Figure 2d shows the grid in the modified vertical coordinate plan. From Fig. 2 it is clear that in order to get a good description of the bathymetry using a cartesian coordinates, a large number of layers is needed. This is especially true to describe the dynamics of the intertidal zones.

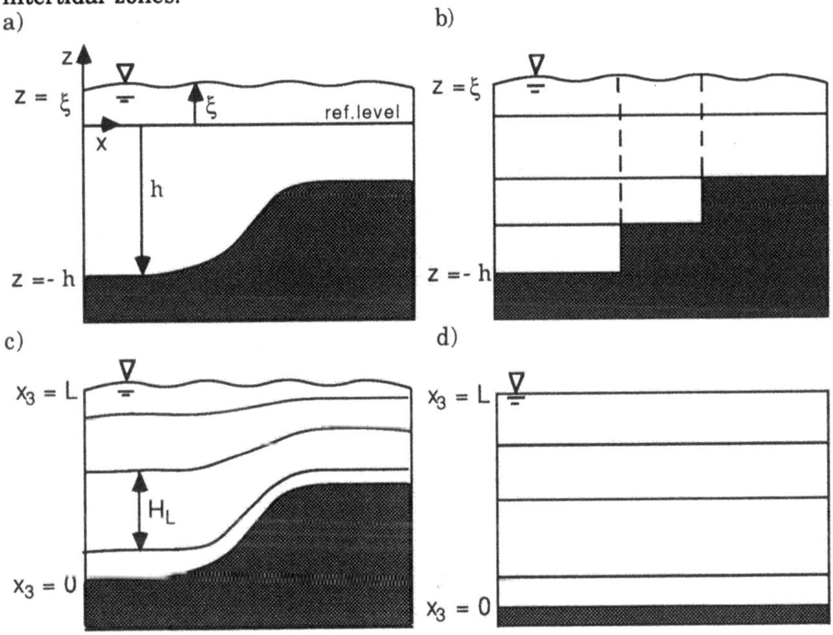

Fig. 2. Representation of an idealized vertical profile a) in a real domain, b) in a cartesian model, c) in a model using the sigma coordinate and d) in the sigma plan

The new vertical coordinate x_3 is expressed by:

$$x_3 = \frac{L}{H}(z + h) \qquad \in (0, L) \tag{8a}$$

$$H = \xi + h \tag{8b}$$

where L is a characteristic length scale of horizontal processes, H is the instantaneous maximum height of the water column, h the local depth, (positive if the bottom lays under the reference sea level), and z the vertical cartesian coordinate.

In the sigma plan the vertical coordinate varies between $x_3 = 0$ at the bottom and $x_3 = L$ at the surface. The horizontal coordinates have the same value in both coordinate systems ($x_1 = x$, $x_2 = y$). The rate of accumulation of matter is calculated as the divergence of the fluxes across the surfaces of the computational cell represented in Fig. 1. So the boundary conditions have been imposed in terms of fluxes as well.

In the sigma coordinate the vertical velocity is nil at the surface and at the bottom, and so is the advective transport. The diffusive fluxes across these surfaces and across the solid boundaries are zero. The mass exchange with the bottom is accounted for by the erosion and deposition fluxes considered as a source/sink term for the layer closer to the bottom. Other than those conditions are needed at the open boundaries. The advective fluxes across the solid boundaries vanishes as a consequence of the normal velocities calculated by the hydrodynamic model.

The most important fluxes across the boundaries are at the bottom. In general at the sea boundary the concentration is small and so is the river flow rate. The computation of the fluxes at the open boundaries were calculated using an upstream formulation.

The sediment erosion algorithm follows the classical approach of Partheniades [22]:

$$dM_E / dt = E(\tau / \tau_E - 1.0) \qquad : \tau > \tau_E \tag{9a}$$

$$dM_E / dt = 0 \qquad : \tau \leq \tau_E \tag{9b}$$

where the left hand sides of Eqs. 9a and 9b are the erosion fluxes; τ - bed shear stress; τ_E - critical shear stress for erosion and E is the erosion constant.

Critical shear stress for erosion is a limiting value and can be estimated as a function of the density of bed sediments (O'Connor, Nicholson,[14]; Stephens et al,[23]), and of the local salinity (Hayter, Mehta,[5]). In this model the formulation proposed by Delo [24] is followed:

$$\tau_E = A1(\rho_d)^{E1} \tag{10}$$

where ρ_d is the dry density of bed sediments and A1 and E1 are empirical coefficients incorporating the influence of mud type and salinity. This author estimated A1=0.0012 and E1=1.2 using different muds. Through the dependence on salinity and on bed consolidation, the critical shear stress becomes a time dependent property.

Deposition algorithm is based upon the approach of Krone [25]. The deposition rate (Eq. 11) is computed as a function of sediment concentration and fall velocity at the bottom layer, and depends on a critical shear stress for deposition, which is assumed in the model as a constant value.

$$dM_D / dt = C_B (W_S)_B (1.0 - \tau / \tau_D) \qquad : \tau < \tau_D \qquad (11a)$$

$$dM_D / dt = 0 \qquad : \tau \geq \tau_D \qquad (11b)$$

where dM_D / dt is the mass deposited per unit area per unit time; τ_D is the critical stress for deposition; $(W_S)_B$ is the deposition velocity and C_B is evaluated near the sediment-water interface. Deposition occurs only when the ambient shear stress is smaller than τ_D

For very fine particles the settling velocity is small and the vertical transport due to Brownian motion can increase the deposition velocity $(W_S)_B$. For larger flocs this effect is negligible $(W_S)_B = W_S$ (Sanford and Halka,[26]).

If τ_E is greater or equal to τ_D there is an intermediate range where neither erosion nor deposition occur. The cohesive sediment transport paradigm of mutually exclusive erosion and deposition processes does not have a general consensus and models assuming continuous deposition can also describe the data with good results (Sanford and Halka,[26]. This formulation corresponds to a classical approach with nil τ_E and τ_D and so the erosion constant (E) and C_B must be different in both formulations to produce similar results. The experimental methodology must be different in each approach, since in the former case only the net exchange between the water column and the bottom must be measured.

The vertical transport across the water column is a result of vertical advection due to the vertical component of flow velocity, of the sediment fall velocity and of vertical diffusion. Vertical flow velocity is computed by the hydrodynamic model, the diffusion coefficient is calculated using a mixing length approach and the fall velocity is a function of the concentration.

In the classical Prandtl formulation the mixing length vertical distribution is proportional to the distance from the bottom, being maximum at the surface. Montgomery (1943; cited by Robert and Ouellet,[27], based upon some laboratory experiments in particular flow), verified that this is not realistic and that the mixing length at the surface must be zero as well, suggesting that the mixing length should be proportional to z(1-z/h). This approach gives a maximum vertical distribution of the mixing length at mid-depth. Later on, Yalin [28] found that the vertical distribution should not be symmetric and suggested that the turbulent vertical viscosity, ε_z should be:

$$\varepsilon_z = Kz\sqrt{(1 - z / h)} \qquad (12)$$

where K is the Von Karman coefficient (assumed as K=0.4).

These three formulation can be synthesized as:

$$\varepsilon_z = Kz(1 - z/h)^{1/n} \qquad (13)$$

where n= 0 in Prandtl .formulation, n= 1 in Montgomery's and n= 2 in Yallin's. In this model n is a calibration parameter. The value of ε_z must be corrected in stratified flows using the local Richardson number.

Settling velocity is a function of the concentration throughout the flocculation processes. Since flocculation depends on the salinity by screening of the electrical forces, the effect of salinity on the settling velocity, bed shear stress and on the erosion and deposition rates has been investigated from laboratory ressuspension tests. Laboratory experiments with suspended matter from the Scheldt (Wollast,[29]) show that an intensive flocculation occurs since 1 psu is reached and is complete for 2.5 psu. Other laboratory tests with mud from Lake Francis (Hayter and Metha,[5]) found that the rate of erosion decreases with increasing salinity up to 2 psu, beyond which it becomes apparently invariant. The deposition rate was found to depend slightly on salinity and a power relationship of the form $W_S \alpha S^{0.13}$ was determined from tests at salinity ranges from 0 to 35 psu. Nevertheless there is not general consensus as to the importance and the parametrization of salinity in flocculation processes and further experimental investigation should be made.

At concentrations lower than a typically value of about 0.3 gl^{-1} (Mehta,[30]) the particles settle without mutual interference and the free settling occurs as a result of the balance between drag and submerged weight. For suspensions with concentration higher than the settling limit, the increase of inter-particle collision due to Brownian motion, internal fluid shearing and/or differential sedimentation and subsequent increase in aggregation leads to the formation of larger flocs with higher settling velocities. In this range settling velocity is simulated as a power function of the concentration, depending on particle size and shape. In presence of very large concentrations (order of tenths of gl^{-1}) the interaction between the downward movement of flocs and the upward movement of the water decreases the fall velocity. The concentration at which the fall velocity reaches its maximum value is called the hindered concentration C_{HS}.

The general expression for concentrations lower than C_{HS} , (Dyer,[13])

$$w_S = k1C^m \qquad :C \leq C_{HS} \tag{14a}$$

For larger values of the concentration Maude and Whitmore,[31] proposed:.

$$w_S = w_0 (1.0 - C)^{m1} \qquad :C > C_{HS} \tag{14b}$$

where W_0 is the fall velocity of a single grain and m, m1 are parameters depending on particle size and shape.

In these equations the flocculation processes are simulated throughout parameters k1, m and m1. Typical values for m are between 1.1 and 2.2, and for m1 between 2.3 and 4.7 (Maude and Whitmore,[31]; Migniot,[32]; Owen,[12]; Odd,[3]). In rigour the Eq. 14a leads to a nill W_0 when the concentration tends to zero while it should lead to the free settling velocity. Nevertheless this is not as significant as in estuaries. The settling on these areas is not relevant.

3. APPLICATION TO THE WESTERN SCHELDT ESTUARY

The Western Scheldt estuary, in The Netherlands, (Fig. 3) is a mesotidal well-mixed estuary. A systematic classification of the tide using the evaluation of the form ratio as the relative magnitude of the two main diurnal and semi-diurnal constituents shows a value less than 0.25, which characterizes a semi- diurnal tidal type (Pond and Pickard,[33]). The upper estuary extends over about 100 Km in a channel 600 m wide, being the dynamic tidal limit located in Gent (Belgium). The lower estuary, approximately 70 Km long and 8 km wide is composed of extended sand banks along the estuary axis, forming flood and ebb channels with a deeper navigation channel of about 20 m deep. The mean river discharge is about 100 m^3 s^{-1} with a residence time in the brackish zone of about 1- 3 months.

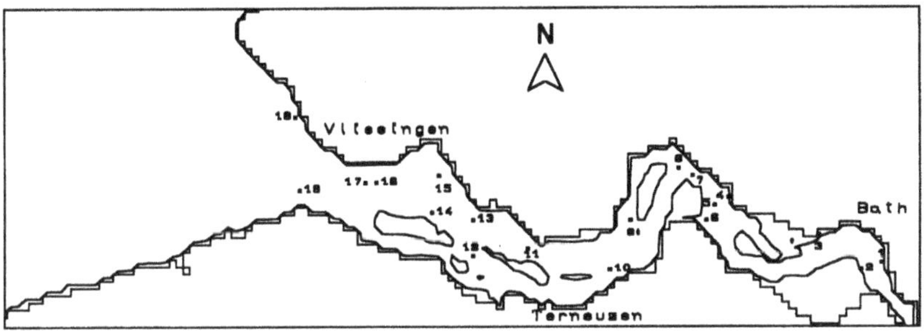

Fig. 3. The Western Scheldt Estuary. Location of the sampling stations (from Mulder & Udink ,[14]) and indication of the intertidal areas.

For the simulation of tidal flow a 600 m grid spacing and 4 vertical layers have been used. The vertical viscosity is calculated at each time step as a function of the local depth and velocity, using a mixing length approach. This is particularly important in the intertidal areas when dry, since the rate of change of the water depth is strong. The maximum time step is related to this effect and is closely related to the minimum depth value considered in the intertidal areas, since when a cell is becoming dry or starts to be flooded vertical viscosity must be often evaluated. In this model a cell is considered dry when the water height is less than 5 cm and the time step used is 20 s. As boundary conditions the 17 major tidal constituents, obtained by interpolation of field data at 3 points, were imposed at the open sea boundary. A mean constant inflow of 100 m^3 was imposed at the river boundary.

In order to validate the numerical models the results were compared with the available field data, as well as with the results of the 2D hydrodynamic and transport models formerly developed. For the relevant harmonic components a comparison was made between observed and computed tidal data at selected points.

The cohesive sediment transport model uses the same space grid and time step of the hydrodynamic model. In order to specify an open boundary concentration characteristic of the open sea and independent of the tidal situation, this boundary was located sufficiently far away from the estuary mouth. Using the same rationale the river boundary was located in the limit of the tidal propagation. Constant values of 0 gl^{-1} and 0.5 $g\,l^{-1}$ were imposed respectively at the sea and river boundaries.

4. RESULTS AND DISCUSSION

To evaluate the variability of suspended cohesive sediment concentration with the tidal situation, were simulated 2 tidal cycles (Fig. 4). The dynamics of suspended matter in a shallow mesotidal estuary can be different during flood and ebb tides. In case of the Western Scheldt estuary the ebb currents are stronger than the flood ones, this asymmetry of the tidal curve being stronger towards the mouth. As a consequence, the slope of the time evolution of the concentration in station 16 (Fig. 4b) is steeper during the ebb period.

a)

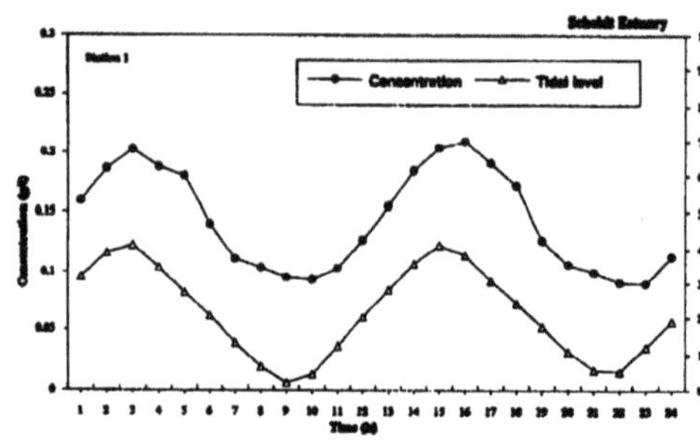

b)

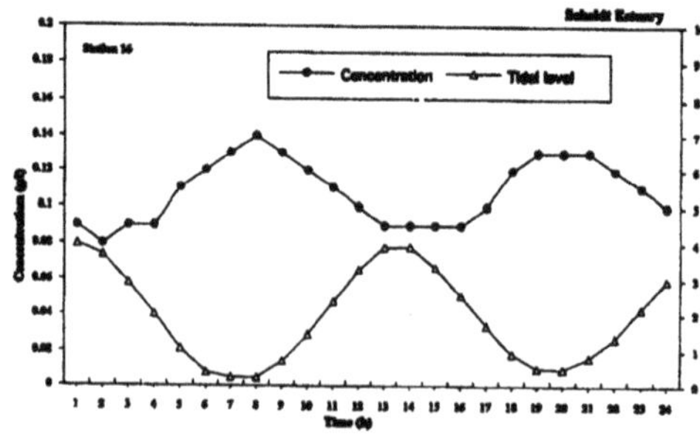

Fig. 4. Tidal evolution of sea level and suspended sediment concentration at two stations in the Western Scheldt estuary at a) station 1 and at b) station 16.

Comparison of figures 4a and 4b shows a phase lag between evolution of concentration in stations 1 and 16. In the former station concentration is maximum during high water while in the latter is maximum during low water. This shows that between station 1 and station 16 there is a maximum

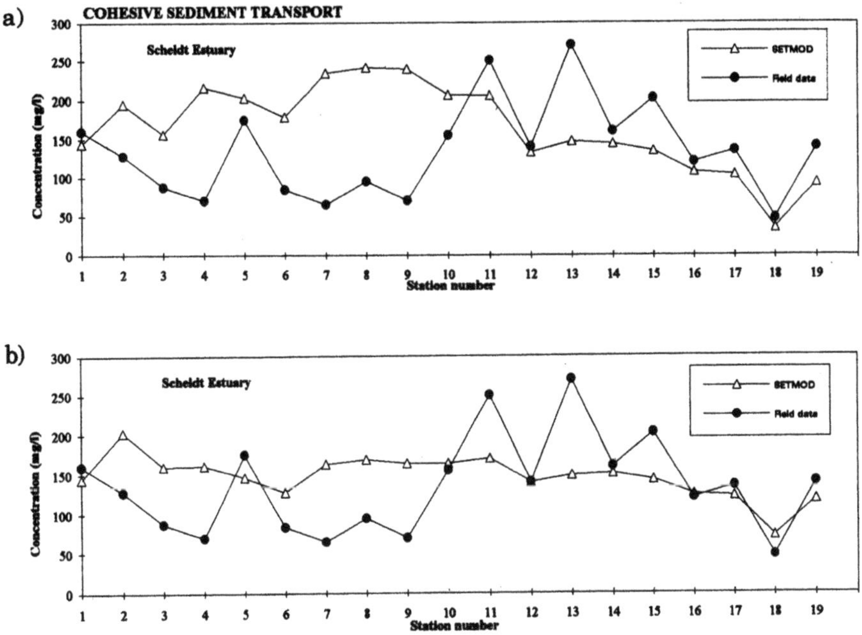

Fig. 5. Comparison between the results of the transport sediment model (SETMOD) and field data from 19 sampling stations in the Western Scheldt estuary (from Mulder & Udink,[14]) with a) E= 5 x 10^{-5} Kg m^{-2} s^{-1} and m = 1.9 throughout the estuary and b) with different spatial distribution of E.

of concentration and puts into evidence the role of horizontal advection in time evolution of concentration in the water column. Figure 5 shows that the location of the maximal concentration is about stations 10 and 15.

In a first attempt to perform sensitivity analysis and verification of the numerical model a set of simulations was performed varying key parameters, and the results from model SETMOD were compared with field data at 19 sampling stations and with the model DELWAQ presented by Mulder and Udink,[34] (Figs. 5-6).

Different values of erosion constants were tested (Eq. 9), showing distinguishable differences in the behaviour of sediment particles. Fig. 5a shows the results of a simulation using E= 5 x 10^{-5} Kg m^{-2} s^{-1} throughout the estuary. The agreement is good except between stations 6 and 10. Different choices with reasonable values were tested in this region and Fig. 5b shows the results obtained reducing the erosion constant to 5 x 10^{-6} Kg m^{-2} s^{-1} between these stations, which provided a rather improved fit with the field data. The improvement obtained could be due to the different composition of the bottom sediments in this region characterized by a lower percentage of silt and higher concentration of fine sand (Looff,[35]). However, these results point out that the question should not only be related with the erosion flux but also with the sedimentation. To verify this hypothesis the exponent m on settling velocity (Eq. 14a) has been reduced from 1.9 to 1.0 providing the results shown in Fig. 6, with a much better agreement between predictions and field data. The need to decrease the coefficient m in the calculation of the settling velocities may show that the flocculation is an important process in the sediment dynamics of the Scheldt estuary.

It can be remarked in Fig. 3 that stations 11, 13 and 15 are located in the same channel (northern channel) and the measured concentrations (Fig. 5a, 5b) show similar values, whereas stations 12 and 14, located in the southern channel, present smaller concentrations but nearly the same streamline line. Modelling results in the northern channel are in good agreement with the field data while in the northern channel (stations 13, 15) the calculated concentrations are considerably lower than the measured ones, except for station 11 where the fit was good.

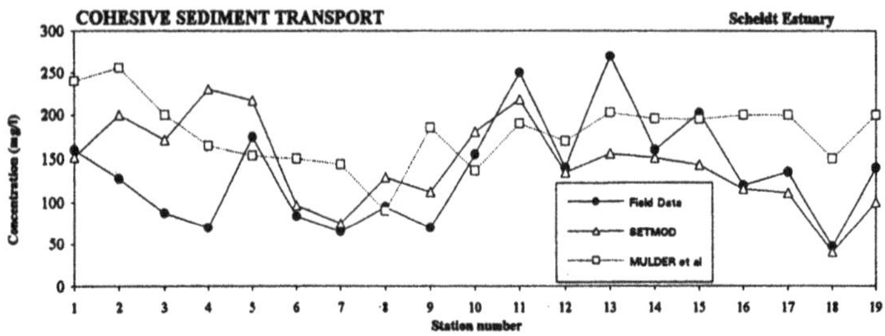

Fig. 6. Comparison between the results of two different sediment transport models (SETMOD and DELWAQ) and field data from 19 sampling stations (from Mulder & Udink, [14]).

This discrepancy can be explained by a different composition and sediment dynamics in the channels and the use of the same parametrization in the numerical simulation for both channels. From the analysis of modelling results of the flow field it is evident that in the northern channel the transient velocities during a tidal period nearly double the ones in the southern channel, leading to higher bottom shear stress and, consequently, to a higher probability of eroding material from the bed and increasing the concentration in the channel. In a posterior analysis of sediment maps (Looff, [35], [36]) it was found that the composition of the bottom sediments is slightly different in both channels. In the northern channel the sediments are characterized by a median diameter of particles, D_{50}, varying between 150-210 μ at stations 13, 15 and 210-420 μ at station 11, while in the southern channel the bottom sediments are characterized by a D_{50} of 300-420 μ. The composition of the sediments in the northern channel presents a lower percentage of sand and of calcium carbonate and a higher percentage of silt and of organic content than the sediments in the southern channel. These different mineralogic and granulometric compositions reflect a distinct behaviour on sediment dynamics and different values of E should be chosen according to the properties of bottom sediments.

5. CONCLUSIONS

Results of a 3D cohesive sediment transport model applied to the Western Scheldt estuary were presented. The verification of the model was done using tidal averaged concentrations in 19 point along the estuary. The results have shown to be sensitive to the erosion and flocculation parameters and, throughout the former to the tidal asymmetry of the velocity profile.

The available field data allowed the choice of a longitudinal distribution of the erosion coefficient and a global choice of the flocculation exponent. In this salinity range the flocculation is a function of the concentration and of the mineralogical composition of the sediments and a constant value can be accepted as a reasonable hypothesis. The same is not the case for the erosion coefficient that must reflect the properties of bottom sediments.

For a more accurate simulation of the sediment transport detailed data is needed for the bottom sediments in order to define a finer distribution of the erosion and deposition coefficients. An important information for the verification of the model are the rates of erosion/sedimentation along the estuary that give a time integral solution of the model, while the concentrations of suspended matter give local instantaneous values.

6. ACKNOWLEDGMENT

Part of this work was undertaken as a part of the CEC Environment Programme project MATURE (Biogeochemical of the MAximum TURbidity zone in Estuaries), under the contract nº 910357. Further investigation will be support by this research project. The first author gratefully acknowledges the financial support of the Programa Ciência JNICT, BD/BIC/M/57/92-IC.

7. REFERENCES

1. H. A. Einstein, *The Bedload Function for Sediment Transportation in Open Channel Flows* **No. 1026** (Soil Cons. Serv. U. S. Dept. Agric. Tech. Bull., 1950) p.78.

2. R. Ariathurai and R. B. Krone, *Finite Element Model for Cohesive Sediment Transport,* **Vol. 102** (J. Hydr. Div.,ASCE, No.HY3, 1976) p. 323-338.

3. N. V. M. Odd, *The Feasibility of Using Mathematical Models to Predict Sediment Transport in the Severn Estuary, The Severn Barrage* (Thomas Telford Ltd, London, 1982)p. 195- 202.

4. J. M. C. Nicholson, *Three- Dimensional Models of Particulate and Cohesive Suspended Sediment Transport,* Thesis for the degree of Doctor of Philosophy (Univ. of Manchester,

England, 1983) p.321.

5. E. J. Hayter and A.J.Mehta, *Modelling Cohesive Sediment Transport in Estuarine Waters*, **vol.10** (Appl. Math. Moddeling, 1986)pp.294-303.

6. Y. P. Sheng, *Modeling Bottom Boundary Layer and Cohesive Sediment Dynamics in Estuarine and Coastal Waters*, ed A. J. Mehta, **vol. 14** (Estuarine Cohesive Sediment Dynamics, Springer, Berlin, 1986) p. 360-400.

7. A. J. Mehta, E. J. Hayter, W. R. Parker, R.B. Krone and A. M. Teeter, *A. Cohesive Sediment Transport.I: Process description*, **vol. 115-8** (J. Hydraul. Eng., 1989) p. 1076-1093.

8. A. J. Mehta, W. N. McAnally, E. J. Hayter, A. M. Teeter and al., *Cohesive Sediment Transport. II: Application*, **vol. 115-8** (J. Hydraul. Eng.,1989)p.1094-1112 .

9. K. R. Dyer and E. M. Evans, *Dynamics of Turbidity Maximum in a Homogeneous Tidal Channel* , **vol. 5** (J. Coastal Research, Special Issue, Fort Lauderdale, Florida, 1989)p. 23-30.

10. L. C. Van Rijn, *State of the Art in Sediment Transport Modeling*, ed. Sam S. Y. Wang (Sediment Transport Modeling, ASCE, 1989) p. 13-32.

11. L. Cancino and R.Neves, *3D- Numerical Modelling of Turbidity Dynamics in Scheldt Estuary* (accepted for publication in vol. 28 of Nether. Journal of Aquatic Ecology, 1993) p.10.

12. M. W. Owen, *A Detailed Study of Settling Velocities of an Estuary Mud*, **Report INT 78** (Hydraulics Research Station,1970) .

13. K. R. Dyer, *Coastal and Estuarine Sediment Dynamics* (Wiley-Interscience, New York, 1986) p. 342.

14. B. A. O'Connor and J. A. Nicholson, *Three-dimensional Model of Suspended Particulate Sediment Transport*, **vol. 12** (Coastal Eng. 1988) p.157-174.

15. P. Le Hirt, P. Bassoullet and J. L'Yavanc, *New Developments About Mud Transport Models. Application to a Macrotidal Estuary,*, ed. Sam S. Y. Wang (Sediment Transport Modeling, ASCE, 1989) p. 94-99.

16. J. E. Mackin and K. T. Swider, *Organic matter decomposition pathways and oxygen consumption in coastal marine sediments*, **vol. 47** (J. of Marine Research, 1989) p. 681-716 .

17. A. J. Santos and R. Neves, *Radiactive Artificial boundaries in Ocean Barotropic Models.* (Computer Modelling in Ocean Engineering, Barcelona, 1991) p. 373- 383.

18. L. H. Thomas, *Elliptic Problems in Linear Difference Equations Over a Network* (Watson Sci. Comput. Lab. Rept., Columbia University, New York, 1949).

19. N. A. Phillips, *A Coordinate System Having Some Spacial Advantages for Numerical Forecasting*, **vol. 14** (J. Meteorol., 1957) p. 184- 186 .

20. J. M. Beckers, *Application of the GHER 3D General Circulation Model to the Western Mediterranean* , **vol.1** (J. of Marine Systems, Elsevier Science Publishers B.V., Amsterdam, 1991) pp. 315-332.

21. R. W. Lardner and Y. Song, *An Algoritm for Three- Dimensional Convection and Diffusion with Very Different Horizontal and Vertical Length Scales*, **vol. 32** (Int. Journal for Num. Meth. in Eng., 1991) pp.1303- 1319.

22. E. Partheniades, *Erosion and Deposition of Cohesive Soils*, **vol. 91, No. HY 1** (Journal of the Hydr. Div., ASCE, 1965) p.105-139.

23. J. A. Stephens, R.J. Uncles et al., *Bulk Properties of Intertidal Sediments in a Muddy Macrotidal Estuary*, **vol. 103** (Mar. Geol., 1992) p.15.

24. E. A. Delo, *Estuarine Muds Manual* , **Report nº SR 164** (Hydraulics Research, Wallingford 1988) p. 64.

25. R. B. Krone, *Flume Studies of the Transport in Estuarine Shoaling Processes* (Hydr. Eng. Lab., Univ. of Berkeley, California, USA 1962) p.110.

26. L. P. Sanford and J. P. Halka, *Assessing the Paradigm of Mutually Exclusive Erosion and Deposition of Mud, with Examples from Upper Chesapeake Bay*, **vol. 114** (Marine Geology, 1993) p. 37- 57 .

27. J. L. Robert and Y.Ouellet, *A Three- Dimensional Finite Element Model for the Study of Steady and Non- Steady Natural Flows*, ed. Nihoul J. C. J. and Jamart B. M, **vol. 45** (Elsevier Oceanography Series, 1987) p. 359-372.

28. M. S. Yalin, *Mechanics of Sediments Transport* (Second ed., Pergamon Press, Oxford, U. K., 1977) p.290.

29. R. Wollast, *The Scheldt Estuary*, ed W. Salomon (Pollution on the North Sea. An Assessment, Springer-Verlag, 1986) p. 183-193.

30. A. J. Mehta, *Hydraulic Behaviour of Fine Sediment*, ed M. B. Abbot and W. A. Price (Coastal, Estuarial and Harbour Engineer´s, Reference Book, Chapman &Hall, 1993) p. 577-584.

31. A. D. Maude and R. L. Whitmore, *A Generalized Theory of Sedimentation*, **vol. 9** (Brit. J. Appl. Phys.,1958) p. 477-482.

32. C. Migniot, *Étude des Propriétes Physiques de Différents Sédiments Très Fins et de leur Comportement Sous des Actions Hydrodinamiques*, **vol. 23** (La Houille Blanche, 1968)p.591-620.

33. S. Pond and G. Pickard, *Introductory Dynamical Oceanography* (Pergamon Press, 1983) p. 329.

34. H. P. J. Mulder and C. Udink, *Modelling of Cohesive Sediment Transport. A Case Study: The Western Scheldt Estuary*, **vol.3** (Coastal Engineering, 1990) p. 3012-3023.

35. D. de Looff, *Kaartering van de Bobemsamenstelling van Het Oostelijk Gedeelte van de Westerschelde. Methode en Resultaten*, (Rijkswaterstaat, 1978) p. 10.

36. D. de Looff, *Kaartering van de Bobemsamenstelling van Het Westelijk Gedeelte van de Westerschelde. Methode en Resultaten*, (Rijkswaterstaat, 1980) p. 9.

A NUMERICAL MODEL FOR SUSPENDED SEDIMENT IN COMBINED CURRENTS AND WAVES

S. HUYNH THANH

Presently at the HOCHIMINHCITY POLYTECHNIC UNIVERSITY - VIETNAM

T. TRAN THU, and A. TEMPERVILLE

Laboratoire des Écoulements Géophysiques et Industriels.

B.P. 53X-38041-GRENOBLE-FRANCE.

ABSTRACT

The suspension of fine non-cohesive sediments is investigated using a numerical turbulent boundary layer model with K-L type closure of order 2. The hydrodynamic equations are coupled with the continuity equation for concentration. The influence of stratification due to sedimentation is introduced into the equations of turbulence. Comparisons are made using several experiments, taking into account the effect of the stratification and the importance of certain input data in the model. In the colinear wave-current case the stratified model yields good results. In the case of oblique interactions, possible explanations are suggested for the apparently contradictory results obtained.

1. Introduction

Sediment transport processes in coastal regions under the action of waves and currents are complicated and at present not fully understood. For some years laboratory experiments on the transport of suspended sediments by waves have been undertaken (Staub 1984, Horikawa 1982, Ahlilan and Sleath 1987, Ribberink and Al-Salem 1991). Improvements in measurement techniques yield results that are now sufficiently accurate to test sophisticated models. In the case of wave-current interaction, only a few experimental measurements are available, except for the recent measurements of Van Rijn and Kroon (1992). At the same time the dynamics of the turbulent boundary layer in the wave-current interaction have been investigated using different turbulent closure models (K-ε, K-L) (Davies 1988, Soulsby et al. 1993, Huynh Thanh 1991, Fredsoe 1984).

Here a numerical model of the suspension mechanism of sediments (fine sand) is presented and some results are compared with the experiments.

In Paragraph 2 the turbulent boundary layer model of K-L type developed by Huynh Thanh (1991) is presented, which was used in the G6M-MAST1 programme for a mutual comparison among 8 models (Soulsby et al.1993). The model is coupled with the continuity equation for the concentration. To solve this equation, the reference concentration C_a at a reference height z_a is specified. The formula of Engelund and Fredsoe (1976) is thus chosen. In Paragraph 3 the experiments of Staub (1984), Horikawa (1982) and Ribberink and Al-Salem (1992) are modelled. The results are in good agreement with the experiments, provided a correct choice is made for the size of the modelled region . In the presence of wave-current interaction the phenomenon is more complex. Van Rijn and Kroon (1992) point out that the concentration is maximum when the waves are perpendicular to the current. It is possible to reproduce this result by taking into account the increase in shear at the bottom caused by the presence of ripples.

2. Equations of the model

The numerical model used in this investigation is a modified version of a second order turbulent closure model originally developed by Lewellen (1977), Sheng and Villaret (1989) and Huynh Thanh (1990). It is applied to the case of a horizontal flow over a flat bottom. The system is composed of dynamic equations for the average flow (u,v) coupled with the conservation equation for the volumetric concentration C of the sediments in suspension. For the closure of the system two equations for the turbulent kinetic energy K and the macro scale L of the eddies were added. This is expressed as:

$$\frac{\partial u}{\partial t}=-\frac{1}{\rho}\frac{\partial P}{\partial x}+\frac{\partial}{\partial z}\left(\nu_t\frac{\partial u}{\partial z}\right) \quad ; \qquad (1)$$

$$\frac{\partial v}{\partial t}=-\frac{1}{\rho}\frac{\partial P}{\partial y}+\frac{\partial}{\partial z}\left(\nu_t\frac{\partial v}{\partial z}\right) \quad ; \qquad (2)$$

$$\frac{\partial C}{\partial t}=\frac{\partial(w_s C)}{\partial z}+\frac{\partial}{\partial z}\left(\gamma_t\frac{\partial C}{\partial z}\right) \; ; \qquad (3)$$

$$\underbrace{\frac{\partial K}{\partial t}}_{\text{evolution}}=\underbrace{\nu_t\left[\left(\frac{\partial u}{\partial z}\right)^2+\left(\frac{\partial v}{\partial z}\right)^2\right]}_{\text{production}}+\underbrace{C_1\frac{\sqrt{2K}}{L}K}_{\text{dissipation}}+\underbrace{C_2\frac{\partial}{\partial z}\left(\sqrt{2K}.L.\frac{\partial K}{\partial z}\right)}_{\text{diffusion}}+\underbrace{\frac{g}{\rho_f}\gamma_t\frac{\partial \rho}{\partial z}}_{\text{buoyancy}} \quad ; \qquad (4)$$

$$\underbrace{\frac{\partial L}{\partial t}}_{\text{evolution}}=\underbrace{C_3\frac{\nu_t}{2K}\left[\left(\frac{\partial u}{\partial z}\right)^2+\left(\frac{\partial v}{\partial z}\right)^2\right].L}_{\text{production}}+\underbrace{C_4\sqrt{2K}}_{\text{dissipation}}$$

$$+\underbrace{C_2\frac{\partial L}{\partial z}\left(\sqrt{2K}.L.\frac{\partial L}{\partial z}\right)+\frac{C_5}{\sqrt{2K}}\left[\frac{\partial}{\partial z}(\sqrt{2K}.L)\right]^2}_{\text{diffusion}}+\underbrace{C_6\frac{L}{2K}\frac{g}{\rho_f}\gamma_t\frac{\partial \rho}{\partial z}}_{\text{buoyancy}} \quad . \qquad (5)$$

The assumption of local equilibrium for the turbulence allows the turbulent viscosity ν_t and the turbulent diffusivity γ_t to be expressed in the following form (Villaret 1987; HuynhThanh 1991):

$$\nu_t=\frac{\sqrt{2K}.L}{4}\frac{1+\left(\frac{1}{A}+\frac{2}{C_1+C_7}\right).\Omega}{\left[1-2\left(1-\frac{1}{C_1C_7}\right).\Omega\right](1-\Omega)} \quad ; \qquad (6)$$

$$\gamma_t=\frac{\nu_t}{A}\frac{1-\Omega}{1+\left(\frac{1}{A}+\frac{2}{C_1C_7}\right).\Omega} \quad , \qquad (7)$$

where Ω is given by:

$$\Omega=\frac{1}{A}\frac{g}{\rho_o}\frac{L^2}{2K}\frac{\partial\rho}{\partial z} \quad . \qquad (8)$$

Coupling between the hydrodynamic and the concentration equations is obtained through the equation for the density:

$$\rho=\rho_f+(\rho_s-\rho_f)C \quad , \qquad (9)$$

where ρ_f and ρ_s are the densities of water and sediment , respectively. C is given by Eq. 3 as a function of z and t. Using Eq. 9 the buoyancy terms containing ρ in Eqs. 4 and 5 may be expressed in terms of C.

The coefficients C_i were determined experimentally by isolating the effect of one or two terms in the equations during calibration (Lewellen 1977). Their values then remain invariant in all model applications in order to build a so-called invariant model. Numerous successful model simulations have validated this choice, although fine tuning of these coefficients could produce better results for a particular experiment (Sheng and Villaret 1989). Empirical values for such coefficients are:

$$C_1 = -0.25 \quad C_2 = 0.3 \quad C_3 = -0.35 \quad C_4 = 0.075$$
$$C_5 = -0.37 \quad C_6 = 0.8 \quad C_7 = 0.45 \text{ and } A = 0.75 \quad . \tag{10}$$

In equations (1) and (2) the horizontal pressure gradient is given by:

$$-\frac{1}{\rho}\vec{\nabla}P = \frac{\partial \overrightarrow{u(t)}}{\partial t} \quad , \tag{11}$$

where $\overrightarrow{u(t)}$ is the velocity of the potential flow outside the inertial layer.
The settling velocity w_s in Eq. 3 is taken to be constant.

The boundary conditions are defined below:

a) At the bottom:

The bottom is considered to be a fixed wall with equivalent roughness k_n. Regardless of the thin viscous sublayer, the calculation begins at a level z_o, called roughness length or zero-velocity level according to the logarithmic law of the wall. At this level the boundary conditions are:

$$u(z_o) = v(z_o) = 0 \quad ; \tag{12}$$

$$\left.\frac{\partial K}{\partial z}\right|_{z_o} = 0 \quad ; \tag{13}$$

$$L(z_o) = \alpha z_o \quad , \tag{14}$$

in which α is a numerical constant taking the value of 0.67 , as given by the analytic solution for a steady current (Villaret 1987).

For the reference concentration at the bottom C_b, the formula of Engelund and Fredsoe (1976) expressing C_b as a function of the Shield parameter Ψ is used:

$$C(z_a) = C_b(\psi) \quad ; \tag{15}$$

where

$$\psi = \frac{\tau_b(t)}{(\rho_s - \rho_f).g.d} \quad \text{and} \quad z_a = 2.d \quad .$$

It is supposed that this formula is valid instantaneously. $\tau_b(t)$ is the instantaneous shear stress at the bottom, calculated using the velocity and turbulent viscosity profiles. d is the grain diameter and g the gravity acceleration.

In the rough hydraulic regime, z_0 is set as $k_n/30$, where the equivalent roughness k_n = 2.5 d_{50} . In the transitory regime, z_0 and k_n are calculated from the curve given by Sleath (1984) , which expressed z_0/k_n as a function of the Reynolds number based on k_n.

b) At the upper boundary: $z = z_{max}$

As a general rule, the upper boundary is set at the limit of the bottom boundary layer. Unfortunately, the boundary layer thickness is also a non uniquely defined variable. In numerical modelling, this quantity takes a constant value independent of time (this is not true for unsteady currents). Such a value may be obtained either by an iterative procedure to find the level defined as boundary layer thickness (for example, the level at which the shear stress vanishes), or by an empirical or analytical formula. Moreover, all these definitions and formulas are etablished with the assumption of a free surface flow and no stratification effect of sediments. In the case of oscillatory flows produced in a wave tunnel as in the following experiments, there is no criterion applicable to the boundary layer thickness. A calculation layer thickness z_{max} satisfying the experiment boundary conditions must therefore be chosen. In principle:

$$\tilde{u}(z_{max}) = \tilde{u}(t) \quad , \tag{16}$$

where $\tilde{u}(t)$is the velocity measured at height z_{max}, outside the boundary layer. $\tilde{u}(t)$ contains the constant component U_c of the current and the oscillatory components of the wave.

In the case of a wave alone, the turbulence vanishes outside the boundary layer, whence:

$$K(z_{max}) = L(z_{max}) = C(z_{max}) = 0 \quad . \tag{17}$$

In the case of the wave-current interaction the following conditions are set:

$$\left.\frac{\partial K}{\partial z}\right|_{z_{max}} = 0 \quad ; \tag{18}$$

$$L\left(z_{max}\right) = \alpha z_{max} \quad ; \tag{19}$$

$$w_s C + \gamma_t \frac{\partial C}{\partial z} = 0 \quad \text{at } z_{max} \quad . \tag{20}$$

The purely hydrodynamic part of this model was used in the framework of the European programme G6M-MAST1 for an investigation of the maximum and average shear stress under the combined action of waves and currents (Soulsby et al.1993). The present article investigates the concentration dependence. The stratification effect of the suspended sediments is taken into account by the density-velocity correlation terms which are written as a function of $\frac{\partial \rho}{\partial z}$ in equations (4), (5) and (8). Neglecting this vertical density gradient, namely the variation of ρ with C , a decoupled flow-sediment version of the model is obtained . The equations for u,v,K, L are independent of C, i.e., the flow is unaffected by the concentration of suspended sediments. Results obtained from the model using the coupled and decoupled versions are then shown and discussed.

3. Experimental comparisons

3.1. *Case of a wave alone*

Huynh Thanh (1990), using the decoupled model, simulated test n° 6 of Staub et al. (1984) and tests (1-1), (1-2) and (1-4) of Horikawa et al. (1982). The parameters of these tests are listed in Table I . U_w is the orbital velocity amplitude of the wave, T the period, d the grain diameter and w_s the settling velocity of sediment.

TABLE I

	Test	U_w (cm/s)	T (s)	d (cm)	w_s (cm)
1	N°6 Staub	128	9.1	0.009	2.1
2	1-1 Horikawa	127	3.6	0.020	2.6
3	1-2 Horikawa	108	4.2	0.020	2.6
4	1-4 Horikawa	87	1.4	0.020	2.6
5	C10 Ribberink	170	7.2	0.021	2.6
6	C8 Ribberink	see text	6.5	0.021	2.6

For the Staub test, Figure 1 displays the instantaneous concentration at height z=1.8 cm, and Figure 2 the vertical profile of the average concentration. Figure 3 shows the vertical profiles of the average concentrations from the Horikawa tests. The results are in good agreement with the experimental data, and one might expect that the coupled model yields concentrations that are too small due to turbulence damping by stratification. In reality, this depends on the chosen size of the modelled region. In these four tests the experimental data are given in a small range of z from 0 to 3 cm. There is no information about what happens beyond this range. It was attempted to vary the upper boundary z_{max} to examine the sensibility of the model on the choice of z_{max}. With the decoupled model, for the first four tests in Table I, a reasonable concentration is obtained with the thickness z_{max}= 6, 8, 7 and 5 cm. In the case of the coupled model, these values are z_{max}= 20, 20, 15 and 10 cm, respectively.

Figure 4 shows a comparison between the two cases (coupled and decoupled) for the Staub test, where the value z_{max}= 9.5 cm, corresponding to the boundary layer thickness for a pure wave as given by the formula of Huynh Thanh (1990), was chosen, i.e.,

$$z_{max} = \delta_K = 0.246\ (a_h/k_n)^{0.81} \quad ,$$

where k_n and a_h are respectively the bed roughness and the orbital amplitude of the wave. The results obtained with the decoupled model are too high, while those for the coupled model are too small.

These results show the great importance of the choice of the upper boundary. Hence, to test the predictibility of the model, further experiments with a more complete set of data are needed.

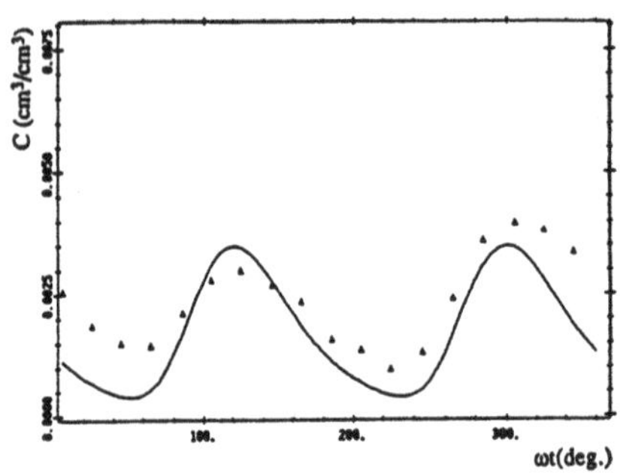

Figure 1: *Test N°6 of Staub et al.: Instantaneous concentration at height z = 1.8cm (▲) measurement, (—) present model*

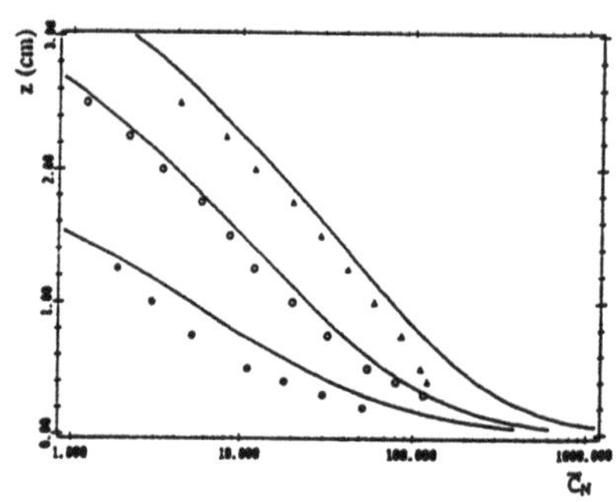

Figure 3: *Horikawa tests - Mean concentration $\overline{C}_N$ concentration normalised by 167 mg/l (▲) Test 1-1, (o) Test 1-2 (•) Test 1-4, (—) present model*

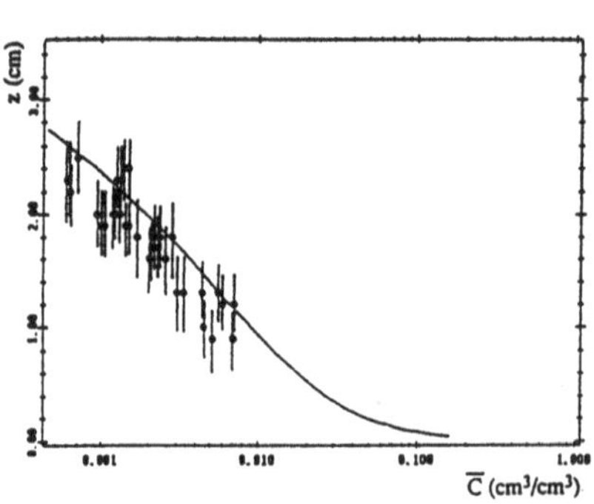

Figure 2: *Test N°6 of Staub et al.: Mean concentration (♦) measurement, (—) present model*

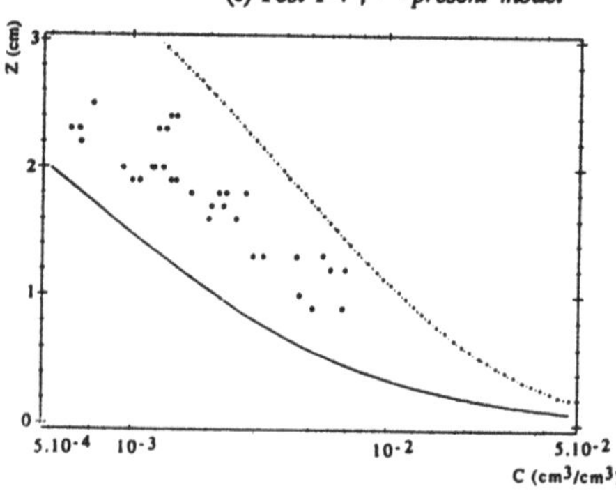

Figure 4: *Test N°6 of Staub et al.: Comparison between mean concentration profiles calculated with damping effect (—) and no damping effect (.....) for the case of z_{max}=9.5 cm. Measured values (•)*

Experiments have been made recently by Ribberink and Al-Salem (1992) using three different methods to measure the concentrations. The C_{10} test involves a sinusoidal wave with U_w=170 cm/s, T = 7.2 s, d = 0.021 cm and w_s = 2.6 cm/s. The thickness of the modelled layer is z_{max}=15 cm., corresponding to the position of the highest point to be measured. At this level the measured concentration is non zero. The condition (17) is therefore replaced by :

$$\left.\frac{\partial K}{\partial z}\right|_{z_{max}} = \left.\frac{\partial L}{\partial z}\right|_{z_{max}} = 0 \qquad ; \qquad (22)$$

and

$$w_s C + \gamma_t \frac{\partial C}{\partial z} = 0 \quad \text{at } z_{max} \quad . \qquad (23)$$

Figure 5 shows that the results obtained with the decoupled model are too high compared with the experiments, while Figure 6 indicates that the results from the coupled model are in good agreement.

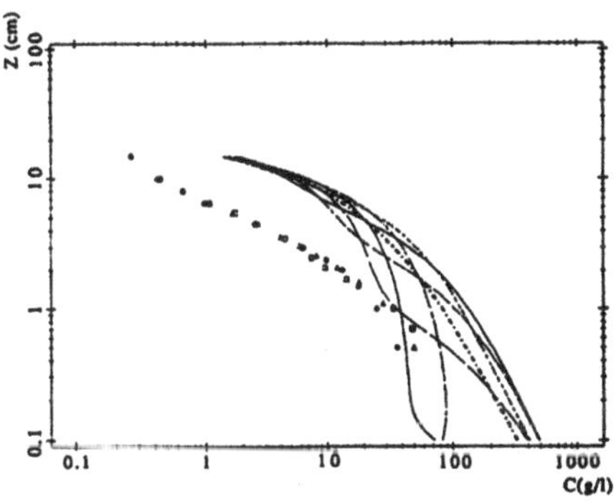

Figure 5: *Test C10 of Ribberink and Al-Salem. Concentration profiles at different phases and their mean value, calculated without damping effect. Comparison with measured mean values.*

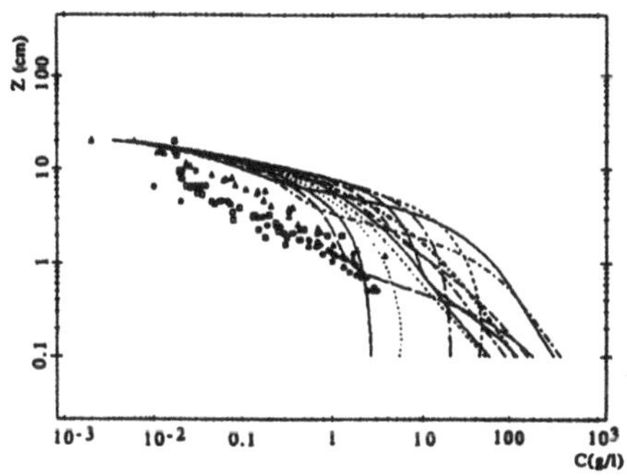

Figure 7: *Test C8 of Ribberink and Al-Salem. Concentration profiles at different phases and their mean value, calculated without damping effect. Comparison with measured mean values.*

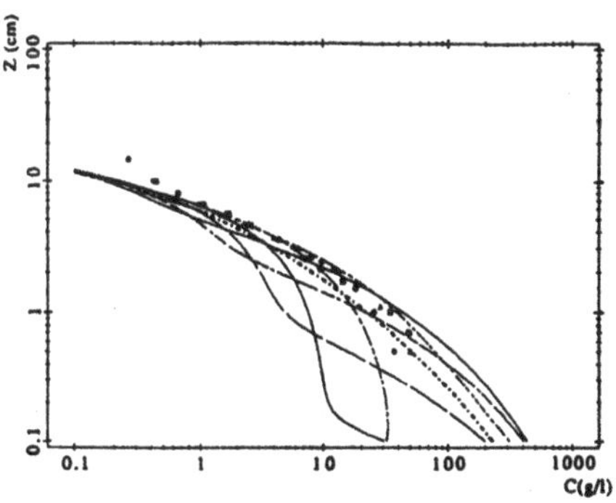

Figure 6: *Test C10 of Ribberink and Al-Salem. Concentration profiles at different phases and their mean value, calculated with damping effect. Comparison with measured mean values.*

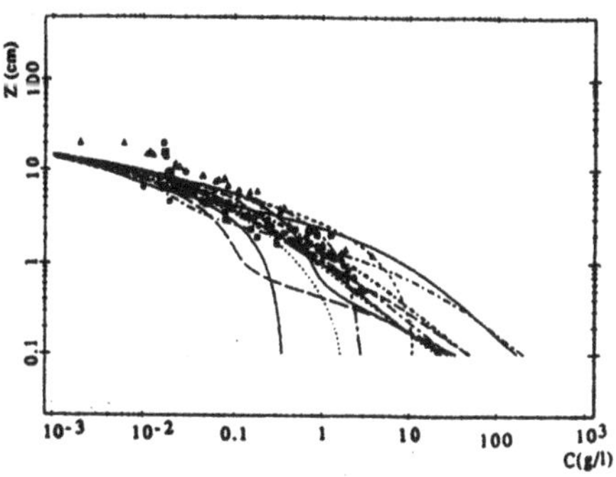

Figure 8: *Test C8 of Ribberink and Al-Salem. Concentration profiles at different phases and their mean value, calculated with damping effect. Comparison with measured mean values.*

For the case of the Ribberink and Al-Salem C_8 test, the wave is assymmetrical. The measured velocity at height z = 20 cm is given by:

$$U(t) = 2.5 + 84.5 \cos \omega t + 26.5 \cos 2\omega t \quad \text{(cm/s)} .$$

with a period $T = 2\pi/\omega = 6.5$ s; d = 0.021 cm and w_s=2.6 cm/s. The thickness of the modelled layer is z_{max} = 20 cm, with the same conditions at the upper boundary as previously. Figures 7 and 8 display the results of the decoupled and coupled models. It can be seen once again that the full model gives results that are in good agreement with the experiments, except for very small values of the concentration (smaller than the measurement errors).

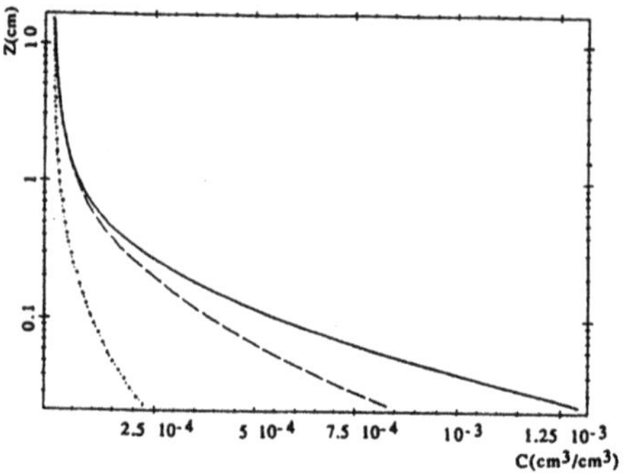

Figure 9: *Time-averaged suspended sediment concentration profiles for the cases where U_c=30 cm/s, U_w=20 cm/s, T=2.3 s, d=0.013 cm, w_s=0.65 cm/s and ϕ = 0° (—) ; 60° (- - -); 90° (. . .). The bed roughness is k_n=0.023 cm.*

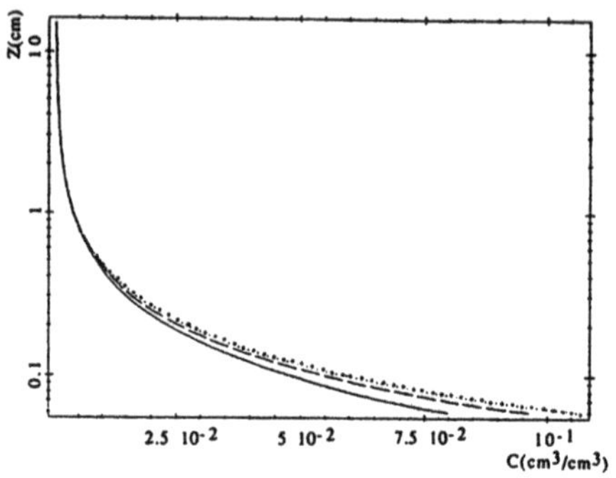

Figure 10: *Time-averaged suspended sediment concentration profiles for the cases where U_c=30 cm/s, U_w=20 cm/s, T=2.3 s, d=0.013 cm, w_s=0.65 cm/s and ϕ = 0° (—) ; 60° (- - -); 90° (. . .). The bed roughness is increased to k_n=1.8 cm to take into account the ripple form roughness effect.*

3.2. *Case of wave-current interaction*

In the case of wave-current interaction, the processes are more complex. The Ribberink C_8 test discussed above can be considered as a case of colinear interaction with a weak current. The case of oblique interaction is now considered.

Van Rijn and Kroon (1992) performed experiments with irregular waves making an angle ϕ with the current. The velocity of the current varies between 12 and 30 cm/s, and the amplitude of the orbital velocity of the waves ranges from 15 to 30 cm/s. They observed that:

- the concentration is maximum when the waves are normal to the current (ϕ=90°);

- the case of opposing interaction (ϕ= 180°) generates a greater concentration than in the colinear case (ϕ=0°);

-ripples are present at the bottom.

Since this model is neither designed for irregular waves nor for the rippled bed, a regular wave is assumed. Figure 9 shows the vertical profile of the mean concentrations calculated for the following cases:

U_c = 30 cm/s; U_w = 20 cm/s; T = 2.3 s; d = 0.013 cm; w_s = 0.65 cm/s and ϕ = 0°, 60°, 90° .

The thickness of the modelled layer is z_{max}=15 cm, and conditions (16),(18) through (20) are applied at the upper boundary. The bed is considered to be flat, the grain roughness is dominant and equal to 0.023 cm in this case. The resulting minimum at ϕ = 90° for the mean concentration is in disagreement with the experimental results. To provide an explanation for such apparent contradiction, Figure 10 shows the results obtained for identical values of the current and waves with a rippled bed. A ripple height of 0.6 cm is assumed and the value for the grain roughness is replaced by that of the ripple form roughness, which is of the order of 1.8 cm. In this case the friction at the bottom and the reference concentration are both larger, and the concentration is maximum at $\phi = 90°$.

The present model does not allow to take into account real waves and the influence of bed-forms. It nevertheless reveals the large effect of friction at the bottom upon the vertical distribution of concentration, without allowing a systematic investigation into the variation of the mean concentration as a function of the interaction angle between wave and current.

4. Conclusions

The hydrodynamic part of the model was tested by comparing with different experiments, as given by Huynh Thanh (1990), and with other models, such as that by Soulsby et al.(1993). For the hydro-sedimentary part, it was shown that the full model can correctly reproduce the variations in concentration in the wave-current interaction, provided that both the size of the modelled layer and the boundary conditions are correctly estimated. Indeed, the reference level of the bottom in formula 15 is selected differently by different authors: for example, Huynh Thanh S. (1990) used with the present model the value $z_a = 2.5$ d; Brors and Eidsvik (1993) adopt $z_a = 4z_0 = 0.32$ d, and observe that the value $z_a = 5z_0 = 0.40$d yields better results. Soulby (1993) suggests calculating z_a as a function of the parameter Ψ. For the thickness of the modelled layer and the upper boundary conditions, flexible choices according to the experimental conditions are proposed . To avoid the arbitrary adjustment, it is preferable to use more complete experiments with whole concentration profiles to validate the global predictibility of different models of the same type .

In the case of the oblique wave-current interaction, the importance of the chosen bed roughness, which is strongly influenced by the existence of bed-forms, on the profiles of the average concentrations was shown. The model thus confirms that the results obtained by Van Rijn et al. (1992) are valid in the regime with ripples, as pointed out by these authors. A more complete investigation should be carried out with a broader range of flow conditions and bed shapes .

Acknowledgements

This work was carried out as part of the G8 Coastal Morphodynamics research programme. It was funded jointly by the French project PNOC and the Commission of the European Communities, Directorate General for Science, Research and Development under contract N° MAS2-CT92-0027.

References

-AHILAN R.V. and SLEATH J.F.A. (1987). *Sediment transport in oscillatory flow over flat beds*. J. Hydraul. Eng. 113(3) pp.308-322.

-BRORS B. and EIDSVIK K.J. (1993). *Sediment concentration in oscillatory flow predicted by a dynamic Reynolds stress model*. MAST II-G8 Coastal Morphodynamics. Grenoble Overall Worshop 6-10 september 1993.

-DAVIES A.G., SOULSBY R.L., KING H.L. (1988). *A numerical model of the combined wave and current bottom boundary layer*. Journal of Geophysical Research, vol .93(C1), pp. 491-508.

-ENGELUND F. and FREDSOE J. (1976). *A sediment transport model for straight alluvial channels*. Nordic Hydrology 7 . pp. 293-306.

-FREDSOE J. (1984). *The turbulent boundary layer in wave-current motion*. Inst. Hydrodyn. Hydraul. Eng. Tech. Univ. Denmark. Series paper 35.

-HORIKAWA K., WANATABE A., KATORI S. (1982). *Sediment transport under sheet flow condition*. Proc. 18th Conf. Coastal Eng. pp. 1335-1352.

-HUYNH THANH S. (1990). *Étude numérique de la couche limite turbulente oscillatoire générée par l'interaction houle-courant en zone côtière.* Thèse de doctorat de l'INPG, France.
-HUYNH THANH S. and TEMPERVILLE A. (1991). *A numerical model of the rough turbulent boundary layer in combined wave and current interaction.* Euromech 262. Sand transport in rivers, estuaries and the sea. pp. 93-100.
-LEWELLEN W. S. (1977). *Use of invariant modeling.* In FROST W. (Ed.); *Handbook of turbulence,* vol.1, Plenum Publishing Corp., pp. 237-280.
-RIBBERINK J.S. and AL-SALEM A. (1991). *Bedforms, sediment concentrations and sediment transport in simulated wave condition.* Euromech 262. Sand transport in rivers, estuaries and the sea. pp. 141-148.
-RIBBERINK J.S. and AL-SALEM A. (1992). *Time-dependent sediment transport phenomena in oscillatory boundary-layer flow under sheet flow conditions.* Data report H 840.20, Part VI. December 1992. Delft Hydraulics.
-SHENG P. and VILLARET C. (1989). *Modeling the effect of suspended sediment stratification on bottom exchange processes.* Journal of Geophysical Research, vol.94(C10) pp. 14.429-14.444.
-SLEATH J.F.A. (1984). *Sea bed mechanics.* Eds. Wiley-Interscience.
-SOUSLBY R.L., HAMM L., KLOPMAN G., MYRHAUG D., SIMONS R.R. and THOMAS G.P.(1993). *Wave-current interaction within and outside the bottom boundary layer.* Coastal Engineering, 21(1993) pp.41-69. Elsevier Science Publishers B.V., Amsterdam.
-SOULSBY R.L. (1993). *The reference concentration for suspended sand in a steady current.* MAST II-G8 Coastal Morphodynamics. Grenoble Overall Worshop 6-10 september 1993.
-STAUB C., JONSSON I.G. and SVEDSEN I.A. (1984). *Variation of sediment suspension in oscillating flow.* Proc. 19th Conf. Coastal Eng. pp. 2310-2321.
-VAN RIJN L.C. and KROON A. (1992). *Sediment transport by currents and waves.* Coastal Engineering . Venise
-VILLARET C. (1987) *Étude expérimentale et numérique des lois d'érosion pour des sédiments cohésifs ..* Thèse de doctorat. de l'USTM Grenoble, France.

MEASUREMENT AND MODELING OF NEAR-BED SEDIMENT TRANSPORT

STEPHEN R. MC LEAN
Department of Mechanical and Environmental Engineering
University of California, Santa Barbara CA 93106, USA

and

JONATHAN M. NELSON
U.S. Geological Survey, Water Resources Division
MS413, Lakewood, CO 80225, USA

ABSTRACT

Much of the uncertainty in sediment-transport estimations arises because of our inadequate understanding of the processes that govern the dynamics of fluid-sediment interaction, especially very near an erodible bed. Comprehensive, physically based, fluid-sediment interaction models, guided by detailed, high-quality, observations in the near-bed region, are needed to advance our capabilities to predict sediment movement. Modeling near-bed flow requires a multi-phase approach wherein the fluid and sediment particles are treated separately in conservation equations for both mass and momentum, and the statistical aspects of the flow are incorporated. In this paper we present new techniques for observing sediment motion very near the bed, typically a region of high sediment concentration and very large gradients in concentration and velocity. Sediment motion illuminated by a laser light sheet can be captured in detail using an imaging fiber-optic probe coupled with a high-speed video camera system. The fiber-optic imaging conduit provides minimal scattering and absorption of the light by the ambient flow and minimal flow disturbance. Using observations collected with this methodology, it is possible to quantify processes such as fluid-particle and particle-particle interactions, which allows characterization and parameterization of the momentum transfer between the constituents of the coupled fluid-sediment system.

1. Introduction

Sediment-transport calculations are characterized by large uncertainty even for steady, uniform flows. Much of this uncertainty arises because of our inadequate understanding of the processes that govern the dynamics of fluid-sediment interaction, especially very near an erodible bed. Unfortunately, for many flows this complex near-bed region is precisely where most of the transport occurs. Therefore, comprehensive, physically based, fluid-sediment interaction models, guided by detailed, high-quality, observations, are needed to enhance our capability to predict sediment movement. Because of the high concentrations of sediment in motion and the very large gradients, accurate measurements of flow and sediment velocity and sediment concentration in the near-bed region are extremely difficult. Consequently, there is a paucity of data with sufficient detail and accuracy to allow the development of insight into the active physical processes.

For steady, uniform flows, Meyer-Peter and Mueller (1948), using extant data, suggested that bedload transport rates were well characterized empirically by

$$\Phi = 8\,(\tau_* - 0.47)^{3/2} \tag{1}$$

where

$$\Phi = \frac{q_b}{[(\rho_s - \rho_w)/\rho_w g D^3]^{1/2}} \tag{2}$$

is the nondimensional transport rate and

$$\tau_* = \frac{\tau_b}{(\rho_s - \rho_w)\, gD} \tag{3}$$

is the nondimensional shear stress. Here q_b is the volume rate of sediment transport per unit width, τ_b is the boundary shear stress, ρ_s and ρ_w are the sediment and water density, respectively, g is the acceleration of gravity, and D is the mean sediment diameter. Others such as Fernandez-Luque and van Beek (1976) have tried to improve on this result empirically by using more recent data, and many models ranging from stochastic (Einstein, 1950), energetics (Bagnold, 1973) and particle trajectory approaches (Yalin, 1963; Wiberg and Smith, 1989) have been developed. But all these yield essentially the same result: for high transport stages the mean transport rate increases with the mean boundary shear stress approximately to the power 3/2. The mean boundary shear stress (or momentum flux) is a measure of the mean force per unit area acting on the bed; however, there is not necessarily a significant correlation between instantaneous momentum flux and instantaneous forces acting on a sediment grain moving near the bed. Instantaneous motion of individual sediment particles is most closely related to near-bed fluid velocity fluctuations. Given the high degree of nonlinearity of the transport process, instantaneous transport rates are likely dominated by a few high-velocity events. Because the velocity statistics, and hence the expected frequency and magnitude of such events, are scaled by the shear velocity for steady uniform flow, this nonlinearity can be included easily in empirical formulae such as Eq. 1. However, for nearly all sediment-transporting flows of interest, both the flow velocity and turbulence are changing in time and/or space in such a way that equilibrium relations do not hold. For example, Nelson et al. (1993) and McLean et al. (1994) found anomalously low values of the correlation coefficient over most of the stoss side of dune-shaped bottom features. Indeed any separated flow, where a wake-like structure interacts with an underlying, developing boundary layer, leads to nonuniform turbulence statistics. Furthermore, the investigations over dune shapes indicate that the exact nature of the turbulence structure is very sensitive to the topographically induced accelerations. In light of this sensitivity in steady, nonuniform flows and given the magnitude of accelerations in oscillatory flows, it is reasonable to assume that local turbulence statistics, and hence sediment transport relations, are strongly affected by acceleration in unsteady flows. Therefore, it is imperative that we understand the basic mechanisms by which sediment is moved in such flows and use that knowledge to develop physically based models that are coupled to detailed measurements of the relevant parameters. In this paper we present a brief formulation of the general, coupled fluid-sediment equations for the case of unsteady flow, and identify the terms in those equations (primarily related to fluid-sediment interaction) that cannot be parameterized accurately using current understanding. We then go on to describe an experimental methodology that can be used to characterize these poorly understood processes for the purpose of modeling sediment transport.

2. Theoretical Framework

In a multiphase approach, the fluid and sediment are treated separately in conservation equations for both mass and momentum; however, in order for such an approach to succeed, it is necessary to include interactions between the phases (sediment and fluid). This is particularly difficult in the near-bed region where sediment grains make frequent contact with the bed and interact with each other, exchanging momentum and energy. For example, the concept of "collisions" is much more complex than presently assumed in particle dynamics approaches. The water between two approaching sediment grains must be displaced, but this is impeded by the presence of other grains. High pressure gradients in the intergranular region can result, thus reducing the likelihood of actual collisions occurring. As a consequence, the grain-grain interaction is in fact mediated by fluid forces arising from the reaction of the interparticle fluid to the relative grain motion. In order to model these and other similar mechanisms, it is necessary to begin with a development of generally applicable statements of conservation of mass (volume) and momentum.

2.1 Conservation of Sediment and Fluid Volume

Conservation of sediment volume is expressed in the following equation:

$$\frac{\partial C_s}{\partial t} + \nabla\cdot(C_s\vec{V}_s) = 0 \tag{4}$$

where C_s is the volume concentration of the sediment (sediment volume per volume of fluid + sediment) and $\vec{V}_s$ is the velocity of the sediment. For simplicity it is assumed here that the sediment is all of one size; in general, however, this is not the case and a distribution of sizes must be considered. Defining the volume concentration of the water phase to be $C_w = 1 - C_s$ a similar conservation equation for water yields

$$-\frac{\partial C_s}{\partial t} + \nabla\cdot(1-C_s)(\vec{V}_\Delta + \vec{V}_s) = 0 \tag{5}$$

where the velocity of the water relative to the sediment is $\vec{V}_\Delta = \vec{V}_w - \vec{V}_s$. Adding Eqs. 4 and 5 yields

$$\nabla\cdot\vec{V}_w = \nabla\cdot(C_s\vec{V}_\Delta) \tag{6}$$

By assuming that for a turbulent flow the velocities and sediment concentrations can be decomposed into mean and fluctuating components, and restricting ourselves to uniform or quasi-uniform conditions such as exist under typical wave conditions where the boundary layer thickness is much less than the wavelength, averaging yields

$$\frac{\partial}{\partial z}\overline{w_w} = \frac{\partial}{\partial z}\left(\overline{C_s}\,\overline{w_\Delta} + \overline{C'_s w'_\Delta}\right) \tag{7}$$

where w_w and w_Δ are the absolute and relative vertical velocity components of the water, respectively. Integrating Eq. 7 in the vertical with the the further assumption that the mean vertical water velocity is zero if the mean concentration of sediment is zero yields

$$\overline{w_w} = -\frac{\overline{C_s}\,\overline{w_s}}{1-\overline{C_s}} + \frac{\overline{C'_s w'_\Delta}}{1-\overline{C_s}} \tag{8}$$

Here the primes denote turbulent fluctuations and the overbar denotes mean quantities. In unsteady flows like waves, this is an ensemble mean, which averages over turbulent fluctuations, but retains the time variation associated with wave motion. Assuming horizontally uniform conditions, averaging Eq. 4 yields

$$\frac{\partial\overline{C_s}}{\partial t} + \frac{\partial}{\partial z}\overline{C_s}\,\overline{w_s} + \frac{\partial}{\partial z}\overline{C'_s w'_s} = 0 \tag{9}$$

These last two equations represent the conservation of water and sediment in the system; analysis of the dynamics of the near-bed flow requires momentum-conservation equations for both the water and sediment.

2.2 Conservation of Fluid Momentum

The momentum equation governing the water is

$$\frac{\partial\vec{V}_w}{\partial t} + \vec{V}_w\cdot\nabla\vec{V}_w = -\frac{1}{\rho_w}\nabla p + \frac{1}{\rho_w}\nabla\cdot\bar{\tau} - \frac{C_s}{\rho_w V}\vec{F} \tag{10}$$

where p is the pressure, ρ_w is the water density, $\bar{\tau}$ is the viscous deviatoric stress tensor, V is the volume of a sediment grain, and the force exerted by the flow on a sediment particle is

$$\vec{F} = \int_S \tilde{\sigma} \cdot \vec{n}\, dS \tag{11}$$

where this represents the integral over the entire surface area S of the particle and $\tilde{\sigma}$ is the full fluid stress tensor including the pressure. Averaging this vector equation and assuming horizontally uniform conditions leads to the following two equations for the horizontal and vertical momentum:

$$\frac{\partial \overline{u}_w}{\partial t} + \overline{w}_w \frac{\partial \overline{u}_w}{\partial z} = -\frac{1}{\rho_w}\frac{\partial \overline{p}}{\partial x} - \frac{\partial \overline{u'_w w'_w}}{\partial z} - \frac{\overline{C_s F_x}}{\rho_w V} \tag{12}$$

and

$$\frac{\partial \overline{w}_w}{\partial t} + \overline{w}_w \frac{\partial \overline{w}_w}{\partial z} = -\frac{1}{\rho_w}\frac{\partial \overline{p}}{\partial z} - g - \frac{\partial \overline{w'_w w'_w}}{\partial z} - \frac{\overline{C_s F_z}}{\rho_w V} \tag{13}$$

where g is the acceleration of gravity and the viscous deviatoric stress has been neglected in Eqs. 12 and 13.

2.3 Conservation of Sediment Momentum

The momentum equation governing the motion of a sediment particle is

$$\rho_s V \frac{D\vec{V}_s}{Dt} = \vec{F} + \rho_s V \vec{g} + \vec{F}_{col} \tag{14}$$

where D/Dt is the substantial derivative (following the particle) and $\vec{F}_{col}$ is the force arising from direct collisions with other particles. For the case of fairly low transport rates, when the near-bed concentration is not that large, saltation is the dominant mode of transport. This is characterized by grains rolling, sliding, and hopping along the bed, making frequent contact with the bed. The consequence of this type of motion is that the motion of near-bed particles is strongly bimodal. There are upward-moving grains that have just collided with the bed and have low horizontal velocities and large upward velocities, and there are grains that are moving downward at approximately the settling velocity and are moving horizontally at nearly the fluid velocity. Such a distribution of particle trajectories makes it difficult to specify $\vec{F}$ without dividing the sediment into classes that characterize their direction of movement (see McLean, 1992). If uniform flow is assumed, Eq. 14 can be averaged to yield

$$\frac{\partial \overline{u}_s}{\partial t} + \overline{w}_s \frac{\partial \overline{u}_s}{\partial z} = -\frac{\partial \overline{u'_s w'_s}}{\partial z} - \frac{\overline{F}_x + \overline{F}_{x,col}}{\rho_s V} \tag{15}$$

and

$$\frac{\partial \overline{w}_s}{\partial t} + \overline{w}_s \frac{\partial \overline{w}_s}{\partial z} = - \frac{\partial \overline{u'_s w'_s}}{\partial z} - \frac{\overline{F}_z + \overline{F}_{z,col}}{\rho_s V} - g \tag{16}$$

It should be noted that the normal approach that is taken in evaluating $\vec{F}$ is to introduce lift, drag, and added mass coefficients, and write the force in terms of the "undisturbed" fluid velocity (eg. Maxey and Riley, 1983; Wiberg and Smith, 1985), defined as the fluid velocity that would exist at the location of the center of mass of a sediment particle if it were not present. This requires the implicit assumption that the pressure and velocity fields about any given particle are the same as those that are found about an isolated particle, which is not generally true. For example, in the case of sheet flow, the volume occupied by sediment is of the same order as that of the water phase, so the pressure and velocity fields around a given particle will be significantly different than those in the isolated particle case. This has the unfortunate consequence that the system of equations that is represented by Eqs. 8, 9, 12, 13, 15, and 16 contains more unknowns than equations because $\vec{F}$ cannot as yet be written in terms of the other unknowns. Unfortunately, $\vec{F}$ and $\vec{F}_{col}$ are functions of many parameters such as the relative velocity, the sediment concentration, the shear in both the water and the sediment velocity, and the gradient of the sediment concentration. To make any further progress in developing a general model for sediment transport that incorporates strong sediment-fluid and grain-grain interactions, it is necessary to develop an accurate parameterization for $\vec{F}$ and $\vec{F}_{col}$. The first step in that development must consist of direct measurement of near-bed flow velocities and sediment concentrations and velocities. From such observations, $\vec{F}$ can be indirectly determined and the governing equations can be closed. At present, however, there is no methodology that allows such measurements, which is a fundamental barrier to further progress on these problems. In the remainder of this paper, we propose a new experimental technique for investigating the details of the flow-sediment and grain-grain interactions and describe a simple preliminary application of the technique.

3. High-Speed Imaging of Near-Bed Sediment Transport

Progress in understanding near-bed sediment dynamics requires simultaneous measurements of instantaneous flow velocity and sediment motion very near an erodible bed. Visualization techniques may provide clues to the nature of this complex environment. Because of the relative opacity of the fluid in this region, however, the field of view must be very small and, because of the relatively high speeds at which fluid and sediment can move even near the bed, the imaging system must operate at high frame rates. Furthermore, the sometimes rapid erosion or accretion rates, coupled with the large gradients near the bed, require a system that can adjust in real time. Video systems with speeds of 250 frame per second or more and over 400 lines of resolution are now available. The use of video provides real-time observations, and analysis can be carried out computationally by using image-processing hardware and software. Because of the need to resolve sediment motion very near the bed, a fiber-optic imaging conduit can be used so that there is minimal scattering and absorption of the signal by the ambient flow and minimal flow disturbance. To satisfy the need for high-intensity illumination for high-speed video systems, a sheet of laser light can be used. The creation of a sheet of light effectively limits the observed motion to a plane (although the "plane" in fact has finite thickness), essentially resulting in a depth-of-field that is determined by illumination rather than by the characteristics of the optics.

Demonstration of a prototype system was carried out in the Ocean Engineering Laboratory at the University of California, Santa Barbara. The flow was generated by oscillating a plate in a small wave tank, 3 m long, 0.3 m wide, and 0.3 m deep. The water depth was approximately 0.15 m, and the period of oscillation was selected to coincide with the 4.95-s seiche period of the tank. Because of gaps between the moving plate and the walls and floor of the tank, the resulting waves were not particularly sinusoidal; the fundamental frequency was contaminated by higher frequency free waves.

An imaging conduit was used to observe the motion of 0.5-mm-diameter sand in the tank. Illumination was provided by a laser that delivered 0.8 W of power in a sheet of light by passing it through a cylindrical lens placed in a wing-shaped plexiglass sting that just penetrated the water surface. The vertical light sheet, about 3-4 grain diameters in thickness, was oriented parallel to the water motion. The best images were found when the viewing angle was about 45° from horizontal. At smaller angles, grains moving between the light sheet and the imaging fiber obscured the illuminated field. The field of view was about 7x7 mm; at this resolution, it was possible to distinguish particles that differed by more than an order of magnitude in diameter. A high-speed video camera with a frame speed of 250 frames per second was used to record the images. The slowest available shutter speed was 1/2500 s. The light provided by the laser was only marginally adequate for this shutter speed and thus produced low-contrast images, making processing more difficult than it would be with a more sensitive camera. Also, only crude calibration of positions within the field of view was possible because orientation of the fiber-optic probe could not be fixed. Nevertheless, the results indicate the promise for such a system.

Presently, details of the sediment motion can only be assessed with relatively high operator input, but computational motion-analysis techniques can be progressively developed to automate the procedure. Here we present some relatively primitive techniques and results that demonstrate some of the potential of high-speed imaging. The image-processing hardware digitizes an image by assigning a grey scale intensity value between 0 and 255 for each pixel: 0 for black and 255 for white. Taking the absolute value of the difference between two consecutive images, pixel by pixel, produces a "difference" image. If there were no motion, then this image would be black; if there is motion, only the changes appear as shades of grey. The averaging of the absolute values of the difference on a pixel-by-pixel basis for the entire image, defined here as β, is plotted in Fig. 1.

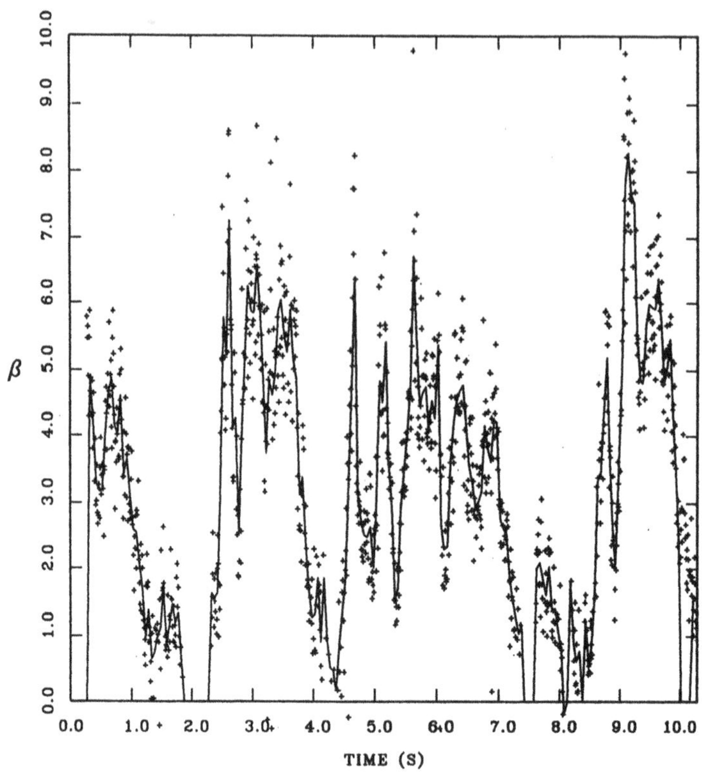

Figure 1: β (absolute value of the difference between two images separated by 1/250 s) versus time. A background value of about 3 has been subtracted.

The physical significance of β and its relation to sediment transport can be considered as follows. Defining Δx_i to be the distance the ith grain (having diameter D_i) in an image moves in time Δt between images, if $\Delta x_i > D_i$, the difference image will contain two images of the same particle and the result is independent of the speed of the grain $u_i = \Delta x / \Delta t$. For this case, the speed of the grains can not be ascertained and β is more a measure of concentration than flux. On the other hand, if $\Delta x < D$, the difference image of the particle appears as two overlapping images of the same particle. The brightness of the overlap region is highly variable depending on the variations in brightness of the particle in the original image. The degree of overlap (and hence the area appearing in the difference) depends on the distance that a given particle has moved (and hence the speed). If the particles were all spheres and if the illumination of all particles were uniform, the average pixel value of the entire difference image is

$$\beta = \frac{1}{M_{tot}} \sum_{i=1}^{N} \begin{cases} 2M_i\, I & \text{if } \Delta x_i / D_i > 1 \\ 2[M_i - (M_i)_{int}]\, I + (M_i)_{int}\, I_{int} & \text{if } \Delta x_i / D_i < 1 \end{cases} \tag{17}$$

where M_{tot} is the total number of pixels in the image, N is the number of particles that moved, M_i is the number of pixels in the projected area of the ith particle, $(M_i)_{int}$ is the number of pixels in the intersection (overlap) region, I is the average illumination intensity of the particles, and I_{int} is the mean intensity of the overlap region. Note that

$$M_i = \frac{S_i\, M_{tot}}{A_{tot}} \tag{18}$$

and

$$(M_i)_{int} = \frac{(S_i)_{int}\, M_{tot}}{A_{tot}} \tag{19}$$

where A_{tot} is the total area covered by the image,

$$S_i = \pi \frac{D_i^2}{4} \tag{20}$$

is the projected area of the particle, and

$$(S_i)_{int} = \frac{D_i^2}{2} \left\{ \cos^{-1}\left(\frac{\Delta x_i}{D_i}\right) - \left(\frac{\Delta x_i}{D_i}\right) \left[1 - \left(\frac{\Delta x_i}{D_i}\right)^2 \right]^{1/2} \right\} \tag{21}$$

is the area of overlap. For the case where $\Delta x_i / D_i$ is less than 1, and in particular is fairly small, and if I_{int} is assumed to be zero (corresponding to an assumption of uniform illumination intensity for every particle), then

$$\beta \approx \sum_{i=1}^{N} \frac{2D_i \Delta x_i I}{A_{tot}} \tag{22}$$

The bedload-transport rate is given by

$$q_b = \sum_{i=1}^{N} \frac{\pi D_i^3 u_i}{6b} \tag{23}$$

where u_i is the velocity of the ith particle and b is the effective thickness of the light sheet; thus bA_{tot} is the volume of the fluid that is sampled by the video system. If the diameter of the particles is nearly the same (ie. $D_i \approx D$) then

$$q_b \approx \frac{A_{tot}\pi D^2}{12bI\Delta t}\beta \tag{24}$$

and the transport is proportional to β. For the case when there is a significant distribution of sizes, Eqs. 22 and 23 can be used along with the measured size distribution to compute the relationship between q_b and β.

Figure 1 is a plot of β for about 10 s during the demonstration. Even when there was no motion, β had a value of about 3 due to inherent noise in the video image. This background noise level was removed from β. The oscillatory nature of the sediment motion is obvious, with direction changes occurring at about 2.0, 4.5 and 7.5 s. In addition to the basic 5-s wave period, there are also fluctuations that are less than a second in duration. These presumably are due to the high frequency waves that were produced by the wavemaker. The fluctuations are very large and represent very rapid changes in the sediment motion over relatively short time periods. In order to determine whether these fluctuations were associated with real events or were artifacts of the image processing, short segments within this record were analyzed in more detail.

It is possible to calculate particle velocities by locating the same particle in two consecutive images and dividing the difference in position by the time between images. Without good positional calibration, the quantitative accuracy of the results shown here is compromised; therefore only crude calculations were made to demonstrate the potential for such a technique. For example, the field of view was assumed to have a uniform scale even though it was viewed at an oblique angle. The grain diameter, in pixels, was estimated for many particle images and the average assumed to be 0.5 mm, thus establishing the pixel size in real units. The velocities calculated using these assumptions are shown in Figs. 2 and 3 for two short time intervals when there were rapid changes in β during the accelerating portions of the wave motion. Note that in these figures sediment motion stops almost completely between about 4.75 and 4.95 s and again between 8.85 and 8.95 s. The pluses in the bottom plots in these figures are individual particles, whereas the circles in the top plots are averages over an entire image. The trends in the two plots are the same, but the variation in the velocities is generally less pronounced than that in β. In Fig. 4 a plot of β versus velocity indicates a degree of correlation, but there is much scatter. This is probably because β becomes more a measure of concentration at high velocities than of velocity, as discussed above. Unfortunately, because of the poor image quality under these conditions and because the frame speed was such that particles frequently moved several diameters (thus making tracking difficult when there were many particles in motion), it was often impossible to obtain an accurate count of the number of grains in motion. Thus, although velocity measurements were possible, it was difficult to accurately estimate transport rates because of the difficulty in counting grains at high transport rates. In spite of these difficulties, a crude estimate of transport is found by summing the different grain velocities for each image (Fig. 5).

The streamwise velocity versus distance from the bed is plotted in Fig. 6. There is a clear trend to these data, but the scatter is again large. Much of this scatter may

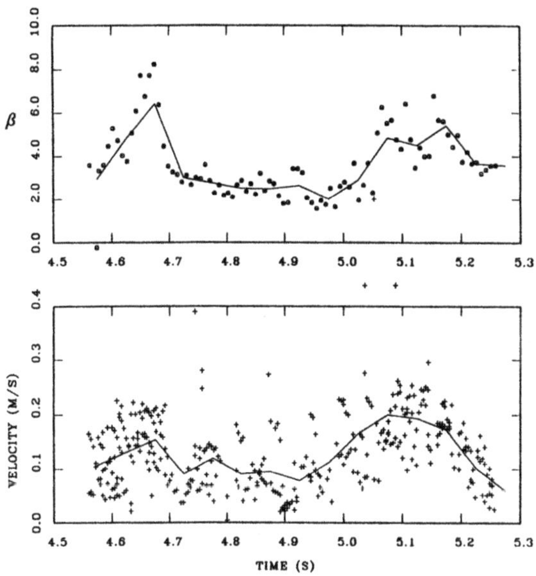

Figure 2: β and particle velocities versus time for a short time interval from Fig. 1. The values of β are integrated over an entire image, whereas the velocities are for individual particles.

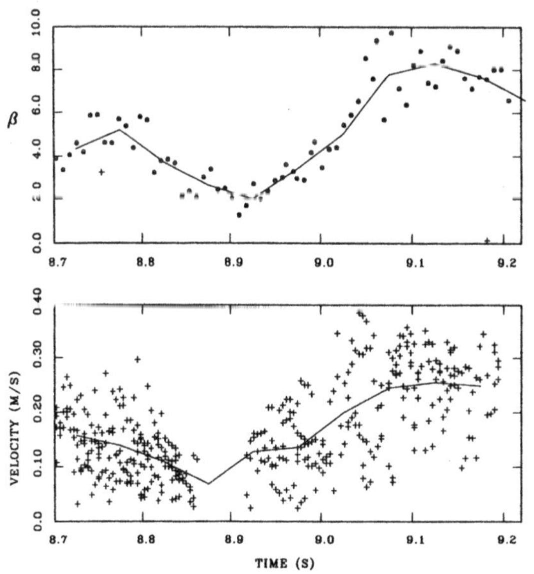

Figure 3: β and particle velocities versus time for a short time interval from Fig. 1.

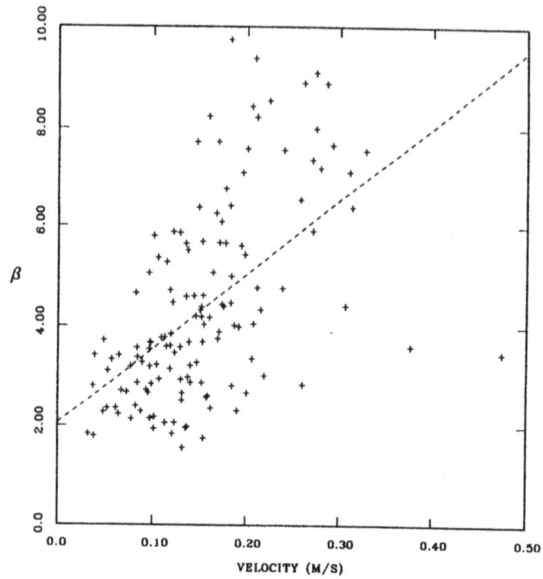

Figure 4: β versus average velocity for each image in the short time segments shown in Figs. 2 and 3. The dashed line is the linear regression, $r = 0.37$.

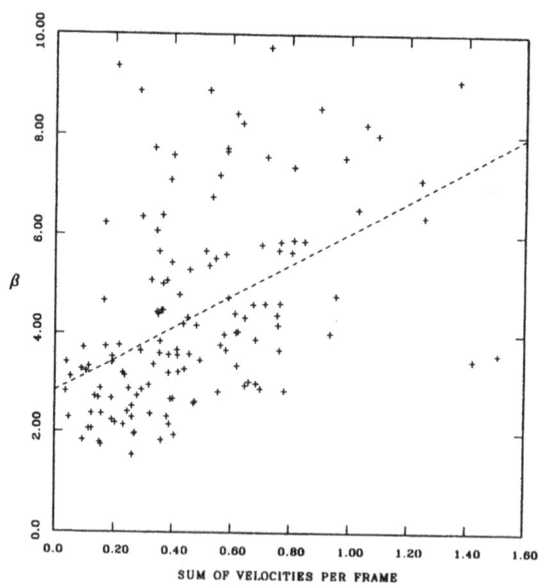

Figure 5: β versus the sum of the velocities from each image from Figs. 2 and 3. The dashed line is the linear regression, $r = 0.47$.

be due to the difficulty in making accurate position estimates because of parallax. The thickness of the light sheet was estimated to be 3-4 grain diameters; therefore, the estimated position depends on whether the particle is near the front or the back of the sheet. Consequently, both vertical positions and vertical velocities are prone to error without more information. The parallax problem can be resolved by adding a second viewing angle, but this doubles the processing. For the case here, the maximum error in absolute particle positions was on the order of 1 mm, but errors in relative particle positions between images was a small fraction of a millimeter.

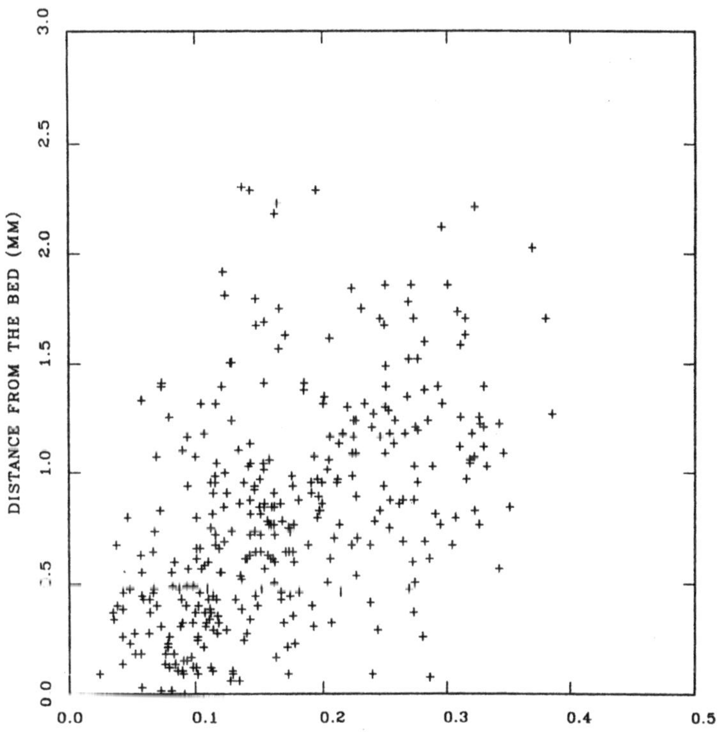

Figure 6: Velocity versus distance from the bed.

With improvements such as better illumination, higher frame rates, and better positioning techniques, a high-speed video system coupled with a fiber-optic imaging system will allow observation of near-bed sediment transport and provide quantitative estimates of rates and sediment particle velocities, resolving gradients at the millimeter scale. The fluid velocity may also be determined by seeding the flow with very small, highly reflective, neutrally buoyant particles, and tracking these smaller particles on the video images. With velocity determined at many locations within an image, correlation between sediment motion and fluid motion can be computed on a reasonable spatial scale.

4. Conclusions

Sediment-transport processes are highly nonlinear, with instantaneous transport rates linked to near-bed velocity fluctuations in a complex way. Understanding this linkage is crucial to our understanding of problems such as beach evolution, bedform dynamics, channel stability and scour around obstacles. Progress in modeling or even understanding many of the phenomena occurring in the near-bed region is hampered by the difficulty of making measurements of flow and sediment motion. Through such observations it is possible to deduce the forces arising from fluid-particle and particle-particle interactions.

Using a high-speed video system, coupled with imaging fiber technology to minimize intrusiveness, yields observations that provide valuable qualitative insight into the mechanisms that are active near the bed and, in conjunction with motion-analysis techniques, the images can provide detailed, quantitative information about both the flow and the sediment transport field that is unparalleled. With a field of view that is only a few millimeters it is possible to measure gradients that are unseen by more traditional measurement techniques. However, because of the tediousness of the analysis, it is not expected that such a system will ever be commonly used; rather it will provide as yet unavailable data that can guide us in the development and testing of realistic theoretical models.

5. Acknowledgements

We would like to express our appreciation to Florence Beaudlet Cayocca for the many hours she spent in data acquisition and analysis. We would also like to thank Don Stafford of NAC Visual Systems for the use of the equipment employed in the demonstration described herein. This work was supported by National Science Foundation grant # CTS92-17804.

6. References

Bagnold, R.A., The nature of saltation and of "bed-load" transport in water, *Proc. Royal Soc. London, A*, 332, pp 473-504, 1973.

Einstein, H. A., The bed-load function for sediment transportation in open channel flows, U. S. Dept. of Agric. Tech. Bull. No. 1026, 1950.

Fernandez-Luque, R. and R. van Beek, Erosion and transport of bed-load sediment,

J. Hyd. Res., **14**, no. 2, pp 127-144, 1976.

Maxey, M.R. and J.J. Riley, Equation of motion for a small rigid sphere in a nonuniform flow, *Phys. Fluids*, **26**(4), pp 883-889, 1983.

McLean, S.R., Modeling the transport of noncohesive sediment, Proc. OCEANS '92, Mastering the Oceans through Technology, IEEE, pp 69-73, 1992.

McLean, S.R., J.M. Nelson, and S.R. Wolfe, Turbulence structure over two-dimensional bedforms: Implications for sediment transport, to appear in *J. Geophys. Res.*, 1994.

Meyer-Peter, E. and E. Mueller, Formulas for bed-load transport, Proc. of 2nd Meeting IAHR, Stockholm, 39-64, 1948.

Nelson, J. M., S.R. McLean and S.R. Wolfe, Mean flow and turbulence fields over two-dimensional bed forms, *Water Resources. Res*, **29**, pp 3935-3953, 1993.

Wiberg, P.L. and J.D. Smith, A theoretical model for saltating grains in water, *J. Geophys. Res.*, **90**(C4), pp 7341-7354, 1985.

Wiberg, P.L., and J.D. Smith, Model for calculating bed load transport of sediment, *J. Hydraul. Eng.*, **115**(1), pp 101-123, 1989.

Yalin, M. Selim, An expression for bed-load transportation, *J. Hyd. Div.*, *ASCE*, **89**, HY3, pp 221-250, 1963.

Settlement : a sedimentary process

Martin SANCHEZ
Laboratoire de Mécanique et Géomécanique de Nantes,
Université de Nantes - 2, rue de la Houssinière, 44072 NANTES, CEDEX

and

Alain GROVEL
Laboratoire de Mécanique et Géomécanique de Nantes,
Université de Nantes - 2, rue de la Houssinière, 44072 NANTES, CEDEX

ABSTRACT

The self-weight settlement of mud is a sedimentary process acting on bed solids concentration and consequently on bed erosion rates. The paper is concerning the settlement in aqueous medium. A proposed model takes into account the mechanical phenomena with the typical properties. This approach simplifies a multi-parameters hydro-sedimentary modelling. The variability of permeability coefficient with solids concentration is the main factor considered.

The validity of the presented model has been proved by a great number of applications on varied materials [Sanchez, 1992]. Finally, this model offers two possibilities : firstly to appreciate the solids concentration above which the sediments is gathering near the bed to possibly give a lens-shaped body of pseudo-plastic mud, and secondly to detect the limit cases where the settlement is occuring quickly.

1. The phenomenon description.

The settlement particularities are varying from one material to another [Migniot, 1989]. The study of this process for a given material requires laboratory experiments and among them, the batch-sedimentation test is widely used.

The curve of the typical surface settlement versus time for a soft material is divided into five parts (Fig. 1) :

- a section from O to A corresponding to the period floculation phenomenon, the settling velocity increases gradually towards the limit velocity associated with the initial solids concentration,
- a section from A to B where the slope of the surface settlement versus time curve is constant. This slope is the velocity associated with the initial solids concentration,
- a section from B to C where the slope of the surface settlement curve decreases gradually in relation with the increment of superficial solids concentration,
- a section from C to D, the surface settlement curve is a straight line because the pulp just below the surface passed from settling to compression [Fitch, 1979],
- a section from D to ∞, the effective stresses play into settling velocity in surface. This is the consolidation phenomenon.

The three first detailed parts occur without effective stresses in surface. The settlement is therefore identified as a sedimentation phenomenon and the settling velocity depends mainly on material's permeability. From C, the effective stresses partially bear all the material, the phenomenon is thus the consolidation.

The increasing of the solids concentration by sedimentation is, in sedimentary problems, more important than increasing by consolidation, because the materials lightly concentrated are more mobile.

In the aim to propose a settlement model compatible with hydro-sedimentary modelling, the settling velocity is determined subsequently in neglecting the effective stresses. However, in order to consider the material behaviour, the settlement is considered only possible if the self-weight consolidation stresses is larger than effective stresses.

2. The material behaviour.

If the consolidation stresses applied to a material is lower than 1000 kPa its behaviour is plastic [Terzaghi, 1961], thus the settlement is not reversible. In sedimentary problems, the consolidation stresses are weak (≈ 1 kPa), and so a plastic behaviour is considered afterwards.

The effective stresses and solids concentration are linked to each material by its behaviour law. This law can be adjusted by self-weight settlement tests. There are two possible ways :

a) the first, requiring two or more batch-sedimentation tests, is consisting in a linear regression which minimizes the square difference between the measured final level and the level resulting from the behaviour law [Sanchez and Grovel, 1993] ;

b) the second way is elaborated in fonction of the solids concentration final profile, thus the effective stresses are calculated by integrating the following equation [Tan et al., 1990] :

$$\frac{\partial \sigma'}{\partial z} = - g\, C_F \left(1 - \frac{\rho_f}{\rho_s} \right) \qquad (1)$$

where σ' is the effective stress, g the acceleration of gravity, C_F the final solids concentration, ρ_s the density of solids phase, ρ_f the density of fluid phase and z the vertical cordinate. This equation has as limit condition in surface $\sigma' = 0$.

3. The velocity of the solid phase.

If the effective stresses are neglected in self-weight settlement, the velocity of the solid phase w_s, in saturated medium on an impermeable bottom is given by the following equation [Been, 1980] :

$$w_s = - \frac{k}{\rho_f} \left(1 - \frac{\rho_f}{\rho_s} \right) C \qquad (2)$$

where k is the permeability coefficient assumed as a function of only solids concentration C, then w_s is indeed a unique function of solids concentration.

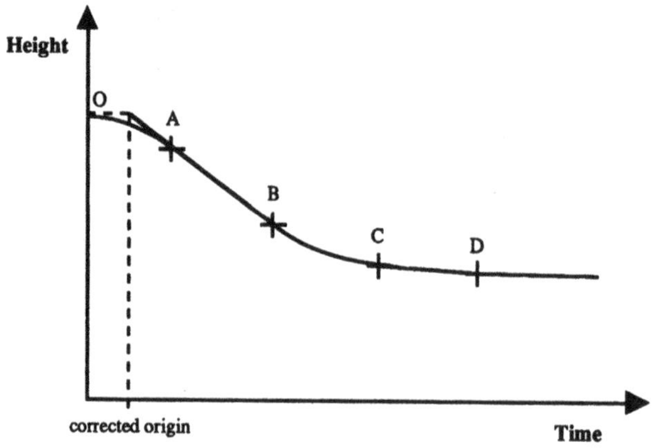

Figure 1. Typical surface settlement versus time.

Thus, the velocity of the solid phase and permeability coefficient can be determined from the surface settlement curve in using a modified Kynch's theory [Tan et al., 1990]. This approach allows the adjustment of an exponential law for the variation of permeability coefficient with solids concentration. This law used has been checked by Sanchez (1992) and can be written as follows :

$$k = A_1 \exp\left(-A_2 \frac{C}{\rho_s}\right) \qquad (3)$$

where A_1 and A_2 are parameters belonging to each material. These parameters values for an estuarine Loire sediment (liquid limit = 150 %) are : $A_1 = 0{,}005$ m s^{-1} et $A_2 = 110$; corresponding to a range of solids concentration $108 < C < 330$ kg m^{-3} [Sanchez and Grovel, 1993]. The typical values of these parameters, calculated from the Laboratoire Central d'Hydraulique de France tests [Migniot, 1989], are given in reference 5, the table n° 1 gives the parameters values corresponding to some estuarine sediments.

4. The "iso-solids concentration" settlement model.

A multi-layer discretisation (Fig. 2) allows a formulation of the settlement model valids with $w_s = f$ (C). The time variation of solid mass in each sublayer is governed by :

$$\frac{\partial M_i}{\partial t} = Q_{s/C\,i+1} - Q_{s/C\,i}\,,\ \text{avec : } M_i \geq 0 \ \text{ pour : } i = 1,2,3,\ldots,n \qquad (4)$$

where M_i is the solid mass in the sublayer i which has a solid concentration between two neighbouring values C_i and C_{i+1} ($C_i > C_{i+1}$) and $Q_{s/C\,i}$ is the solid flux through the interface i with solid concentration C_i constant in time.

If the effect of effective stresses on velocity of the solid phase is neglected, the flux $Q_{s/C\,i}$ is given by :

$$\text{for } \sigma > \sigma' : \quad Q_{s/C\,i} = C_i \times \left. \frac{d[\, w_s(C) \times C \,]}{dC} \right|_{C=C_i} - w_s(C_i) \times C_i \qquad (5)$$

$$\text{for } \sigma \leq \sigma' : \quad Q_{s/C\,i} = 0 \qquad (6)$$

where σ is the self-weight consolidation stresses. The equation 5 is written in using the characteristics of the equation of continuity of the solids for w_s fonction only of solids concentration [Kynch, 1952]. The characteristics linked to each solids concentration C_i are straight lines whose the slopes W are given by :

$$W(C_i) = \left. \frac{d[\, w_s(C) \times C \,]}{dC} \right|_{C=C_i} \qquad (7)$$

Along each characteristic line, solids concentration is constant [Kynch, 1952]. Eq. 4 can be integrated to obtain a linear variation in time of M_i. It is right for all sublayers except the upper sublayer which can be affected by a sedimentation rate Q_D from the superficial flow.

$$Q_{s/C\,n+1} = Q_D \qquad (8)$$

Finally, the equation 6 implies that the settlement is not possible if the self-weight consolidation stresses are smaller than effective stresses.

5. Applications of the model of settlement.

5.1. Formation of a pseudo-plastic suspension near the bed.

The solids concentration C_d above which the sediments are gathering near the bed to possibly give a lens-shaped body of pseudo-plastic mud depends on the sedimentation rate Q_D. The following equation linkes C_d and Q_D :

$$Q_D = C_d \left[W(C_d) - w_s(C_d) \right] \qquad (9)$$

The equation solution for each sedimentation rate, shows that an increase of Q_D value brings about a decrease of solids concentration above which the sediments are gathering near the bed.

5.2. The granular materials.

The settlement phenomena concerns all materials, however this process is usually neglected in granular material sedimentary modelling. This is justified because the permeability coefficient value of these materials is great and then, settlement is almost instantaneous. Morever, the granular material behaviour is generally incompressible which implies the non variability of solids concentration versus consolidation stress.

5.3. Weak sedimentation rate.

If rate sedimentation is weak beside solid flux through all the interfaces of bed multi-layer model, settlement is also almost instantaneous. In this case, solids concentration in the bed is linked to the highest consolidation stress applied before. A criterion allowing to detect this case can be made out if the solid concentration C_{FF} immobile beside the shear stress which is beared on by superficial flow, is known. This condition is :

$$Q_D = Q_{s/C}(C_{FF}) \tag{10}$$

This case is frequent in the weak turbidity zones (e.g. oceanic medium). The suspended matter in superficial flow settles to straight form a bed highly concentrated. The material behaviour law is sufficient to know the time variation in the bed solids concentration profile.

6. Conclusions.

The multi-layer "iso-solids concentration" model (Eq. 4) is compatible with multi-parameters hydro-sedimentary modelling. This model presents the following advantages :

a) in absence of effective stresses, this model agrees with the Kynch's sedimentation theory and for a range of weak effective stresses, the model approaches the one-dimensional consolidation theory [Been, 1980 ; Tan et al., 1990]. A good agreement with experiments confirms the validity for a great number of cohesive sediments [Sanchez, 1992],

b) on the one hand, the solid flux limit condition in silt layer surface is the sedimentation rate Q_D, on the other hand, Q_D is a term in the turbulent mass transfer equation,

c) the possible discontinuity in solids concentration profile on the superior boundary of the silt layer is the direct consequence of the application of the model, in agreement with experiments,

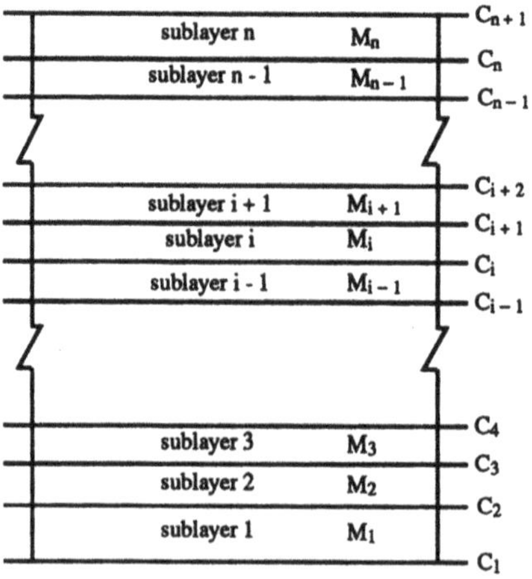

Figure 2. Used scheme in multi-layer "iso-solids concentration" model.

	parameters of k variation law A_1	A_2	correlation coefficient r	range of solids concentration C_{min}	C_{max}	floculation time factor
ESTUARINE SEDIMENTS						
MAHURY	1,8969	194,30	0,990	200	300	125000
DOUALA	0,2431	166,11	0,987	200	300	125000
OWENDO	0,2567	163,66	0,977	200	350	200000
ESCAUT	0,002968	104,44	0,983	200	350	0
LOIRE	0,006591	108,60	0,997	200	350	0
GIRONDE	0,01743	93,73	0,963	200	500	18000
LA VILAINE	0,01675	84,04	0,975	200	500	12000
ORNE	0,02000	71,61	0,990	200	510	16000
DIVES	0,01459	66,81	0,990	200	530	6000
LARACHE	0,02422	59,21	0,990	200	675	1000
TANCARVILLE	0,03099	46,54	0,990	200	675	1250

Table n° 1

d) the linearity of the equation 4 makes the settlement modelling easier,

e) the material properties linked to the solids concentration as cohesion or critical erosion shear stress can be assumed constant in each bed sublayer because $C_i \approx C_{i+1}$.

Finally, this model offers the possibility to appreciate the limit cases when the settlement has a secondary importance in hydro-sedimentary problems.

7. References.

1. K. Been, "Stress strain behaviour of a cohesive soil deposited under water", *Thesis presented to the University of Oxford*, at Oxford, England, in partial fulfillment of the degree of Doctor of Philosophy (1980).
2. B. Fitch, "Sedimentation of floculent suspensions : state of the art", *AIChE Journal*, Vol. 25, No. 6, (1979) 913-930.
3. G.F. Kynch, "A theory of sedimentation", *Faraday Society Transactions*, Vol. 48, (1952) 166-176.
4. C. Migniot, "Tassement et rhéologie des vases", *La Houille Blanche*, No. 1 et 2 (1989).
5. M. Sanchez, "Application du modèle iso-concentration de tassement dans la simulation du tassement des vases molles étudiées au Laboratoire Central d'Hydraulique de France", *Rapport de recherche No. CFL-92-006*, Nantes, octobre, (1992).
6. M. Sanchez and A. Grovel, "Modélisation du tassement sous poids propre des couches de vase molle et saturée, sur un fond imperméable", *La Houille Blanche*, No. 1, (1993) 29-34.
7. T.S. Tan, K.Y. Yong, E.C. Leong and S.L. Lee. "Sedimentation of clayey slurry", *Journal of Geotechnical Engineering*, Vol. 116, No. 6, (1990) 885-898.
8. K. Terzaghi and R.B. Peck. *Mécanique des sols appliquée*, (Dunod, Paris, 1961).

APPLICATION OF THE METHOD OF CHARACTERISTICS TO A 2-DH 2-LAYER MODEL FOR SEDIMENTATION IN RESERVOIRS

C.J. SLOFF
Faculty of Civil Engineering, Delft University of Technology, P.O.Box 5048, 2600 GA Delft, The Netherlands

ABSTRACT
Reservoirs are loosing an important part of their storage capacity due to sedimentation by turbidity currents. To simulate flow and morphology in such reservoirs a 2-D depth-averaged 2-layer model is presented. This hyperbolic eulerian model is analyzed by means of the method of characteristics, which shows how disturbances develop in the solution as physically relevant wave fronts (for example surface waves, internal waves, and bed waves). Hence the behaviour of numerical and analytical solutions depends greatly on these results. The analysis proves how for turbidity currents (or other types of density currents) with low velocities (internal subcritical) the model can be significantly reduced by decoupling flow and morphology. For this type of flow Monge equations are derived which express the characteristics geometrically. However, for swift (internal supercritical) turbidity currents is found that flow and morphology fully interact, and hence they can only be simulated with a fully coupled model. From the characteristics also the required number of boundary conditions, and a simple stability criterion for the two-layer flow is derived. The model is found to be more stable in deep reservoirs. The presented elaborations and general results can easily be applied to other types of (hyperbolic) 2-DH models.

1. Introduction

Reservoirs, such as hydropower and water supply reservoirs, are loosing their storage capacity due to accumulation of sediments from inflowing rivers. Under certain conditions a riverine inflow with large suspended-sediment concentration can plunge below the quiescent reservoir water and continue as a dense underflow. These turbidity currents are able to transport sediments up till the dam where they may settle or may be sluiced out. The nature of these turbidity currents allows us to simulate the processes by means of a two-layer stratified-flow model. For instance a one-dimensional (cross-sectionally averaged) two-layer approach for turbidity currents in submarine canyons was proposed by Parker et al. (1986).

To simulate the evolution of flow and bed morphology in a non-prismatic reservoir we have derived a two-dimensional (depth-averaged) two-layer model (2-DH model). It corresponds to an extended form of the model for sediment-laden shallow-water flow on a mobile bed. Furthermore it forms a hyperbolic system of first-order partial differential equations characterized by wave-like solutions. A useful tool in analyzing the structure of these complicated solutions, and those of other types of 2-DH hydraulic models, is *the method of characteristics*, which concerns the study of propagation of physical disturbances. Disturbances with infinitely small amplitude propagate along the characteristics in the solution space (t,x,y-space) as wave fronts, carrying information from initial and boundary conditions. As we show herein, this method supplies us with some essential information for simplifying, and analytically or numerically solving of the equations. It gives us rules for imposing boundary conditions to make the system well posed, and it can be used to formulate a stability criterium for two-layer flow.

The mathematical treatment of the characteristic theory for a general system of partial differential equations is given by Courant and Hilbert (1962). Often their treatment has been used to analyze and solve the 2-D (depth-averaged) shallow-water equations for flow on a fixed bed (without stratification), for example by Daubert and Graffe (1967) and Hirsch (1988). De Vriend (1985,1987a,b) added the influence of a mobile bed into the characteristic analysis of 2-D shallow-water flow. More general applications of the method to 2-D depth-averaged hydraulic models are given by Sloff (1992), where the theory is elaborated starting from simple shallow-water flow to fully coupled two-layer flow on a mobile bed. A general treatment of 1-D models is given by Abbott (1966).

Application of the method of the method of characteristics is pursued herein for our reservoir model. It shows how, and under which conditions, the mobility of the bed and the high sediment concentrations affect the solutions, by comparison with fixed-bed models for conservative density currents (e.g. saline density underflows). Various simplifications and rules are obtained for further elaboration of the mathematical model, explained from a physical point of view. Presented theory can easily be applied to other type of 2D models.

2. Model formulation

A definition sketch for the chosen model is given in fig. 1.

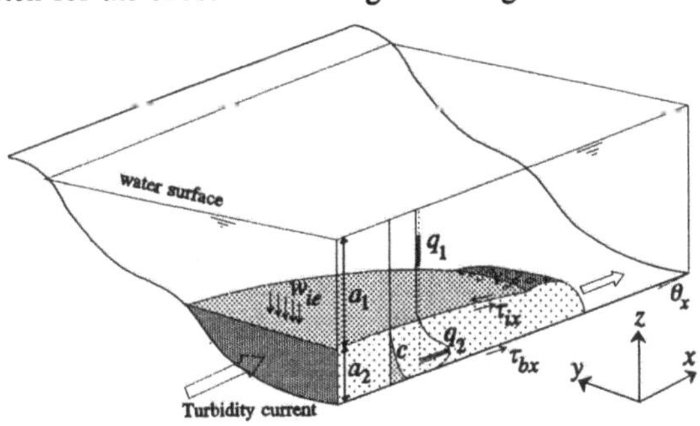

Fig.1 Definition sketch

A coordinate system is defined along the average bottom slope with x and y the horizontal coordinates and z the vertical coordinate (tilted coordinate system to account for steep bottom slope, average angles θ_x and θ_y). We assume a two-layer approach where the upper layer (layer 1) is quiescent ambient fluid (clear water), while the lower layer (layer 2) is a swift turbidity current (density excess only caused by suspension). At the interface between the layers interfacial shear stresses (τ_{ix},τ_{iy}) and interfacial mixing (entrainment velocity w_{ie}) is defined, while at the bottom bed-shear stresses (τ_{bx},τ_{by}) and bed-load transport (s_{bx},s_{by}) are assumed.

By means of depth-integration of the Reynold's equations for each layer, basic eulerian equations for motion and continuity have been derived. These derivations were carried out in an identical way as those for unstratified one-dimensional sediment-laden flow presented by Sloff (1993) neglecting horizontal diffusion. We define discharges per unit of width as q, layer depths as a, volumetric suspended-sediment concentration as C_s, bed levels as z_b, velocities as u,v, porosity of the bed as ϵ_p, and densities as ρ. The density of the lower layer is defined as $\rho_2=\rho_f(1+\sigma' C_s)$ where ρ_f is the density of clear water, $\sigma'=(\rho_s-\rho_f)/\rho_f$, and ρ_s is the density of sediment grains.

The following general system of basic equations has been derived:
Conservation of momentum in upper layer 1 (equations of motion) with $j=1,2$ and $x_1=x$, $x_2=y$, $q_{11}=q_{1x}$, $q_{12}=q_{1y}$:

$$\frac{\partial q_{1xj}}{\partial t} + \sum_{k=1}^{2}\left[\frac{\partial}{\partial x_k}\left(\frac{q_{1xj}q_{1xk}}{a_1}\right)\right] + g_z a_1 \frac{\partial(a_1+a_2+z_b)}{\partial x_j} + \frac{\tau_{i;xj}}{\rho_f} - g_{xj}a_1 = 0 \qquad (1),(2)$$

Conservation of mass in upper layer 1 (continuity equation)

$$\frac{\partial a_1}{\partial t} + \sum_{k=1}^{2}\left[\frac{\partial q_{1k}}{\partial x_k}\right] - w_{ie} = 0 \qquad (3)$$

Conservation of momentum in lower layer 2 (equations of motion) with $j=1,2$ and $x_1=x$, $x_2=y$, $q_{21}=q_{2x}$, $q_{22}=q_{2y}$:

$$\begin{aligned}&\frac{\partial q_{2xj}}{\partial t} + \sum_{k=1}^{2}\left[\frac{\partial}{\partial x_k}\left(\frac{q_{2xj}q_{2xk}}{a_2}\right)\right] + \left(\frac{\rho_f}{\rho_2}\right)g_z a_2 \frac{\partial a_1}{\partial x_j} + g_z a_2 \frac{\partial(a_2+z_b)}{\partial x_j} + \\ &+ \gamma_s \frac{1}{2} g_z a_2^2 \sigma'\left(\frac{\rho_f}{\rho_2}\right)\frac{\partial C_s}{\partial x_j} + \frac{(\tau_{b;xj}-\tau_{i;xj})}{\rho_2} - g_{xj}a_2 + \\ &+ \gamma_s q_{2xj}\sigma'\left(\frac{\rho_f}{\rho_2}\right)\left(\frac{\partial C_s}{\partial t} + \sum_{k=1}^{2}\frac{q_{2xk}}{a_2}\frac{\partial C_s}{\partial x_k}\right) = 0\end{aligned} \qquad (4),(5)$$

Conservation of mass of fluid-sediment mixture in lower layer 2 (continuity equation):

$$\frac{\partial a_2}{\partial t} + \sum_{k=1}^{2}\left[\frac{\partial q_{2xk}}{\partial x_k}\right] + \gamma_s \frac{\partial z_b}{\partial t} + w_{ie} = 0 \tag{6}$$

Sediment balance bed and lower layer (conservation of sediment mass) with $s_{s;xk}$ = suspended-load rate per unit of width = $C_s \cdot q_2$:

$$(1-\epsilon_p)\frac{\partial z_b}{\partial t} + \sum_{k=1}^{2}\left[\frac{\partial(s_{b;xk}+s_{s;xk})}{\partial x_k}\right] + \frac{\partial a_2 C_s}{\partial t} = 0 \tag{7}$$

Suspended sediment concentration lower layer (depth-averaged form of the convection-diffusion equation for suspended sediment)

$$C_{se} - C_s = T_a \frac{\partial C_s}{\partial t} + L_{ax}\frac{\partial C_s}{\partial x} + L_{ay}\frac{\partial C_s}{\partial y} \tag{8}$$

where C_{se} = the equilibrium concentration relating to the transport capacity of the flow. This equation accounts for the spatial and temporal variation of the density (i.e. the concentration) due to deposition or erosion as proposed by Galappatti (1983) and Galappatti and Vreugdenhil (1985), and later generalized by Wang (1989). The redistribution process of the concentration profile (adaptation of C_s to C_{se}) is governed by adaptation lengths L_a and adaptation time T_a.
The gravity vector (for the sloping coordinate system) is approximately equal to

$$\vec{g} = (g_x, g_y, g_z) \approx (g\cdot\sin\theta_x\cos\theta_y\ ,\ g\cdot\cos\theta_x\sin\theta_y\ ,\ g\cdot\cos\theta_x\cos\theta_y) \tag{9}$$

A dummy-coefficient γ_s is defined which is either equal to one if the flow is sediment-laden, or equal to zero if concentrations are low (Boussinesq approximation). In the following analysis we assumed the bed-load transport rate to be proportional to, and directed towards the total flow velocity:

$$s_{b;x} = \frac{u_2}{u_{2tot}} s_{btot}\ ;\ s_{b;y} = \frac{v_2}{u_{2tot}} s_{btot}\ ;\ u_{2tot} = \sqrt{u_2^2 + v_2^2}$$

Now the presented model consists of eight partial differential equations which are non-linear in their dependent variables but quasi-linear in their derivatives. It is fairly general. For instance by neglecting concentrations and density differences the model governs unsteady shallow-water flow on a mobile bed. By neglecting gradients in the sediment-transport rates the model governs flow on a fixed bed. For the characteristic analysis the model is written in the following matrix form

$$A_t \frac{\partial \vec{u}}{\partial t} + A_x \frac{\partial \vec{u}}{\partial x} + A_y \frac{\partial \vec{u}}{\partial y} = \vec{b} \tag{10}$$

where $\vec{u}$ is the vector of independent variables i.e., $(q_{1x}, q_{1y}, a_1, q_{2x}, q_{2y}, a_2, z_b, C_s)$.

3. Method of characteristics

Corresponding to the number of partial differential equations (expressed by Eq.10) eight characteristic surfaces $C{:}\phi(t,x,y)$ exist along which the discontinuities in the solutions propagate. These surfaces follow from the characteristic condition defined as (Courant and Hilbert, 1962, Sloff, 1992)

$$Q \equiv \det \left\| A_t \xi_t + A_x \xi_x + A_y \xi_y \right\| = 0 \tag{11}$$

where each vector $\vec{\xi}$, which satisfies this equation, is the vector normal to the corresponding characteristic surface $C{:}\phi(x,y,t)=0$ which can be expressed as

$$\vec{\xi} = (\xi_t, \xi_x, \xi_y) = \left(\frac{\partial \phi}{\partial t}, \frac{\partial \phi}{\partial x}, \frac{\partial \phi}{\partial y} \right) \tag{12}$$

For each $\vec{\xi}$ which satisfies Eq.11 we can define bicharacteristic directions (i.e. characteristic celerities or propagation velocities) in the t,x,y-space as

$$\frac{dx}{dt} = \frac{\partial Q / \partial \xi_x}{\partial Q / \partial \xi_t} \quad ; \quad \frac{dy}{dt} = \frac{\partial Q / \partial \xi_y}{\partial Q / \partial \xi_t} \tag{13}$$

These celerities are the directions of the bicharacteristic rays in the t,x,y-space which span the characteristic surfaces. By eliminating $\vec{\xi}$ from Eq.11 by means of Eq's.13 the characteristic equation can be expressed in terms of dx/dt and dy/dt and is then called the Monge equation.

The characteristic condition for our model is an algebraic equation in $\vec{\xi}$ of the order eight but it can be factorized into two linear and one sixth-order equation:

1. $Q_1 = 0$; First linear equation gives $dx/dt = q_{1x}/a_1$ and $dy/dt = q_{1y}/a_1$. This is a type of wave, along the stream lines in the upper layer, along which vorticity is transported (Vreugdenhil, 1989). It is a wave only observable in the velocity pattern, and does not exist in 1D theory.
2. $Q_2 = 0$; Second linear equation gives $dx/dt = L_{ax}/T_a$ and $dy/dt = L_{ay}/T_a$. This is a wave (almost located along the stream lines) in the lower layer representing convective transport of suspended sediment.
3. $Q_3 = 0$; The remaining irreducible sixth order algebraic equation in $\vec{\xi}$ cannot

be solved analytically, and therefore cannot be written as a Monge equation. The six characteristic cones expressed by equation $Q_3=0$ make up four cones (Monge cones) in the t,x,y-space.

These cones can be subdivided in two groups: 2 cones generated by circular (or elliptical) wave fronts, and 2 cones generated by star-shaped fronts. An example is given in figure 2 which shows how wave fronts develop in time originating from a point disturbance at the origin and how bicharacteristics envelop the cones.

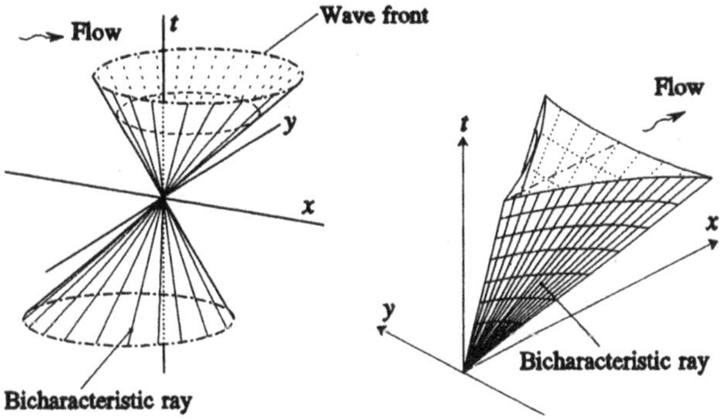

Fig.2 Monge cones for 2-DH mobile bed models

In reality the bicharacteristics are curved since they are dependent on the local values of the dependent variables (depth, discharge, etc.) which also vary in time and space. Due to this curvature wave fronts of different disturbances may intersect and allow for jumps and fronts with finite amplitude (e.g. Courant and Hilbert, 1962). This property plays an evident role in shock-capturing methods for solving the equations. The physical meaning of each cone is discussed in the following section.

We found that one of the star-shaped cones almost coincides with the stream line in the lower layer if concentrations are large so that $\gamma_s = 1$, and fully coincides with the stream line if $\gamma_s = 0$ (then Q_3 factorizes into a fifth order and another linear algebraic equation). This wave is associated to transport of vorticity in the lower layer (cf. $Q_1 = 0$).

An important difference in model-behaviour, as we show next, is found for internal sub- and supercritical flow. The flow is internal critical if the densimetric Froude numbers Fr_1 and Fr_2 satisfy

$$Fr_1^2 + Fr_2^2 = \frac{\left(q_{1x}^2 + q_{1y}^2\right)}{g a_1^3\left[1 - \left(\rho_f/\rho_2\right)\right]} + \frac{\left(q_{2x}^2 + q_{2y}^2\right)}{g a_2^3\left[1 - \left(\rho_f/\rho_2\right)\right]} = 1 \tag{14}$$

If they are smaller than 1 then the flow is internal subcritical, else supercritical.

4. Analysis of wave fronts for the 2-DH 2-layer model

Wave fronts (with infinitely small amplitude) can be obtained from the Monge equation by means of integration to t. By expressing $dx/dt = \Delta x/\Delta t$ and $dy/dt = \Delta y/\Delta t$ we can draw wave fronts developing for each value of Δt. However, for the characteristic condition $Q_3=0$ a Monge equation cannot be formulated and we used Huygen's method (Courant and Hilbert, 1962) to construct the wave fronts. Taking $\Delta t=1$ s, the lines $\mu - (\alpha_x \cdot \Delta x) - (\alpha_y \cdot \Delta y) = 0$ envelop the wave front, where the values for μ (eigenvalues) follow (numerically) from $Q_3(-\mu,\alpha_x,\alpha_y) = 0$ and where $\vec{\alpha}=(\alpha_x,\alpha_y)$ is a unit vector which has to be varied over the unit sphere. Equation $Q_3(-\mu,\alpha_x,\alpha_y) = 0$ gives six roots for μ relating to the four cones expressed by this equation (Sloff, 1992).

4.1 *Internal subcritical flow*

Consider an internal subcritical two-layer flow. These types of turbidity currents have a low sediment-transport capacity so that we may assume $\gamma_s=0$. In figure 3 the wave fronts are plotted for a mobile bed derived from condition Q_3 originating from a point disturbance (in the origin at $t=0$ s) after $\Delta t=1$ s. We used the fictitious values $q_{1x}=-0.3 m^2/s$; $q_{1y}=q_{2y}=0$ m²/s; $q_{2x}=0.5$ m²/s; $a_1=3$ m; $a_2=1$ m; exaggerated $C_s \approx 0.1$; $s_{btot}=0.001$ m²/s; $Fr_1^2+Fr_2^2\approx 0.16$.

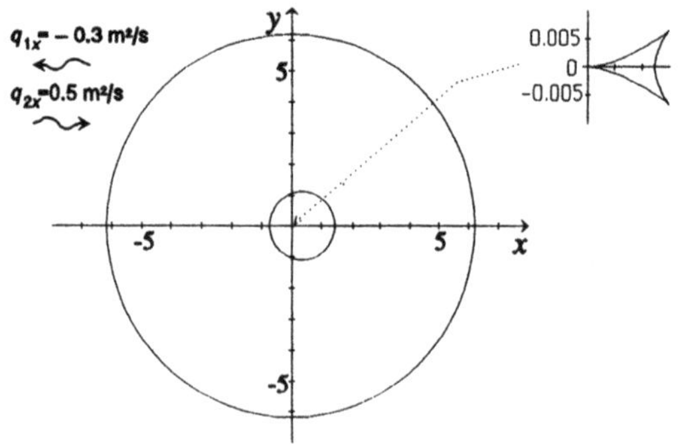

Fig.3 Wave fronts internal subcritical flow.

We found that for a fixed bed model ($\partial z_b/\partial t=0$) the two circular fronts are practically identical to those for a mobile bed in figure 3, but that the star-shaped front cannot be found. The wave fronts in figure 3 can be associated to the following physical phenomena:

- The large circular front is associated to *surface waves*. This external flow phenomenon disappears if a rigid lid model is defined. Surface waves have a relatively large propagation speed.
- The small elliptical front is associated to *internal waves* at the density

interface, which have a much smaller propagation speed than surface waves. These internal waves only occur in multiple-layer models.

- The small star-shaped front is associated to *bed waves* and is typical for all morphological models. Their propagation speed is very low.

The distinct differences in magnitude of the celerities of these waves, notably for $Fr_1^2+Fr_2^2<0.8$, allows us to simplify the model as follows:

- The propagation velocity of disturbances in the bed is small compared to those in flow. Consequently flow and morphology can be *decoupled*: Use a quasi-steady approach for computing the morphology by neglecting all derivatives to t except those for $\partial z_b/\partial t$, but by still allowing for the time variation of discharges. Consider a fixed bed for computing the unsteady flow phenomena.
- Also the difference in magnitude of the internal and external flow propagation velocities allows us to *decouple* the internal and external flow if the density differences between the layers are small.

By simplifying and reducing the characteristic equation $Q_3=0$ according to these considerations we can define Monge equations for each type of wave front separately. For external flow (surface waves) assume fixed bed and neglect density differences. Then the Monge cone practically corresponds to that of open-channel flow on a fixed bed:

$$\left[\frac{dx}{dt}-\left(\frac{q_{1x}+q_{2x}}{a_1+a_2}\right)\right]^2+\left[\frac{dy}{dt}-\left(\frac{q_{1y}+q_{2y}}{a_1+a_2}\right)\right]^2=g_z\left(a_1+a_2\right) \tag{15}$$

For internal flow (internal waves) assume fixed bed and neglect terms of order $(\varepsilon g(a_1+a_2))^2$ where $\varepsilon=1-\rho_f/\rho_2 \ll 1$ is the relative density. Then the following Monge cone can be found

$$\begin{aligned}K_2\left[\frac{dx}{dt}-\gamma_{q1}\right]^2+K_1\left[\frac{dy}{dt}-\gamma_{q2}\right]^2+2\left(\frac{q_{1x}}{a_1}-\frac{q_{2x}}{a_2}\right)\left(\frac{q_{1y}}{a_1}-\frac{q_{2y}}{a_2}\right)\left[\frac{dx}{dt}-\gamma_{q1}\right]\left[\frac{dy}{dt}-\gamma_{q2}\right]+\\+\frac{a_1a_2}{(a_1+a_2)^2}\left[\left(\frac{q_{1x}}{a_1}-\frac{q_{2x}}{a_2}\right)^2\left(\frac{q_{1y}}{a_1}-\frac{q_{2y}}{a_2}\right)^2-K_1K_2\right]=0\end{aligned} \tag{16}$$

where

$$K_1=g_z(a_1+a_2)\varepsilon-(q_{1x}/a_1-q_{2x}/a_2)^2\ ;\quad K_2=g_z(a_1+a_2)\varepsilon-(q_{1y}/a_1-q_{2y}/a_2)^2$$

$$\gamma_{q1}=\left(a_1^2q_{2x}+a_2^2q_{1x}\right)/\left(a_1a_2(a_1+a_2)\right)\ ;\ \gamma_{q2}=\left(a_1^2q_{2y}+a_2^2q_{1y}\right)/\left(a_1a_2(a_1+a_2)\right)$$

For the morphology we assume the flow to be quasi-steady and rewrite the system of equations for coordinates in the flow direction of the underflow (we assume the bed morphology to be practically independent of the quiescent ambient flow):

$$s = \frac{u_2}{u_{2tot}}x + \frac{v_2}{u_{2tot}}y \quad ; \quad n = \frac{-v_2}{u_{2tot}}x + \frac{u_2}{u_{2tot}}y \tag{17}$$

Then the following Monge equation can be found:

$$\eta_R^4 + \frac{\eta_R^2}{4\left(1 - Fr_D^2\right)}\left[8\,\sigma_R^2 - 36\,\Psi\sigma_R + 27\,\Psi^2\right] + \frac{\sigma_R^3\left[\sigma_R - \Psi\right]}{\left(1 - Fr_D^2\right)} = 0 \tag{18}$$

where $\sigma_R = \left(1 - Fr_D^2\right)\left[a_2(\mathrm{d}s/\mathrm{d}t) - S_{btot}/p'\right]$; $\eta_R = \left(1 - Fr_D^2\right)a_2(\mathrm{d}n/\mathrm{d}t)$;

$Fr_D = u_{2tot}\left(\sqrt{\varepsilon g_z a_2}\,\right)^{-1}$; $\Psi = (1/p')\left[f_b u_{2tot} - S_{btot}\left(1 - Fr_D^2\right)\right]$;

$f_b = \mathrm{d}S_{btot}/\mathrm{d}u_{2tot}$ and $p' = 1 - \epsilon_p$

This equation can be solved for η_R, yielding four roots (forming two families):

$$\eta_{R_{1,2,3,4}} = \pm\frac{1}{4}\sqrt{2}\left(\frac{\left(-27\,\Psi^2 - 36\,\sigma_R\Psi + 8\,\sigma_R^2\right)}{\left(1 - Fr_D^2\right)} \pm \sqrt{\frac{\Psi\left(9\,\Psi - 8\,\sigma_R\right)^3}{\left(1 - Fr_D^2\right)}}\right)^{1/2} \tag{19}$$

if the following conditions are satisfied: for the first family (with a positive sign before the secondary square root) must hold that

$$\frac{f_b u_{2tot}}{p' a_2\left(1 - Fr_D^2\right)} \leq \frac{\mathrm{d}s}{\mathrm{d}t} \leq \frac{9}{8}\frac{f_b u_{2tot}}{p' a_2\left(1 - Fr_D^2\right)} - \frac{1}{8}\frac{S_{btot}}{p' a_2} \tag{20}$$

For the second family (with a negative sign before the secondary square root) must hold that

$$\frac{S_{btot}}{p' a_2} \leq \frac{\mathrm{d}s}{\mathrm{d}t} \leq \frac{9}{8}\frac{f_b u_{2tot}}{p' a_2\left(1 - Fr_D^2\right)} - \frac{1}{8}\frac{S_{btot}}{p' a_2} \tag{21}$$

The general characteristic celerity for bed disturbances for the corresponding one-dimensional quasi-steady model is approximately

$$\frac{ds}{dt} = \frac{f_b u_2 \left(1 - Fr_1^2\right)}{a_2 \left(p' - \gamma_s C_s\right)\left(1 - Fr_1^2 - Fr_2^2\right) + f_b \gamma_s \left(1 - Fr_1^2 + Fr_2^2\right)} \tag{22}$$

This value corresponds to the downstream point of intersection of the two-dimensional star-shaped bed-wave front with the s-axis in flow direction. At this point the elementary behaviour of the bottom is identical with the one-dimensional behaviour. Compared to the one-dimensional models it can be noticed that at the edges of the star shaped front the two-dimensional approach yields slightly larger propagation velocities of bed disturbances.

The separate Monge equations presented above can be used to determine the Monge cones for subcritical unsteady density and turbidity currents with small density differences. They are required for numerical methods which are based on the method of characteristics.

4.2 *Internal supercritical flow*

Now consider an internal supercritical turbidity current which can be sediment-laden ($\gamma_s = 1$). In figure 4 the wave fronts are plotted for a mobile bed derived from

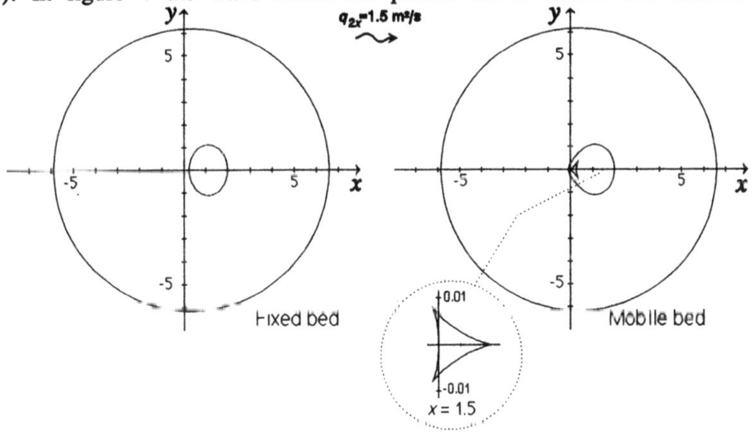

Fig.4 Wave fronts internal supercritical flow, above fixed bed; below mobile bed.

condition Q_3 originating from a point disturbance (in the origin at $t=0$ s) after $\Delta t=1$ s. We used $q_{1x}=q_{1y}=q_{2y}=0$ m²/s; $q_{2x}=1.5$ m²/s; $a_1=3$ m; $a_2=1$ m; $C_s \approx 0.1$; $f_b=0.01$ m; $s_{btot}=0.003$ m²/s; $Fr_2=1.17$.

Clearly the figure for fixed bed resembles that for internal subcritical flow, except for the internal wave front, which is now located fully downstream of the disturbance source (typical for supercritical flow). Furthermore the fixed bed and mobile bed models show a fundamentally different behaviour except for the large circular (external) wave front.

Firstly a small star-shaped front is found near the lower layer streamline (amplified in the dotted circle), which, as previously stated, we may associate to vorticity. Secondly on a mobile bed the small elliptic and star-shaped front near the origin *interact* in such a way that upstream disturbance propagation occurs. This negative propagation is usually associated to the propagation of anti-dunes on the bed.

From the strong interaction between the two wave fronts we may conclude that in internal supercritical flow the wave fronts cannot be associated to either internal waves or bed waves, and consequently the processes are fully coupled. A similar behaviour was found for supercritical shallow water flow on a mobile bed without stratification by Sloff (1992). Solving of the equations for this type of flow requires at least a fully coupled (numerical) solution method for the internal flow and the bed morphology. External flow may still be decoupled for the same reasons as in subcritical flows.

Accounting for large concentrations by taking $\gamma_s=1$ causes, besides the formation of a star-shaped vorticity wave, an increase in propagation velocities in the direction of the turbidity current and a decrease in counterflow direction. This phenomenon decreases the critical value of the densimetric Froude number and decreases the Froude number at which instability occurs (Sloff, 1993). Dependent on the concentration the critical value for the Froude numbers in Eq.14 approximately reduces from 1 to 0.8.

5. Boundary conditions

The knowledge on the development of the wave fronts can be used to determine the number of boundary conditions (b.c.) required to make the system well posed. For hyperbolic problems the number of boundary conditions, needed at any particular point at the boundary equals the number of characteristics entering the region at that point (Daubert and Graffe, 1967). Boundary conditions have to be given at each instant t at the contour of a boundary Ω. Consider the lower-layer flow-velocity normal to the boundary to be u_{nor2}; the upper-layer flow-velocity u_{nor1}. Furthermore we assume $u_{nor2} \gg u_{nor1}$. From the discussed wave fronts (external, internal, bed, sediment convection, and vorticity) the required number of boundary conditions is determined in Table I in the appendix for different flow types on a mobile bed as well as on fixed bed (Sloff, 1992). The type of boundary conditions to be given is usually determined from physical reasoning.

6. Stability of the two-layer model

Under certain flow conditions the two-layer model becomes instable when internal-wave celerities become imaginary. At that point the model becomes elliptic and breaks down. A stability (or hyperbolicity) criterion can be obtained from the approximative Monge equation for internal waves, Eq.16:

$$\left(\frac{q_{1x}}{a_1}-\frac{q_{2x}}{a_2}\right)^2+\left(\frac{q_{1y}}{a_1}-\frac{q_{2y}}{a_2}\right)^2 \leq g\varepsilon\left(a_1+a_2\right) \tag{23}$$

Also for internal supercritical turbidity currents on a mobile bed we found that this condition expresses well the stability of the flow, provided that concentrations are not too large. At large concentrations a small reduction of the right-hand side of Eq.23 is recommended (e.g. by means of trial and error).

The stability condition expressed by Eq.23 practically equals that which can be found by analyzing the Kelvin-Helmholtz instabilities (e.g. Thorpe, 1971) and requiring that the depth of a subsequent developing interfacial mixing layer is less than about 1/3 of the total depth or layer depth. Coherently also a consideration of energy of the mixing process due to Kelvin-Helmholtz instability yields the same criterion. Eq.23 shows that the two-layer model is more stable in deep-water reservoirs.

7. Discussion and conclusions

The method of characteristics is used to analyze a two-dimensional eulerian depth-averaged two-layer model for simulation of morphological processes in a reservoir with turbidity currents. It shows mathematically how a point-disturbance develops into a number of (characteristic) wave fronts (with infinitely small amplitudes) which are related to physical phenomena. Since information from boundaries is transported along characteristics it is well known that these results are important fro numerical and analytical solutions.

By comparing two-layer flow on a fixed bed to that on a mobile bed and by comparing low concentration turbidity currents to sediment-laden ones we found that:

- Besides circular wave fronts which can be associated to surface waves and internal waves, and a star-shaped wave front which can be associated to bed waves, two downstream propagating waves are found associated to transport of vorticity in the upper and lower layer respectively. Furthermore another downstream directed wave is associated to the adaptation of suspended sediment.
- For internal subcritical flow the wave fronts for surface waves (external flow), internal waves (internal flow) and bed waves differ several orders of magnitude and are practically independent. Therefore morphological computations can be carried out by means of a quasi-steady approach, or unsteady flow computations can be carried out with a fixed bed. Also external

and internal flow computation may be decoupled if density differences are small. These adjustments greatly simplify the model, and allow for analytical formulations of the wave fronts.

- For internal supercritical flow the internal and morphological processes are fully coupled and require a fully coupled computation. Still external and internal flow can be decoupled for small density differences.
- At high concentrations the propagation speed of disturbances in flow direction increases, while that in counter-flow direction decreases. For that reason the flow may be become critical (and also instable) at lower Froude numbers.
- The intersection of characteristic wave fronts plays an evident role in computing shocks and jumps in the solutions, because these discontinuities develop due to intersection of these fronts.
- The required number of boundary conditions for numerical or analytical solution directly follow from the number of characteristic entering through the boundary.
- A stability criterion for two-layer flow follows from the condition for hyperbolicity of the model. In deep-water reservoirs the flow remains stable up to higher densimetric Froude numbers than in shallow reservoirs.

The presented elaboration of the theory for characteristic analysis can easily be used for other 2-D models by means of simplification of the model equations and of the results. It is a useful tool in analysing a complicated model and a first step to be made for numerical solution.

8. Acknowledgements

The writer would like to express his gratitude to Prof. Dr. M. de Vries, Delft University of Technology, Department of Civil Engineering, for his encouraging support and guidance. Also Dr. Z.B. Wang, Dr. E. Mosselman, and Dr. H.J. De Vriend are greatfully acknowledged for their help and valuable comments.

9. Notations

a_1	= depth of ambient fluid	[L]
a_2	= depth of turbidity current	[L]
C_s	= depth-averaged suspended-sediment concentration	[-]
C_{se}	= depth-averaged equilibrium	[-]
Fr_1, Fr_2	= densimetric Froude numbers	[-]
$\bar{g}$	= (g_x, g_y, g_z) = gravity acceleration vector	$[LT^{-2}]$
L_{ax}, L_{ay}	= adaptation length for suspended-sediment concentration	[L]
q_{1x}, q_{1y}	= discharge per unit of width of ambient fluid	$[L^2T^{-1}]$
q_{2x}, q_{2y}	= discharge per unit of width of turbidity current	$[L^2T^{-1}]$

s_{bx}, s_{by}	=	bed-load transport discharge per unit of width	$[L^2T^{-1}]$
s_{sx}, s_{sy}	=	suspended-load transport discharge per unit of width	$[L^2T^{-1}]$
s_{btot}	=	total bed-load discharge per unit of width = $\sqrt{(s_{bx}^2+s_{by}^2)}$	$[L^2T^{-1}]$
t	=	time coordinate	$[T]$
T_a	=	adaptation time for suspended-sediment concentration	$[T]$
u_1, u_2	=	depth-averaged velocities in x-direction	$[LT^{-1}]$
u_{2tot}	=	total flow velocity of turbidity current = $\sqrt{(u_2^2+v_2^2)}$	$[LT^{-1}]$
v_1, v_2	=	depth-averaged velocities in y-direction	$[LT^{-1}]$
w_{ie}	=	interfacial entrainment velocity	$[LT^{-1}]$
x, y, z	=	spatial coordinates	$[L]$
z_b	=	bed-level	$[L]$
γ_s	=	dummy variable (= 1 for sediment-laden flow, else: = 0)	[-]
ε	=	relative density difference = $1 - \rho_f/\rho_2$	[-]
ϵ_p	=	porosity of the bed = $1 - p'$	[-]
$\underline{\xi}$	=	(ξ_t, ξ_x, ξ_y) = vector normal to a characteristic surface	$[LT^{-1}, -, -]$
ρ_f	=	density of clear water	$[ML^{-3}]$
ρ_s	=	density of sediment grains	$[ML^{-3}]$
ρ_2	=	density of turbid flow: $\rho_2=\rho_f(1+\sigma' C_s)$	$[ML^{-3}]$
σ'	=	relative density of sediment = $(\rho_s-\rho_f)/\rho_f$	[-]
τ_{bx}, τ_{by}	=	bed-shear stresses	$[ML^{-1}T^{-2}]$
τ_{ix}, τ_{iy}	=	interfacial-shear stresses	$[ML^{-1}T^{-2}]$

10. References

Abbott, M.B. (1966) *An introduction to the method of characteristics*. Thames & Hudson, London, and American Elsevier, New York, 243 pp..

Courant, R. and Hilbert, D. (1962) *Methods of mathematical physics, Vol. II: Partial differential equations*. Interscience publishers, New York, 830 pp.

Daubert, A., Graffe, O. (1967) Quelques aspects des écoulements presque horizontaux a deux dimensions en plan et non permanents application aux estuaires. La Houille Blanche No.8, p.847-860 (in French).

Galappatti, R. (1983) A depth-integrated model for suspended sediment transport. Comm. on Hydr., Report No. 83-7, Delft Univ. of Techn., 114 pp.

Galappatti, R. and Vreugdenhil, C.B. (1985) A depth-integrated model for suspended sediment transport. J. Hydr. Res., IAHR, Vol.23, No.4, p.359-377.

Hirsch, C. (1988) *Numerical computation of internal and external flows*; Vol.1, Fundamentals of numerical discretization. Wiley Series in Numerical Methods in Engrg., J. Wiley & Sons, England, 515 pp.

Parker, G., Fukushima and Y., Pantin, H.M. (1986) Self-accelerating turbidity currents. J. Fluid Mech., Vol.171, oct. 1988, p.145-181.

Sloff, C.J. (1992) The method of characteristics applied to analyze 2DH hydraulic models. Comm. on hydr. and geotechn. engrg., Report No. 92-4, Delft Univ.

of Techn., The Netherlands, 62 pp.

Sloff, C.J. (1993) Analysis of basic equations for sediment-laden flows. Comm. on hydr. and geotechn. engrg., Report No. 93-8, Delft Univ. of Techn., The Netherlands, 54 pp.

Thorpe, S.A. (1971) Experiments on the stability of stratified shear flows: miscible fluids. J. Fluid Mech., Vol.46, p.299-319.

Vreugdenhil, C.B. (1989) *Computational hydraulics, an introduction.* Springer Verlag, Berlin, 182 pp.

Vriend, H.J. de (1987a) Analysis of 2DH morphological evolutions in shallow water. J. Geoph. Res, No.92, p.3877-3893.

Vriend, H.J. de (1987b) 2DH mathematical modelling of morphological evolutions in shallow water. Coastal Engineering, Vol.11, p.1-27

Wang, Z.B. (1989) Mathematical modelling of morphological processes in estuaries. Communications on Hydr. and Geotechn. Engrg., Report No.89-1, Delft Univ. of Techn., Delft, 208 pp (doctoral thesis).

11. Appendix

The required number of boundary conditions (b.c.), for different types of two-layer 2-DH models is presented in Table I.

Table I Number of boundary conditions for 2-DH stratified flow

Internal subcritical density current		b.c. mobile	b.c. fixed	Internal supercritical density current		b.c. mobile	b.c. fixed
lower layer	inflow $u_{nor2}<0$	4	3	lower layer	inflow $u_{nor2}<0$	4	4
	outflow $u_{nor2}>0$	2	2		outflow $u_{nor2}>0$	2	1
	solid $u_{nor2}=0$	2	2		solid $u_{nor2}=0$	2	2
upper layer	inflow $u_{nor1}<0$	1	1	upper layer	inflow $u_{nor1}<0$	1	1
	outflow $u_{nor1}>0$	0	0		outflow $u_{nor1}>0$	0	0
	solid $u_{nor1}=0$	0	0		solid $u_{nor1}=0$	0	0
susp. trans-port	inflow $u_{nor2}<0$	1	-	susp. trans-port	inflow $u_{nor2}<0$	1	-
	outflow $u_{nor2}>0$	0	-		outflow $u_{nor2}>0$	0	-
	solid $u_{nor2}=0$	0	-		solid $u_{nor2}=0$	0	-

PREDICTION OF SHORELINE CHANGE AND GRAIN-SIZE SORTING AT THE SENDAI BAY COAST

HITOSHI TANAKA
Department of Civil Engineering,
Tohoku University, Aoba, Sendai 980, Japan

and

MASASHI SUZUKI
Japan Tokai Railway Company, Nakamura, Nagoya 450, Japan

ABSTRCT

A mathematical model is proposed for predicting shoreline change as well as grain size sorting along a beach. The model consists of three portions: (i) the wave transformation model based on the wave ray method in collaboration with Goda's wave breaking criterion, (ii) the shoreline change model and (iii) the model for predicting the fraction of bed material composition. In the third part of the present model, the conservation equation for each fraction of mixed sediment proposed by Hirano is modified for a computation on a beach. Furthermore, in order to compute the transport rate of littoral drift of each grain size, the longshore sediment transport rate formula proposed by Kamphuis et al. is extended to mixed sand. The threshold angle for incipient motion of longshore sediment is also considered in the formula. In the model applications, an idealized beach consists of mixed sand, being bounded by two groins, is firstly chosen. The difference between the results of the present model and that of the conventional model is discussed. Secondly, the model is used to compute a time-variation of the shoreline and the grain size frequency at the Sendai Bay Coast in Japan.

1. Introduction

Several mathematical models have been proposed for predicting shoreline change due to wave motion (see e.g. Horikawa [1]). Among these models, the one-line theory originated by Pernald-Considere [2] has extensively and successfully been applied to a computation of long-term shoreline change. As an extension based on this predictive model, more sophisticated models such as two-line model and multi-line model have also been proposed in order to deal with both cross-shore and longshore sediment (Bakker et al. [3] ; Perlin and Dean [4]). In these models, in general, the grain size on a beach is not considered or assumed to be represented by a uniform sand grain. In reality, temporal and spatial variation of sand grain can be observed along a shoreline. For example, sorting process on a beach after artificial nourishment is of practical importance for coastal engineers. Another example can be seen in a coastal area where a certain amount of sediment discharge is supplied from eroding rocky coasts or river mouths. However, all the mathematical models mentioned above are unable to provide any information of the grain size sorting along a beach due to wave motion.

In contrast, in the field of river sedimentation, mathematical models have already been proposed for computing grain–size sorting as well as change of river bed elevation along a channel, and it has been confirmed that these models reproduce quite reasonable results (e.g. Ashida and Michiue [5]; Hirano [6]; Proffit and Sutherland [7]; Shen and Lu [8]).

In the present study, a mathematical model is proposed to predict a longshore distribution of the grain size as well as corresponding shoreline change, so that the deficiency of the previous models may be overcome. The model can be constructed by combining the one–line model and a sorting model which was originally proposed by Hirano [6] for an alluvial channel. The present model is first applied to an idealized hypothetical beach bounded by two groins, and secondly to the Sendai Bay Coast in Japan, in which two rivers pour to the ocean, resulting in distinct spatial variation of the grain size.

2. MATHEMATICAL MODEL

2.1. Wave Model

In the computation of wave transformation, the wave ray method proposed by Munk and Arthur [9] is employed in collaboration with Goda's wave breaking criterion (Goda [10]). The equations used are given as:

$$\frac{d\alpha}{d\xi}=\frac{1}{C}(\sin\alpha\frac{\partial C}{\partial x}-\cos\alpha\frac{\partial C}{\partial y}) \tag{1}$$

$$\frac{d^2b}{d\xi^2}-p(\alpha)\frac{db}{d\xi}+q(\alpha)b=0 \tag{2}$$

in which

$$p(\alpha)=\cos\alpha\frac{1}{C}\frac{\partial C}{\partial x}+\sin\alpha\frac{1}{C}\frac{\partial C}{\partial y} \tag{3}$$

$$q(\alpha)=\sin^2\alpha\frac{1}{C}\frac{\partial^2 C}{\partial x^2}-2\sin\alpha\cos\alpha\frac{1}{C}\frac{\partial^2 C}{\partial x\partial y}+\cos^2\alpha\frac{1}{C}\frac{\partial C}{\partial y^2} \tag{4}$$

where α the wave angle, b the width of adjacent two wave rays, C the wave celerity, (x,y) the plane coordinate system and ξ is the coordinate in the direction of wave ray. The wave breaking criterion of Goda is expressed as:

$$\frac{H_b}{L_0}=0.17[1-\exp[-1.5\frac{\pi h_b}{L_0}(1+15m^{4/3})]] \tag{5}$$

where H_b the breaking wave height, h_b the water depth at the breaking point, m the beach slope, and L_0 the deep water wave length.

The wave reflection can also be calculated by use of the same method as the computation method of wave refraction above. Furthermore, the wave diffraction is calculated by using the angular spreading method (Goda [11]).

2.2. *Equations of Conservation of Sediment Mass*

Regarding the computation of shoreline change, we can use the following conservation equation which has been commonly applied in the previous studies (see Fig.1).

$$\frac{\partial x_s}{\partial t} = -\frac{1}{D}\frac{\partial Q_y}{\partial y} \tag{6}$$

where x_s the position of the shoreline from a reference line, t the time, Q_y the longshore sediment transport rate, D the depth of sediment transport zone and y the alongshore coordinate.

To simulate corresponding time-variation of grain size along a beach, a conservation equation for each fraction of sediment should be solved as well. In the field of river engineering, the sorting of mixed sand has extensively been studied by many researchers in connection with a formation of armor coat. Among them, Hirano [6] proposed a concept of a surface bed layer, in which time-variation of the bed material composition takes place, and this theory was applied to the prediction of an armor coat formation in rivers. Hirano's method has been extended to on-offshore sand movement due to wave motion (Tanaka [12]) and that due to wave-current coexistent flow (Tanaka [13]).

For the use of prediction of grain-size sorting on a beach in alongshore direction, the conservation equations proposed by Hirano can be modified as follows (see Fig.1).

(i)In case of shoreline advance:

$$\frac{\partial i_b}{\partial t} = -\frac{m}{AD}\left[\frac{\partial i_B Q_y}{\partial y} - i_b\frac{\partial Q_y}{\partial y}\right] \tag{7}$$

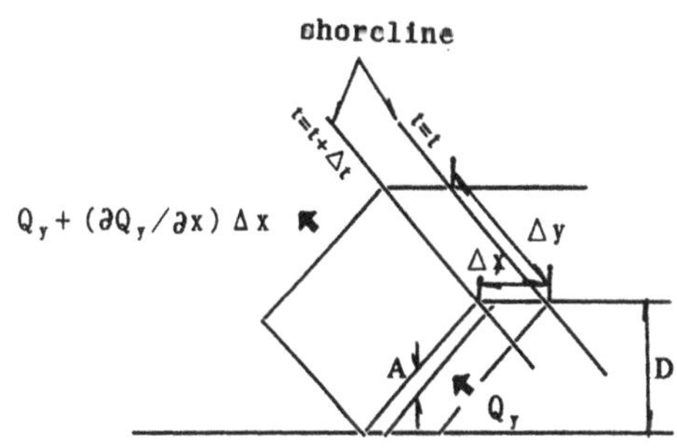

Fig.1 Conservation of sediment mass.

(ii)In case of shoreline retreat:

$$\frac{\partial i_b}{\partial t}=-\frac{m}{AD}[\frac{\partial(i_B Q_y)}{\partial y}-i_{b0}\frac{\partial Q_y}{\partial y}] \tag{8}$$

where i_b, i_{b0} and i_B the fraction of the sand grain of i-th rank contained in the surface bed layer, the bed material beneath the surface bed layer and the transported sand, respectively, and A the thickness of the surface bed layer.

2.3. Longshore Sediment Transport Rate

The sediment transport rate formula which is expressed as a function of the longshore component of the wave energy and the wave angle has been used in most of the previous simulation methods of shoreline change. Most commonly used formula is given by (Komar and Inman [14]; CERC [15]):

$$I_y=K(Ec_g)_b\sin\alpha_b\cos\alpha_b \tag{9}$$

where I_y the immersed weight of sediment transport rate, K an empirical coefficient, $(Ec_g)_b$ the incident wave energy flux evaluated at the breaker line and α_b the breaking wave angle to the shoreline. However, this type of formula cannot be applied in the present study, since the effect of the grain size is not involved in the above equation. Thus, the formula proposed by Kamphuis et al. [16] will be adopted and extended to mixed bed material as follows, by using the similarity hypothesis proposed by Hirano [6] and Ashida and Michiue [5].

Kamphuis et al. [16] obtained a energy-related littoral sand transport rate formula in which grain size effect is explicitly included:

$$Q_y'=0.001\rho(\frac{g}{2\pi})^{1/2}\frac{mH_b^{7/2}}{d}\sin2\alpha_b \tag{10}$$

where Q_y' the sediment mass transport rate in Kg/s, ρ the density of sea water, and d the grain size of uniform sand. Using empirical relationship proposed by Kamphuis and Sayao [17] to relate beach slope m to H_b/d:

$$m=1.8(\frac{H_b}{d})^{-1/2} \tag{11}$$

Eq.10 can be simplified as:

$$Q_y'=2.3\frac{H_b^3}{d^{1/2}}\sin2\alpha_b \tag{12}$$

Here, it should be noted that in the above equation, critical velocity or critical energy for the incipient motion of sediment movement is not contained. In an alluvial channel, it is known that the selective sediment movement takes place due to the difference of critical shear stress for initial motion of each grain size. In the field of coastal sedimentation, however, there have been quite less number of longshore sediment transport rate formula in which such a critical external force for incipient motion is included. Only one exception is the study by Kraus et al. (1988). Thus, incorporating a critical inception for transport presented by Kraus et al., Eq.12 can be modified as:

$$Q_y' = 2.3\frac{H_b^3}{d^{1/2}}\sin 2\alpha_b\left[1 - \frac{1.75m\gamma^{1/2}}{H_b^{5/2}\sin 2\alpha_b}\right] \tag{13}$$

where γ the breaker index $(=(H/h)_b)$.

Finally, the similarity hypothesis proposed by Hirano [6] and Ashida and Michiue [5] for alluvial sedimentation will be applied to the longshore movement of sediment with non-uniform grain size. The basic idea of the similarity hypothesis is as follows.

The sediment transport rate with uniform grain size is usually expressed as:

$$q^* = f(\tau^*, \tau^*_c) \tag{14}$$

where $q^*=q/(sgd^3)$, q the sediment transport rate per unit width per unit time, s the immersed specific gravity of the sediment, g the gravity acceleration, $\tau^*=u^{*2}/(sgd)$, u^* the shear velocity, $\tau^*_c=u^{*2}_c/(sgd)$, and u^*_c is the critical shear velocity for incipient motion of sand particle. According to the similarity hypothesis of Hirano [6] and Ashida and Michiue [5], Eq.14 can be generalized for mixed sediment:

$$q^*_i = f(\tau^*_i, \tau^*_{ci}) \tag{15}$$

where $q^*_i=i_Bq/(i_b(sgd_i^3))$, $\tau^*_i=u^{*2}/(sgd_i)$, $\tau^*_{ci}=u^{*2}_{ci}/(sgd_i)$, d_i the grain size of sand particle of i-th lank, and u^*_{ci} is the critical shear velocity for the incipient motion of sand with grain size d_i.

By use of the above hypothesis, Eq.13 may be extended to sediment mixture.

$$i_BQ_y = i_b\frac{0.0014H_b^3}{d_i^{1/2}}\sin 2\alpha_b\left[1 - \frac{1.75m\gamma^{1/2}}{H_b^{5/2}\sin 2\alpha_b}\left(\frac{d_i}{d_m}\right)^{0.08}\right] \tag{16}$$

here d_m the mean diameter in a surface bed layer. The dependence on grain size in the braces, $(d_i/d_m)^{0.08}$, is based on the laboratory study of Tanaka [19]. Eq.16 can be substituted into Eqs.7 and 8 to compute i_b.

2. Numerical Computaion

The governing equations for wave transformation, Eqs.1 and 2, are solved by use of the Runge-Kutta method, whereas Eqs.6, 7 and 8 are solved simultaneously by means of an explicit finite difference method.

At first, the present model is applied to a hypothetical beach with the length of 2km bounded by two groins as seen in Fig.2. It is assumed that artificial beach nourishment is carried out in the central region of the beach. In this case, the wave transformation process is not computed, but the wave height and the wave angle at the breaking point are assumed to be constant along the beach. Furthermore, the wave diffraction is not considered. In this simplified situation, an analytical solution

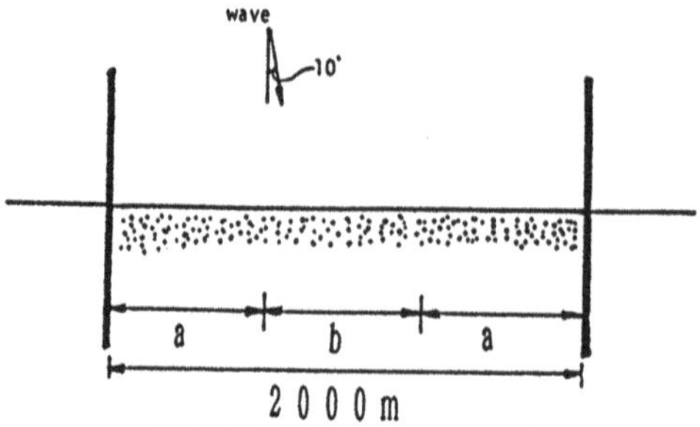

Fig.2 Beach between two groins.

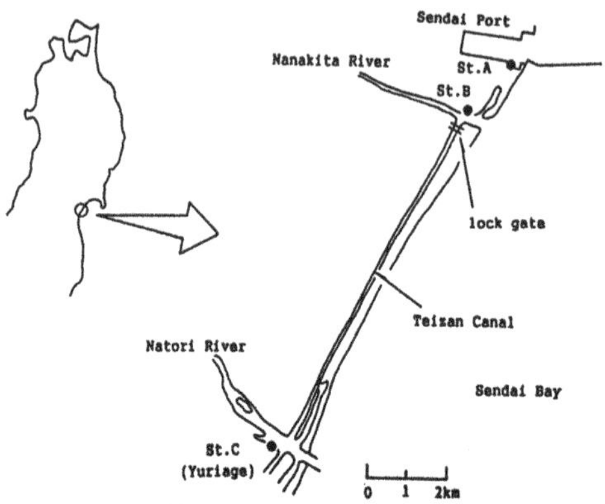

Fig.3 Map of the Sendai Bay Area.

Table 1 The Natori river and the Nanakita river.

	Natoti River	Nanakita River
Catchment Area (km^2)	984	229
Length (km)	55	45
Estimated rate of sediment yield ($m^3/year$)	12,000	3,000

is possible for a beach with uniform sand (Tanaka and Nadaoka [20]).

Secondly, the model is applied to the Sendai Bay Coast (see Fig.3), in which two rivers, the Natori river and the Nanakita river, pour to the Pacific Ocean. As seen in Fig.3, the coastal area considered in this study is about 12km long, and thesouthern and the northern ends of the coast are bounded by the jetties at the Natori river mouth and by the breakwater at the Sendai Port, respectively. The catchment area, the length and the estimated rate of sediment yield of each river are summarized in Table 1. As seen in the table, the sediment supply from the Natori river is more predominant. Detailed description about this study area can be found elsewhere (Sato et al. [21]; Tanaka [22]; Tanaka [23]).

The estimation method of the thickness of a bed surface layer, A, is still unclear even in an alluvial channel flow. In the present computations, the thickness is assumed to be 10cm, referring to the field study of Kraus [24].

4. Results

4.1. Hypothetical Beach

In the first computation, the time-variation of shoreline change and the corresponding grain-size sorting after a beach nourishment will be considered. The beach configuration is already shown in Fig.2, and the initial fraction of sand grain composition in each region is given in Table 2. The other conditions are specified as follows: m=1/40, A=10cm, D=5m, H_b=1m and α_b=10°.

Table 2 Initial composition of grain size.

	d_i				d_m
	0.1mm	0.3mm	0.6mm	1.0mm	(mm)
Region A	30%	50%	20%	0%	0.3
Region B	0%	20%	40%	40%	0.7

Figure 4 shows time-variation of the shoreline change, the distribution of mean grain size and the frequency distribution of bed material in the surface bed layer. The thicker and the thinner lines in the upper figure are the numerical solution based on the present model and the analytical solution for uniform sand derived by Tanaka and Nadaoka [20], respectively. The thicker and the thinner line in the middle figure are computed result at each time step and the initial profile, respectively. And the numbers from 1 to 4 in the lowest figure corresponds to the fraction of sand grain with the diameter of 0.1mm, 0.3mm, 0.6mm and 1.0mm, respectively. The computed shoreline changes more quickly as compared with the analytical solution for uniform sand. It is observed that the finer sand is predominant in the surface bed layer at the downdrift end in Figs.4(b) and 4(c). Thus, it can be concluded that the faster shoreline change in the present computation is due to the higher mobility of the sand grains with smaller diameter.

Comparing Figs.4(c) and 4(d), it is seen that the grain size distribution has already attained to its equilibrium stage at t=115.0 days, while the shoreline is still changing from time to time at this moment. In Fig.4(d), the shoreline has also attained to final profile. There can not be any difference in the final shoreline position between the present result and the solution for uniform sand, because, at the equilibrium stage, the shoreline must be perpendicular to the wave direction, independent on the sand grain property whether mixed or uniform.

As seen in Fig.5, the convex profile of d_m propagates to the downdrift

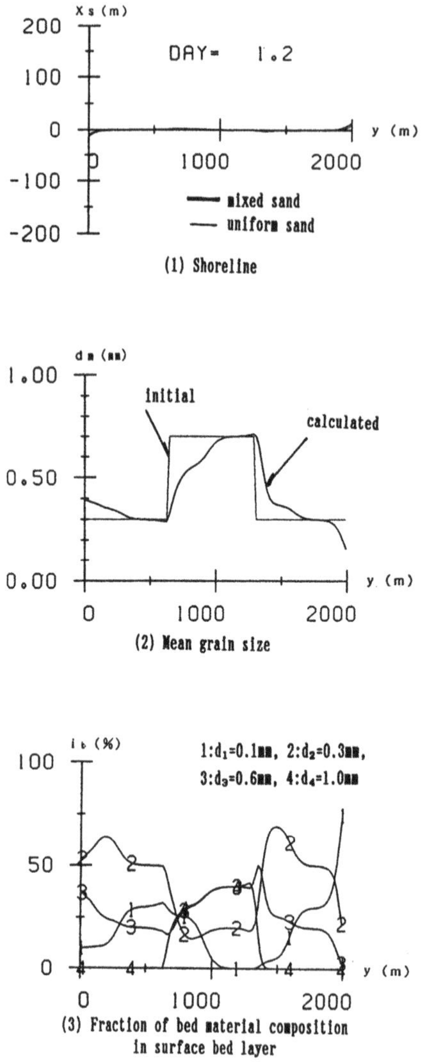

Fig.4(a) Shoreline and grain size change (t=1.2days).

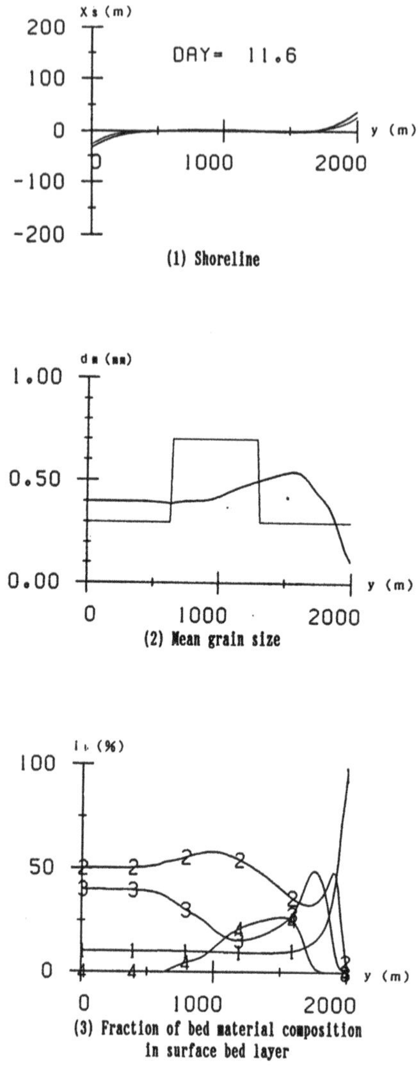

Fig.4(b) Shoreline and grain size change (t=11.6days).

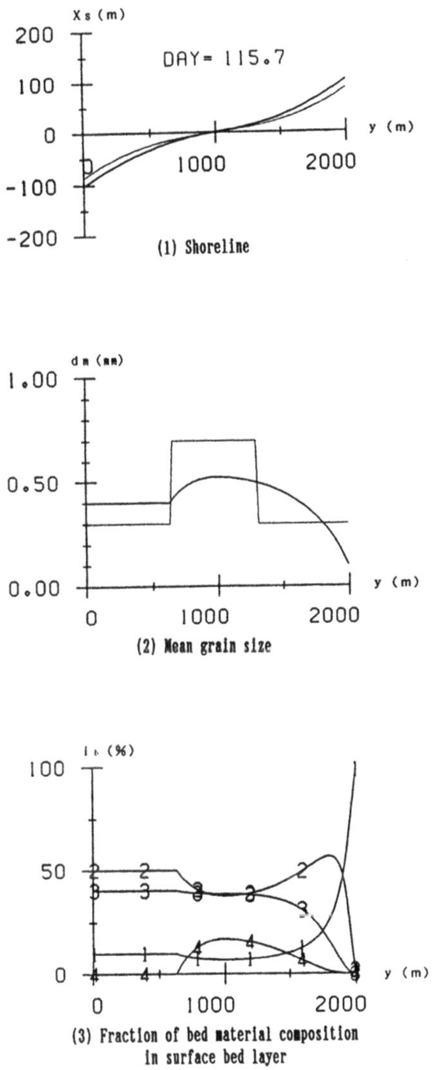

Fig.4(c) Shoreline and grain size change (t=115.7days).

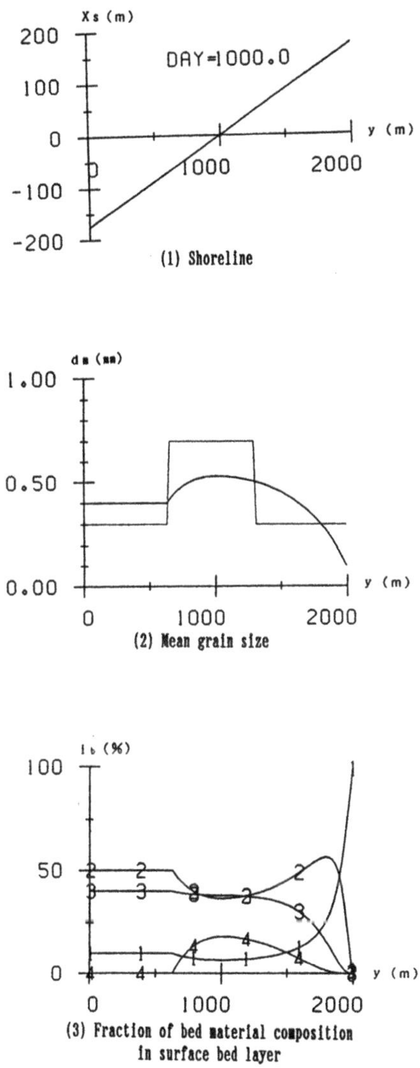

Fig.4(d) Shoreline and grain size change (t=1000days).

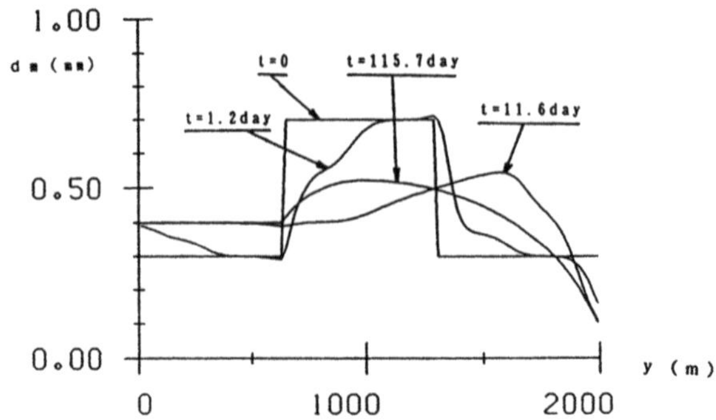

Fig.5 Time-variation of d_m.

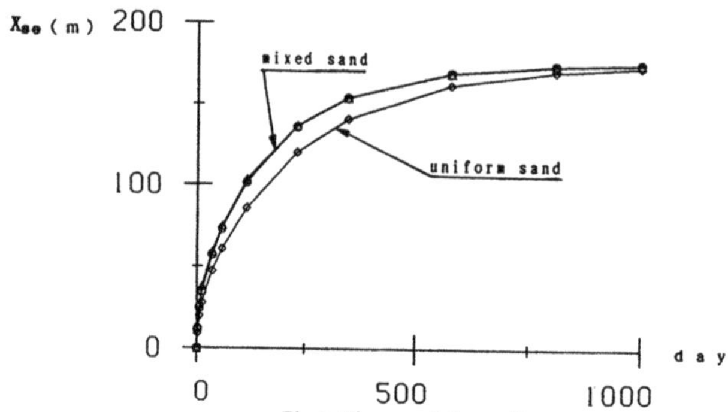

Fig.6 Time-variation of x_{se}.

direction with time. This is due to the fact that the Eqs.7 and 8 are fundamentally one-dimensional wave equations (Tanaka [12]).

The computed time-variation of the position of advanced shoreline at y=2000km is plotted in Fig.6. Furthermore, the result of the analytical solution isshown by open diamonds. As already seen in Fig.4, the computed result of the present study attains to its equilibrium beach profile more quickly as compared with the solution for uniform sand beach.

4.2. Sendai Bay Coast

In Fig.7, the shoreline change, the longshore distribution of mean grain size, and the fraction of bed material composition in surface bed layer are shown for t=1, 3, 5, and 7 years, where the definition of y-axis can be seen at the top of Fig.7(a). In the figure of d_m, the thinner and thicker lines are the initial condition based on Sato et al. [21] and computed result, and in the figure of i_b, the numbers from 1 to 4 mean the fractions of sand grain with the diameter of 0.2mm, 0.4mm, 0.6mm and

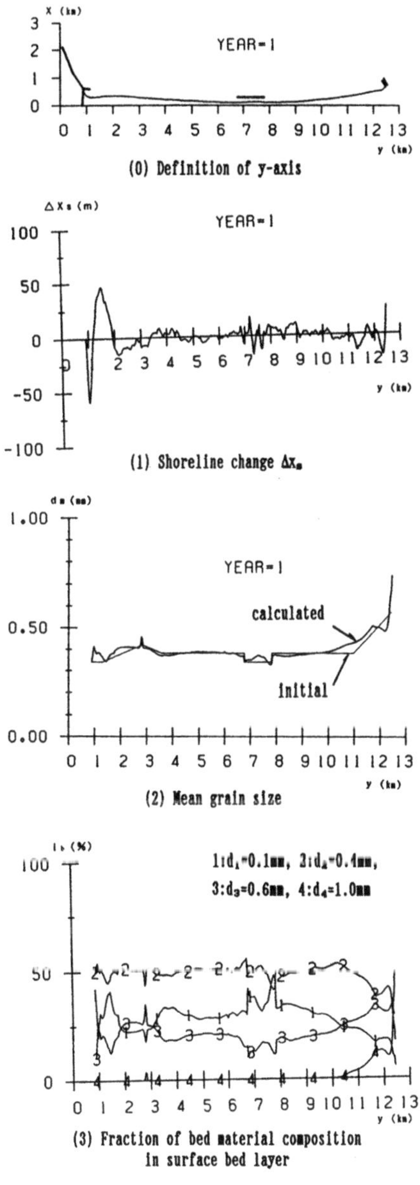

Fig.7(a) Shoreline change at the Sendai Bay (t=1year).

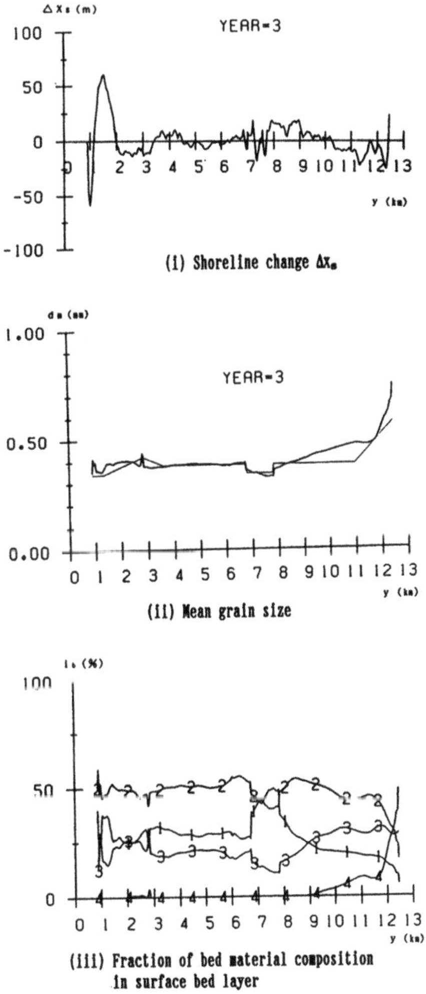

Fig.7(b) Shoreline change at the Sendai Bay (t=3years).

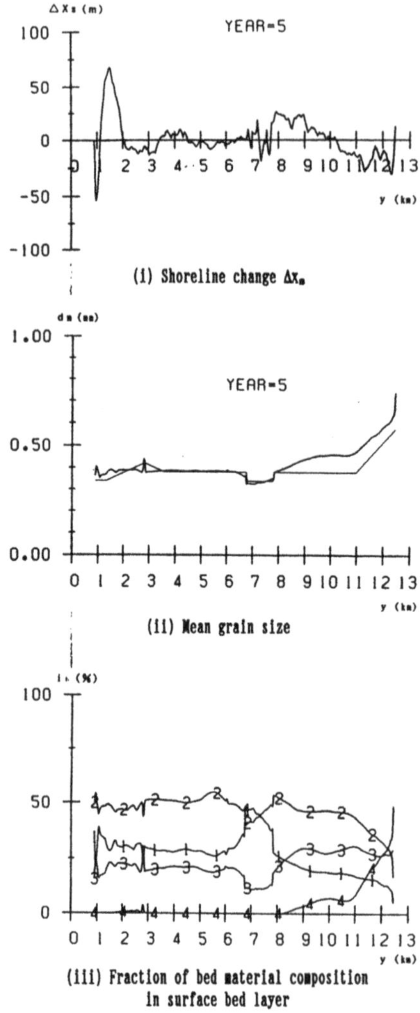

Fig.7(c) Shoreline change at the Sendai Bay (t=5years).

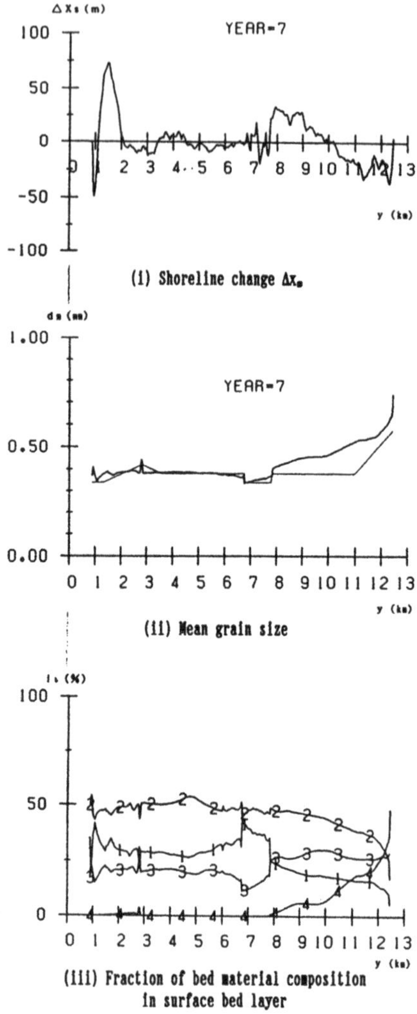

Fig.7(d) Shoreline change at the Sendai Bay (t=7years).

1.0mm, respectively. The computed time–variation of longshore distribution of d_m is summarized in Fig.8.

As reported by Sato et al. [21], the predominant direction of longshore sediment movement along this coast is from the south to the north. Thus, thesediment discharge from the Natori river mouth is transported from the right to the left in Figs.7 and 8. Accordingly, the coarser sand grains propagate in the northern direction from the Natori river mouth as depicted in Fig.8. Another peak is seen in front of the Nanakita River mouth at y=2.8km, induced by sediment supply from the mouth.

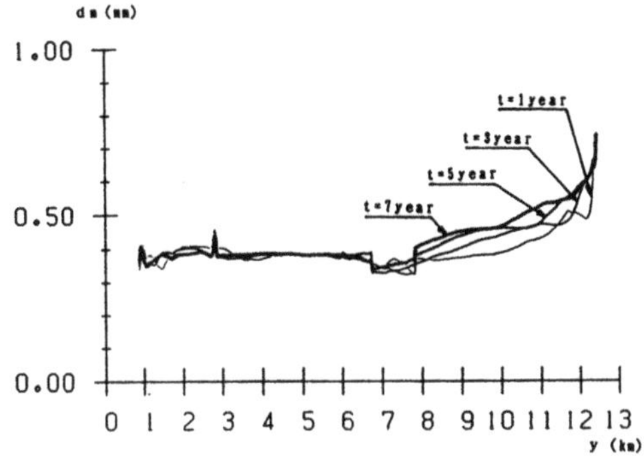

Fig.8 Time–variation of d_m.

In Fig.9, the measured and the computed shoreline change are shown. As seen in the figure at the top of Fig.7(a), the detached breakwaters have been constructed at y=7–8km, resulting in the deposition of sand at the updrift side. This tendency as well as the erosion near the Natori River mouth is predicted very well.

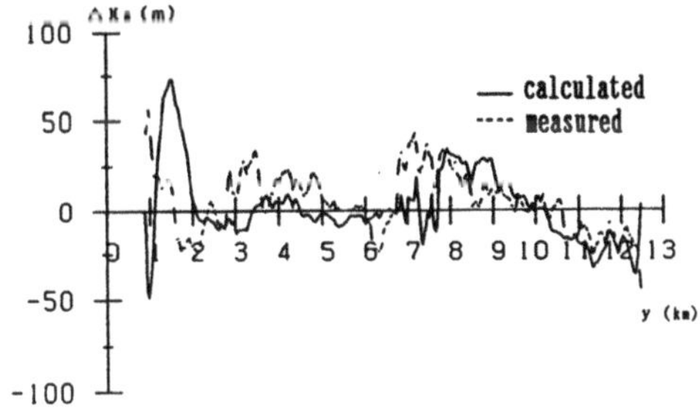

Fig.9 Comparison of shoreline in 1990.

Since the measurement of grain size along the coast was not made in 1990, a quantitative comparison is impossible between the computation and the field observation. For a tentative comparison, the measured grain size distribution in 1991

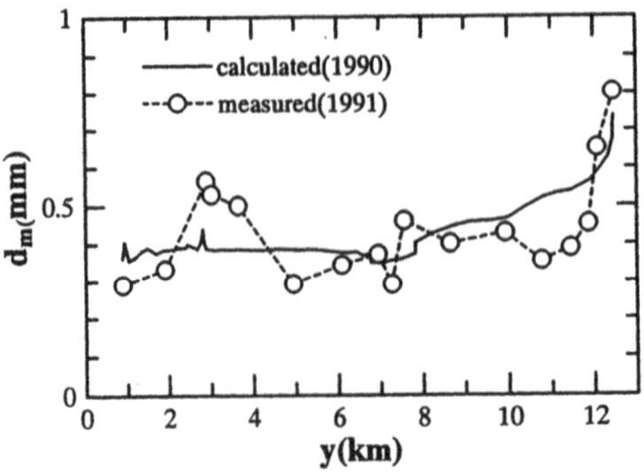

Fig.10 Comparison of grain size in 1990 and 1991.

and computed one in 1990 are shown in Fig.10. The grain size near the Natori river mouth is slightly overestimated, whereas the sand diameter near the Nanakita river mouth is much overestimated.

Computation as a beach with uniform sand was also made using the same conditions, and there was no distinct difference in the shoreline change.

5. Conclusions

Mathematical models proposed up to now for predicting shoreline change can not be applied to compute time-variation of grain-size on a beach. In the present paper, the one-line model and the grain-size sorting modelling for river sedimentation were combined to predict grain size distribution in a longshore direction as well as shoreline deformation due to wave motion. In the computation for a hypothetical beach bounded by groins, the selective transport of finer sand could be observed, resulting in a faster change of the shoreline as compared with the computation for a beach with uniform sand. Furthermore, the present model was applied to the Sendai Bay Coast located in the north-east district in Japan. Although the computed shoreline change showed reasonable agreement with the measurement, a quantitative comparison of the sorting process of sand grain could not thoroughly be made due to a lack of field observation. Further verification of the present model is needed through a comparison with laboratory data as well as field observations.

6. Acknowledgements

The authors would like to express their grateful thanks to Prof. Nobuo Shuto, Tohoku University, and Mr. Katsuhiro Sato, Japan Tetrapod, for their helpful suggestions and encouragement during the course of this study. The appreciation is extended to the Sendai Office, Ministry of Construction, for their kind supply of the valuable field data. A part of this study was financially supported by Grant-in-Aid for Encouragement of Young Scientist, Ministry of Education, Science and Culture.

7. References

1. K. Horikawa, *Nearshore Dynamics and Coastal Processes,* (University of Tokyo Press, Tokyo, 1988)p.522.
2. R. Pelnard-Considere, *IV Journ 'ees de l'Hydraulique* **1** (1956) 289–298.
3. W. T. Bakker et al., *Proc. 12th Coastal Eng. Conf.* (1970) 1001–1020.
4. M. Perlin and R.G. Dean, *U.S. Corps of Engineers, Miscel. Report* **83–10** (1993) p.199.
5. K. Ashida and M. Michiue, *Proc. 14th IAHR Congress* (1971) 247–255.
6. M. Hirano, *Proc. JSCE* **195** (1971) 51–60. (in Japanese)
7. G. T. Proffit and A.J. Sutherland, *Jour. Hydr. Res.* **21/1** (1982) 33–43.
8. H. W. Shen and J.Y. Lu, *Jour. Hydr. Div.* **109/4** (1983) 611–629.
9. W. H. Munk and R.S. Arthur, *U.S. National Bureau of Standards Circular* **521** (1952) 95–109.
10. Y. Goda, *Proc. JSCE* **80** (1970) 39–49.
11. Y. Goda, *Random Seas and Design of Maritime Structure* (University of Tokyo Press, Tokyo, 1988)p.323.
12. H. Tanaka, *Proc. of the 1989 National Conf. Hydr. Eng.* (1989) 95–100.
13. H. Tanaka, *in Sand Transport in Rivers, Estuaries and the Sea,* eds. R. Soulsby and R. Bettes (A.A.Balkema, Rotterdam, 1991)p.291.
14. P. D. Komar and D.L. Inman, *Jour. Geophys. Res.* **75/30** (1970) 5914–5927.
15. CERC, *Shore Protection Manual* (U.S. Govt. Printing Office, 1984)2 Vols.
16. J. W. Kamphuis et al., *Coastal Eng.* **10** (1986) 1–21.
17. J. W. Kamphuis and O.J. Sayao, *Proc. 19th Coastal Eng. Conf.* (1982) 1305–1325.
18. N. C. Kraus et al., *Proc. IAHR Symp. Mathematical Modelling of Sediment Transport in the Coastal Zone* (1988) 117–126.
19. H. Tanaka, *Coastal Eng. in Japan* **31/2** (1988) 265–276.
20. N. Tanaka and K. Nadaoka, *Tech. Note Port and Harbor Res. Inst.* **436** (1982) p.40 (in Japanese)
21. S. Sato et al., *Proc. 13th Japanese Conf. Coastal Eng.* (1966) 176–182.(in Japanese)
22. N. Tanaka, *Coastal Sediments '77* (1977) 1026–1041.
23. H. Tanaka, and N. Shuto, *Proc. Int. Symp. Sediment Transport Modelling* (1989) 480–485.
24. N. C. Kraus, *Jour. Sed. Petrl.* **55** (1985) 3–14.

NUMERICAL APPROACH TO HYDRAULIC JUMP IN NON-PRISMATIC CHANNELS

ALESSANDRO VALIANI
Dipartimento di Ingegneria Civile dell'Università di Firenze,
via di S. Marta, 3 - I 50139 Firenze - ITALY

ABSTRACT

The principal aim of this paper is to show how important is, in numerical modelling of free surface flows, to preserve conservation properties in numerical codes. In fact, when transition from supercritical to subcritical flow conditions occurs, mass and momentum conservation -after numerical integration- is not a trivial goal to achieve.

Shallow Water equations are numerically integrated using a finite differences, McCormack scheme with second order, TVD high resolution technique.

This method is applied to the simulation of the hydraulic jump in a diverging channel. The case is interesting with respect to the prismatic one, because boundary conditions are more complicated, because physical gradients of water surface elevation, velocity and unit width discharge are not zero far from the jump.

The problem is approached for the purely convective case, in order to check the ability of the code to cross over the sharp step in water depth. In such a way, diffusion terms due to momentum exchange by turbulence can be tuned on the basis of the physics only, and not as some necessary artificial term based on convergence or stability requirements.

Good agreement is shown between analytical results and numerical ones. Conservation properties of the mechanical system are well preserved. Stationary solution is found as the asymptotic result of a time varying process, started from an initial non conservative stage; convergence history plots show the stability of the method reaching stationary conditions.

1. Introduction

The present work is part of a research programme, carried out in the Hydraulic branch of the Civil Engineering Dept. of Florence, dealing with the numerical integration of two-dimensional shallow-water equations in fluvial and coastal environments, generally including erosion and deposition processes. A fundamental requirement that is pursued is the ability of the algorithms to maintain convergence, stability and robustness also in presence of flow transitions from supercritical to subcritical flows, in order to be used in common engineering applications.

For the sake of clarity, some limitations inherent the system of 2D Shallow Water Equations are here assumed: physical quantities are vertically averaged; pressure distribution is assumed to be hydrostatic over vertical lines (in coastal environments this hypothesis is appropriate only for long waves); effects of free surface curvature are generally neglected, focusing the attention on the propagation characteristics of bores, and so neglecting some important local aspects of bores generation; turbulence effects are taken into account only by simple "engineering" models, that is, by using some simple closure model for the turbulent stresses between adjacent fluid vertical layers; bed friction is taken into account by simple Coolebrok or Manning-Strickler formulas, that are really valid only for uniform flow conditions.

The methods that are chosen to test the codes consist in checking the results by simple, classical cases for which an analytical solution exist; checking the results by comparison with numerical results by others (Valiani and Montefusco, 1990; Valiani, 1992, a, b), for some classical case study; checking the results by the direct comparison with experiments (at a quite large laboratory scale), in some case of complex geometries (Montefusco and Valiani, 1993).

The present work is an example of the first method, and deals with the test of the numerical code with the hydraulic jump in a diverging open channel, whose width increases linearly in space.

The reasons why such a case is considered of particular interest are the following. First, the capacity to pass over a strong shock, satisfying the fundamental conservation laws, is a severe test for a numerical code. Second, the non prismatic channel case permits: to test the capacity of the method to find the position of the shock; to test the capacity of the method to accept the presence of source terms due to the curvature of the coordinate system (in the prismatic case, if distributed friction losses are neglected with respect to the energy losses in the roller, such terms are zero); to test a not trivial set of boundary conditions (particularly the downstream boundary conditions).

2. Some Remarks on the Physics

The work was inspired by two accurate series of experimental investigations, performed by Rubatta (1963; 1964) in the Hydraulics Laboratory of the University of Bologna. This author made experimental runs, using a glass walls divergent or convergent flume (a complete radial symmetry is realised), to obtain relationships between the fundamental quantities that control the hydraulic jump. The bottom was smooth and horizontal.

A definition sketch of the geometrical configuration is given in fig. 1.

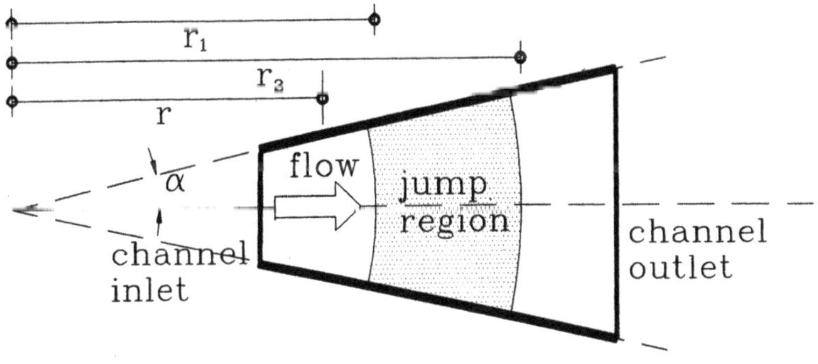

FIG. 1. Definition sketch of the diverging channel considered in present simulations.

The liquid volumetric discharge (Q), the specific energy of the flow at the inlet of the channel (E_1), the specific energy of the flow at the outlet of the channel (E_2) are considered as independent quantities; distributed energy losses are neglected with respect to the energy losses in the jump.

The above mentioned experimental analysis, coupled with dimensional analysis, shows that the independent non dimensional parameters are the following (2α is the angle formed between the idealised intersection of the walls, g is the gravity acceleration).

- An inlet Froude number (with specific energy used as a length scale):

$$\zeta = \frac{Q}{2\alpha\, E_1^2 \sqrt{g\, E_1}} \tag{1}$$

- The energy ratio from outlet and inlet:

$$\eta = \frac{E_2}{E_1} \tag{2}$$

Outside the jump region, continuity equation and energy conservation give the surface profile solution, in terms of non dimensional surface elevation ($\tilde{y}$) as a function of non dimensional distance from the centre of radial symmetry ($\tilde{r}$).

If the level of specific energy, E, is chosen as a typical vertical length scale, and $U_E = \sqrt{g\,E}$ is chosen as a typical velocity scale, the typical horizontal length scale is: $R_E = Q/(2\,\alpha\, E \sqrt{g\,E})$, so that non dimensional depth and non dimensional abscissa are:

$$\tilde{y} = \frac{y}{E}; \tag{3}$$

$$\tilde{\tilde{r}} = \frac{r}{R_E} = \left(\frac{Q}{2\,\alpha\, E^2 \sqrt{g\,E}}\right)^{-1} \frac{r}{E} \tag{4a}$$

that is:

$$\tilde{\tilde{r}} = \left(\frac{Q}{2\,\alpha\, E^2 \sqrt{g\,E}}\right)^{-1} \tilde{r} \tag{4b}$$

Double cedilla is used for lengths that are made non dimensional using R_E as a length scale, while single cedilla is used for lengths that are made non dimensional using E as a length scale ($\tilde{\tilde{r}} = \zeta^{-1}\tilde{r}$).

The above mentioned solution outside the jump region is:

$$\tilde{\tilde{r}} = \frac{1}{\tilde{y}\sqrt{2\,(1-\tilde{y})}} \tag{5}$$

that is, an implicit equation that has two roots, corresponding to the subcritical solution and to supercritical one. Critical values (corresponding to unity for Froude number) of non dimensional water surface elevation and non dimensional radius are:

$$\tilde{y}_{cr} = \frac{2}{3}; \quad \tilde{\tilde{r}}_{cr} = \sqrt{\frac{27}{8}} \tag{6}$$

The series of experiments allow to obtain the non dimensional position of the jump (the location of the beginning and of the end) and the non dimensional amplitude of the jump, as functions of the non dimensional physical parameters, that is:

$$\tilde{r}_1 = \frac{r_1}{E_1} = f(\zeta, \eta) \tag{7}$$

$$\tilde{r}_2 = \frac{r_2}{E_1} = g(\zeta, \eta) \tag{8}$$

$$\frac{\tilde{y}_2 - \tilde{y}_1}{\tilde{r}_2 - \tilde{r}_1} = h(\eta) \tag{9}$$

in the experimental range $0.5 \leq \zeta \leq 2$; $0.3 \leq \eta \leq 0.8$. f, g, h are empirical functions, reported in the original papers by Rubatta. The estimation of the conjugate depths ratio involves the evaluation of the streamwise component of the force exerted by lateral walls, that is (γ is the specific weight of water):

$$F_L = \int_{r_1}^{r_2} \gamma \, Y^2 \, dr \, , \tag{10}$$

even if this estimate is affected to the non negligible error of the deviation from the hydrostatic pressure distribution.

The "shock" solution is obtained if the length of the jump is neglected with respect to the value of the initial position:

$$\left(\frac{r_2 - r_1}{r_1}\right) \ll 1. \tag{11}$$

In that case:

$$\frac{r_1}{E_1} = \frac{r_2}{E_1} = \frac{r_0}{E_1} = \zeta \, \lambda_0(\eta) \, , \tag{12}$$

being λ_0 a function in a closed form (see Rubatta, 1963).

In order to stress mainly the capacity of the numerical method to pass over the shock, this analytical, complete solution is the solution reproduced by the present numerical method (with lengths made non dimensional with upstream energy level, E_1):

$$\tilde{r} \leq \tilde{r}_0 : \quad \tilde{r} = \frac{1}{\tilde{y}\sqrt{2\,(1-\tilde{y})}} \tag{13a}$$

$$\tilde{\tilde{r}} = \tilde{\tilde{r}}_0 : \quad \tilde{\tilde{r}}_0 = \lambda_0(\eta) \tag{13b}$$

$$\tilde{\tilde{r}} \leq \tilde{\tilde{r}}_0 : \quad \tilde{\tilde{r}} = \frac{1}{\tilde{y}\sqrt{2\,(\eta-\tilde{y})}} \tag{13c}$$

3. The Numerical Approach

The present contribution intend to apply to the divergent channel case a shock capturing method of integration, that takes into account of the recent developments in computational fluid mechanics. In fact, while a large amount of literature exists treating shock transitions from supersonic to subsonic flow in compressible fluids (see, i. e., Fletcher, 1988, or Hirsch, 1990 for a detailed list of references), only more recently the scientific community have applied shock capturing methods to shallow water equations, when transition from supercritical flow to subcritical one occurs (Alcrudo, Garcia-Navarro and Saviron, 1992; Alcrudo and Garcia-Navarro, 1993;

Bellos, Soulis and Sakkas, 1990; Fennema and Chaudry, 1989, 1990; Garcia-Navarro, Alcrudo and Saviron, 1992; Glaister, 1988, 1993; Yang, Hsu and Chang, 1993; Yang and Hsu, 1993).

The present method is an application of a mixed explicit-implicit method of a TVD (total variation diminishing) McCormack scheme, that is a second order space-time accurate upwind scheme of integration of shallow water equations.

A cylindrical system of coordinates is considered, as sketched in fig. 1. The bottom is considered to be plane and horizontal, while diffusive terms due to viscosity (that are physically almost negligible) or to turbulence (physically important in the region of the roller) are not considered, with the specific aim to check a method that is able to follow the hyperbolic character of the shallow water eq.ns.

In fact, it is proven that the hyperbolic nature of the convective terms require an 'ad hoc' method of numerical integration. Dispersive terms due to the non uniformity of velocity distribution along the vertical are also neglected, so that only one value of the velocity components is significant on a vertical. Pressure distribution is considered to be hydrostatic on each vertical, and the role of free surface curvature is neglected.

So, the main goal of this work is not to explain the local physics of the phenomenon (that would be not possible by shallow water equations), but to find a method that is able to by-pass strong singularities of the flow field, with particular attention to the conservation properties of the system, that is the respect of the global, integral balances of mass and momentum.

The following symbols are used:

g is the acceleration of gravity;

t, r, θ are respectively time, radial coordinate and tangential coordinate;

$Y(r,\theta,t)$ is the local, instantaneous flow depth;

$U_r(r,\theta,t)$ is the vertically averaged velocity along r - axis;

$U_\theta(r,\theta,t)$ is the vertically averaged velocity along θ - axis;

C is the non dimensional friction Chezy coefficient (the ratio between the mean velocity and the friction velocity), assumed isotropic for simplicity.

Continuity equation may be written as follows:

$$\frac{\partial Y}{\partial t} + \frac{1}{r}\left[\frac{\partial(rU_rY)}{\partial r} + \frac{\partial(U_\theta Y)}{\partial\theta}\right] = 0 \tag{14}$$

Momentum equation in r direction may be written in the form:

$$\frac{\partial(U_rY)}{\partial t} + \frac{\partial(U_r^2Y)}{\partial r} + \frac{1}{r}\frac{\partial(U_rU_\theta Y)}{\partial\theta} + \frac{\partial}{\partial r}\left(\frac{1}{2}gY^2\right) + $$
$$+ \frac{U_r^2Y}{r} - \frac{U_\theta^2Y}{r} + U_r\frac{\sqrt{U_r^2+U_\theta^2}}{C^2} = 0 \tag{15}$$

Momentum equation in θ direction may be written in the form:

$$\frac{\partial(U_\theta Y)}{\partial t} + \frac{\partial(U_rU_\theta Y)}{\partial r} + \frac{1}{r}\frac{\partial(U_\theta^2Y)}{\partial\theta} + \frac{1}{r}\frac{\partial}{\partial\theta}\left(\frac{1}{2}gY^2\right) + $$
$$+ \frac{2U_rU_\theta Y}{r} + U_\theta\frac{\sqrt{U_r^2+U_\theta^2}}{C^2} = 0 \tag{16}$$

Just to put into the largest evidence the role of purely convective terms, the maximum simplification of equations is considered.

A complete radial symmetry is considered, so that, for any (r, θ, t), equalities: $\frac{\partial}{\partial\theta} = 0$; $U_\theta = 0$ hold.

Distributed flow resistances are neglected ($C \to \infty$), just to point out the role of the source terms due to the curvature of the coordinate system; as a consequence, the loss of specific energy $E_1 - E_2$ is concentrated in the shock.

The nature of the problem is purely hyperbolic. It is important to stress the fact that the aim of the work is not to explain the jump (that is not possible without a complete analysis of production-dissipation-diffusion turbulence mechanism in the roller); on the contrary, the aim of the work is to obtain a code able to pass on the jump, maintaining the fundamental conservation properties of the system. In such a way, diffusive terms to to turbulence, and more or less complicated terms such as dispersion terms, resistance terms, etc. may be eventually inserted, and tuned, 'a posteriori' on the basis of physics, not of numerical analysis.

Now, the problem to solve, in a compact form, is:

$$\frac{\partial \mathbb{U}}{\partial t} + \frac{\partial \mathbb{E}}{\partial r} + \mathbb{S} = \mathbb{O} \tag{17}$$

where:

$$\mathbb{U} = \begin{bmatrix} Y \\ U_r Y \end{bmatrix}; \quad \mathbb{E} = \begin{bmatrix} U_r Y \\ U_r^2 Y + \frac{1}{2} g Y^2 \end{bmatrix}; \quad \mathbb{S} = \begin{bmatrix} \frac{U_r Y}{r} \\ \frac{U_r^2 Y}{r} \end{bmatrix}; \quad \mathbb{O} = \begin{bmatrix} 0 \\ 0 \end{bmatrix} \tag{18}$$

The physical boundary conditions are:

$$U_r Y = \frac{Q}{2 \alpha r_{inlet}} \tag{19}$$

given at the entrance of the channel;

$$Y = Y_{inlet} \tag{20}$$

given at the entrance of the channel (a sharp-edged sluice gate was used in quoted experiments to control the depth at the channel inlet);

$$Y = Y_{outlet} \tag{21}$$

given at the end of the channel (an overflow weir was used in quoted experiments to control the depth at the channel outlet).

A deep analogy exists between the structure of the shallow water equations and the Euler equation for compressible flows: the flow depth in the former plays the role of the density in the latter, the Froude transition in the former plays the role of the Mach transition in the latter. As a consequence, numerical techniques are commonly used in classical literature of CFD of transonic flows (see Fletcher, 1988; Hirsch, 1990; AIAA Journal for a large amount of references).

In this frame, the TVD method is selected, as an appropriate tool to obtain a conservation preserving numerical solution.

The TVD (total variation diminishing) methods are based on the principal assumption that if the scalar conservation law (in which x, $x \in D$ is space, t is time, u is a physical quantity, f=f(u)):

$$u_{,t} + f_{,x} = 0 \tag{22}$$

holds, the total variation of any physical admissible solution is:

$$TV = \int_D \left| \frac{\partial u}{\partial x} \right| dx, \tag{23}$$

and this quantity does not increase in time (Lax, 1973).

In discrete terms (indexes refer to space discretization, apexes to time discretization):

$$\mathrm{TV}(u) = \sum_i |u_{i+1} - u_i|. \tag{24}$$

A numerical solution is TVD if:

$$\mathrm{TV}(u^{k+1}) \leq \mathrm{TV}(u^k). \tag{25}$$

If, in time, *i*) no new local extrema in x can be created, and *ii*) the value of a local minimum (maximum) is not decreasing (increasing), the numerical scheme is monotonicity preserving: u^{k+1} remains monotone if u^k is monotone. In other words, monotone profiles are preserved during time evolution, so that the major benefit of these methods is the capacity to avoid spurious oscillations of the numerical solution.

Some important properties of these methods are:

- All monotone schemes are TVD.
- All TVD schemes are monotonicity preserving.
- Any linear TVD scheme is first order accurate.
- High resolution (second order) TVD schemes must be nonlinear.

More detailed (and rigorous) details may be found in Hirsch (1990).

Equation (17) may be written in terms of characteristic variables:

$$\frac{\partial \mathbb{W}}{\partial t} + \Lambda \frac{\partial \mathbb{W}}{\partial r} + \mathbb{P}^{-1}\mathbb{S} = \mathbb{O}, \tag{26}$$

where: $\mathbb{J}_{\mathbb{E}} = \frac{\partial \mathbb{E}}{\partial \mathbb{U}}$ is the Jacobian matrix of the flux vector.

The following splitting technique is used:

$$\mathbb{J}_{\mathbb{E}} = \mathbb{P}\,\Lambda\,\mathbb{P}^{-1}, \tag{27}$$

where $\mathbb{P}$ is the left eigenvectors matrix, Λ is the diagonal eigenvalues matrix, $\mathbb{P}^{-1}$ is the right eigenvectors matrix, and the relation between increments in terms of characteristic variables and independent variables is:

$$\delta\mathbb{U} = \mathbb{P}\,\delta\mathbb{W}. \tag{28}$$

In the specific case:

$$\mathbb{J}_{\mathbb{E}} = \begin{bmatrix} 0 & 1 \\ gY - U_r^2 & 2U_r \end{bmatrix}; \quad \Lambda = \begin{bmatrix} U_r + \sqrt{gY} & 0 \\ 0 & U_r - \sqrt{gY} \end{bmatrix}; \tag{29, 30}$$

$$\mathbb{P} = \begin{bmatrix} 1 & 1 \\ U_r + \sqrt{gY} & U_r - \sqrt{gY} \end{bmatrix}; \quad \mathbb{P}^{-1} = \begin{bmatrix} -\dfrac{(U_r - \sqrt{gY})}{2\sqrt{gY}} & +\dfrac{1}{2\sqrt{gY}} \\ +\dfrac{(U_r + \sqrt{gY})}{2\sqrt{gY}} & -\dfrac{1}{2\sqrt{gY}} \end{bmatrix} \tag{31}$$

The use of characteristic variables allows the method to follow the physics of the propagation phenomena, adapting the finite difference operations to the direction and intensity of the propagating waves. In terms of increments of characteristic and physical variables (being λ_j, δw_j, $\mathbf{r}^{(j)}$, respectively, the jth eigenvalue, the jth increments of characteristic variable, the jth right eigenvector):

$$|\mathbb{J}_{\mathbf{E}}|\,\delta\mathbb{U} = \sum_j |\lambda_j|\,\delta w_j\,\mathbf{r}^{(j)}. \tag{32}$$

The description of the complete algorithm is given by the following relations.

Purely explicit method:

Predictor step:

$$\mathbb{U}_i^p = \mathbb{U}_i^n - \frac{\Delta t}{\Delta x}(\mathbb{E}_{i+1}^n - \mathbb{E}_i^n) - \Delta t\ \mathbb{S}_i^n \tag{33}$$

Corrector step:

$$\mathbb{U}_i^c = \mathbb{U}_i^n - \frac{\Delta t}{\Delta x}(\mathbb{E}_i^p - \mathbb{E}_{i-1}^p) - \Delta t\ \mathbb{S}_i^p \tag{34}$$

Following a suggestion by Hirsch (1990), source terms may be treated with an implicit method; in this case, because of the nature of these terms, a mixed implicit-explicit splitting is performed, in order to reduce instabilities without solving a linear system for each step.

Mixed implicit-explicit treatment of source terms:

$$\mathbb{S} = \begin{bmatrix} \frac{U_r Y}{r} \\ \frac{U_r^2 Y}{r} \end{bmatrix} = \left(\frac{U_r}{r}\right)\begin{bmatrix} Y \\ U_r Y \end{bmatrix}, \text{ that is: } \mathbb{S} = \left(\frac{U_r}{r}\right)\mathbb{U}. \tag{35}$$

In discrete terms, $\mathbb{S}_i^n$ is replaced by $(U_r/r)_i^n\,\mathbb{U}_i^p$, while $\mathbb{S}_i^p$ is replaced by $(U_r/r)_i^p\,\mathbb{U}_i^c$, so that the mixed McCormack procedure become:

Predictor step: $$\left[1 + \Delta t\left(\frac{U_r}{r}\right)_i^n\right]\mathbb{U}_i^p = \mathbb{U}_i^n - \frac{\Delta t}{\Delta x}(\mathbb{E}_{i+1}^n - \mathbb{E}_i^n) \tag{36}$$

Corrector step: $$\left[1 + \Delta t\left(\frac{U_r}{r}\right)_i^p\right]\mathbb{U}_i^c = \mathbb{U}_i^n - \frac{\Delta t}{\Delta x}(\mathbb{E}_i^p - \mathbb{E}_{i-1}^p) \tag{37}$$

Both in the cases of pure explicit method or mixed explicit-implicit (for source terms) method, the second part of the algorithm is the same, i. e. the following.

Second order TVD correction:

$$\mathbb{U}_i^{n+1} = \frac{1}{2}(\mathbb{U}_i^p + \mathbb{U}_i^c) + \left\{\frac{\Delta t}{\Delta x}\left(\mathbb{D}_{i+1/2}^n - \mathbb{D}_{i-1/2}^n\right)\right\}. \tag{38}$$

The term in graphs is the TVD correction to the classical, explicit McCormack scheme, being:

$$\mathbb{D}_{i+1/2}^n = \frac{1}{2}\sum_k \delta W_{i+1/2}^{(k)}\,\varphi\left(c_{a_i+1/2}^{(k)}\right)\left(1 - \frac{\Delta t}{\Delta x}\left|c_{a_i+1/2}^{(k)}\right|\right)\left[1 - \Psi\left(r_{i+1/2}^{(k)}\right)\right]\mathbf{r}_{i+1/2}^{(k)} \tag{39}$$

Quantities at the point $(i+1/2)$ are calculated according the criterion proposed by Roe (1984), and also suggested by Hirsch (1990):

$$Y_{i+1/2} = \sqrt{Y_i\,Y_{i+1}}\ ; \quad c_{r_i+1/2} = \sqrt{g\frac{Y_i + Y_{i+1}}{2}} \tag{40, 41}$$

$$U_{r_{i+1/2}} = \frac{U_{r_i}\sqrt{Y_i} + U_{r_{i+1}}\sqrt{Y_{i+1}}}{\sqrt{Y_i} + \sqrt{Y_{i+1}}} \tag{42}$$

$$c_{a_{i+1/2}}^{(1)} = U_{r_{i+1/2}} + c_{r_{i+1/2}} ; \quad c_{a_{i+1/2}}^{(2)} = U_{r_{i+1/2}} - c_{r_{i+1/2}} \tag{43, 44}$$

$$r_{i+1/2}^{(1)} = \begin{bmatrix} 1 \\ c_{a_{i+1/2}}^{(1)} \end{bmatrix} ; \quad r_{i+1/2}^{(2)} = \begin{bmatrix} 1 \\ c_{a_{i+1/2}}^{(2)} \end{bmatrix} \tag{45, 46}$$

The function φ is introduced to obtain a minimum celerity that cleans out small perturbations:

$$\varphi(v) = \begin{cases} |v| & \text{if } |v| \geq \delta \\ \delta & \text{if } |v| < \delta \end{cases} \tag{47}$$

where δ is a parameter of the model to be tuned (of the order of 10^{-2} m/s).

The function Ψ is the "limiter" nonlinear function, that takes into account the spatial growth of consecutive variations of characteristic variables, in order to avoid spurious numerical oscillations of physical quantities:

$$\Psi\left(r_{i+1/2}^{(k)}\right) = \frac{\delta W_{i+1/2-s}^{(k)}}{\delta W_{i+1/2}^{(k)}}, \text{ being: } s = \text{sign}\left(c_{a_{i+1/2}}^{(k)}\right), \quad k = 1, 2. \tag{48, 49}$$

Different forms of Ψ function have been tested (Van Leer's; minmod; β; Chakravarthy and Osher limiter); presented results are obtained by Roe's superbee limiter (1985):

$$\Psi(r) = \max(0, \min(2r, 1), \min(r, 2)). \tag{50}$$

4. Boundary and Initial Conditions

Regarding the boundary conditions, the physical nature of the problem and the numerical structure of the algorithm make evidence on some facts.

If we suppose that the physical domain is divided in $N-1$ intervals by points $r_1, r_2, \ldots\ldots, r_N$, the method requires, for each time, the values of the state variables in the fictitious points r_0 and r_{N+1}.

In the predictor step, 2 conditions are needed in r_{N+1}, but only one is physical:

$$Y_{N+1} = Y(r_{N+1}) \tag{51}$$

given by outlet condition on free surface elevation.

In the corrector step, 2 conditions are needed in r_0, and both are physical:

$$Y_0 = Y(r_0) \tag{52}$$

given by inlet condition on free surface elevation;

$$U_{r_0} Y_0 = \frac{Q}{2\alpha r_0}, \tag{53}$$

given by inlet condition on liquid discharge free surface elevation; in (53) all quantities of right end side are known.

In the predictor step, a first order linear extrapolation in time on characteristic variables is used to obtain the value of U_{rN+1}, that is the second, mathematical boundary condition; this estimation is performed according a suggestion by Hirsch (1990) (Δ is the difference operator in time):

$$\Delta W_{N+1}^{(n+1)} = \Delta W_{N+1}^{(n)} \tag{54}$$

Initial conditions are proven not to be relevant for the final solution; in performed runs they have been assigned in a very simple way, that is supposing the position of the shock in a random index s, far from the boundaries, neglecting the kinetic part of specific energy downstream the shock, and neglecting the potential part of specific energy upstream the shock.

Different initial conditions are tested, but the influence of these is forgotten after some hundred of calculation steps.

In any case, a Courant number of the order of 1 is used in the method.

5. Results

Preliminary results are promising, in the sense that mass and momentum relative errors are found to be of the order of 0.1% even near the shock, so that conservation properties of the system may be maintained by numerical integration.

In the asymptotic case of prismatic channel, source term vanishes and the accuracy of the method increases.

In fig. 2 the solution of a test-case from Rubatta (1963) is presented. The plotted free surface profile is obtained using artificial dissipation terms by Jameson and others (1981). All the quoted experimental data may be reasonably reproduced by this technique, with an accuracy that may seem quite encouraging.

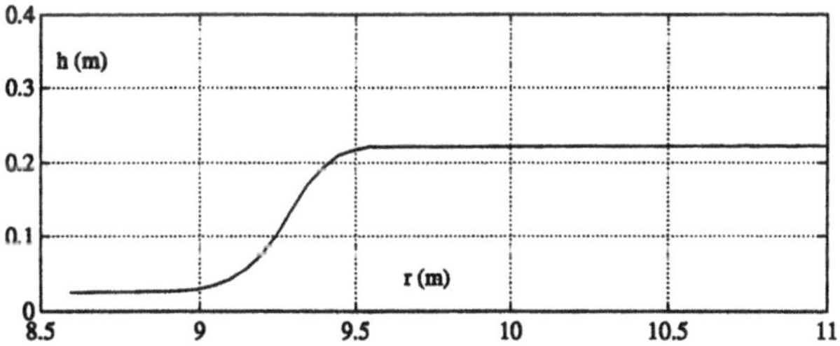

FIG. 2. Free surface profile versus radial coordinate, obtained with artificial dissipation terms. This solution does not preserve conservation of mass.

Nevertheless, a simple analysis of conservation properties of the solution puts into evidence that mass and momentum balances are not satisfied by such a solution.

In fact, looking at fig. 3 (referred to a quite similar test case), in which the ratio between liquid discharge and liquid discharge at the entrance are plotted versus space step index, when asymptotic solution is reached, it is evident how the conservation of mass is not satisfied across the shock. Besides, in other cases, a net loss of mass occurs after the shock, so that conservation of mass is not satisfied not only locally, but even globally.

Looking at momentum balance, the quantity that is used here to check the conservation properties of the method is the specific force of the flow, that increases with respect to the radial coordinate, r , due to the widening of the flow:

$$F = \frac{1}{2} r Y^2 + \frac{U_r^2}{g} r Y \tag{55}$$

corresponding to an increasing ratio of: $\frac{dF}{dr} = \frac{1}{2} Y^2 .$ (56)

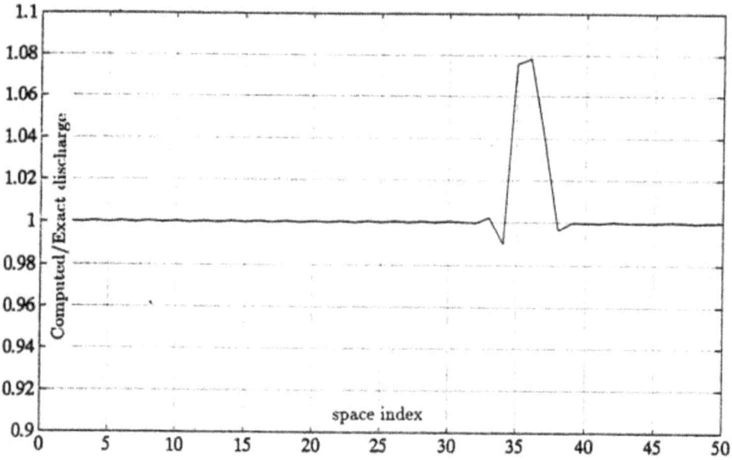

FIG. 3. Ratio between computed discharge and exact discharge. Mass conservation is not satisfied by methods using artificial dissipation.

In fig. 4 (referred to the same case reported for mass balance), the ratio between the computed specific force and the exact specific force is plotted versus space step index, again when the steady-state solution is reached.

Linear momentum conservation is preserved with relative errors quite less than mass conservation. In any case, the momentum balance is not satisfied only in a limited region near the shock, while global momentum balance (in the quoted range of data) is not affected by the local not conservative behaviour.

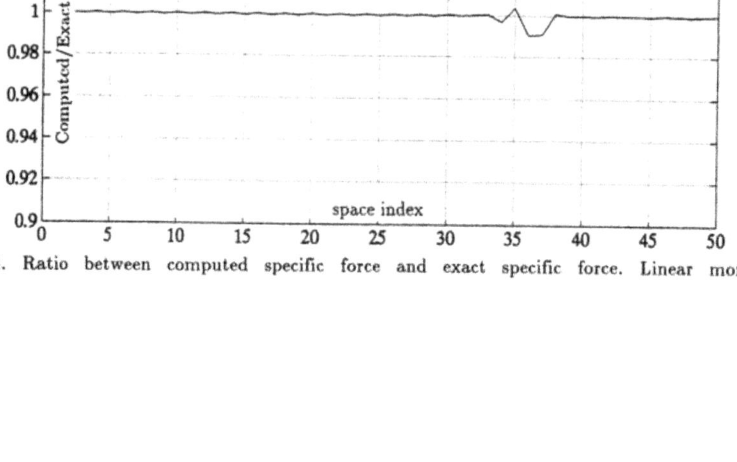

FIG. 4. Ratio between computed specific force and exact specific force. Linear momentum

conservation is satisfied by methods using artificial dissipation, with relative errors across the shock of about 1%.

On the contrary, the following results are obtained dropping any artificial dissipation term, and considering the purely convective case, as described in the preceding paragraph.

The time marching technique was stopped after about 200 seconds of simulation, because this time was proven to be sufficient to reach steady state conditions for all experimental cases here analysed.

Time histories of the values of the state variables in a section located downstream of the jump are considered significant as steady-state indicators. The location of the cross-section is chosen at a distance of about 15% of the total length of the channel upstream of the weir at the end of the channel.

A typical behaviour in time is shown in figs. 5-6, where the above mentioned depth and unit discharge are plotted versus time. Final values are reached after a transient whose duration is about 30 sec.; steady state is very stable in time, and any significant oscillation is completely dropped after about 40 sec.. The complete absence of oscillations is maintained until the final time of 200 sec. .

The maximum amplitude of oscillations is about of $\pm 25\%$ of final value for the depth of the channel and about of $\pm 300\%$ of final value for the unit discharge (velocity times depth).

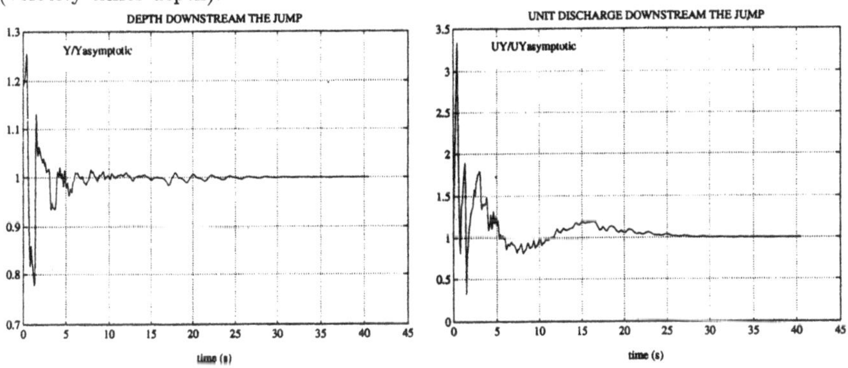

FIGG. 5, 6. Ratio between depth and final depth (left), and ratio between unit discharge and final unit discharge (right), computed in a cross section downstream the jump. The final, steady state value is very stable in the simulate range of physical quantities.

Typical behaviour of steady state solution is shown in figs. from 7 to 11. All these figures are referred to the case I.12 of the quoted experiments by Rubatta (1963), but analogous results may be obtained for the entire set of data. In the specific case the numerical values of independent parameters are:

$$\eta = 0.517; \qquad \zeta = 0.5.$$

In all the following figures, supercritical exact solution is plotted as a continuous line, while subcritical exact solution is plotted as a dashed line. Computed numerical solution is plotted as a dashed line, with open circles where computed points take place.

Lengths are made non dimensional by upstream energy level E_1, kinematical quantities by their value at the entrance of the channel (U_{r_0}, $Q_0 = Q$), and specific force also by its value at the entrance (F_0).

Fig. 7 shows the free surface profile. Its easy to verify how, for practical values of physical parameters, the subcritical profile is not far from being horizontal, so that, if the specific force of the supercritical flow is large enough, the approximation obtained neglecting kinetic height with respect to pressure height is acceptable from a technical point of view.

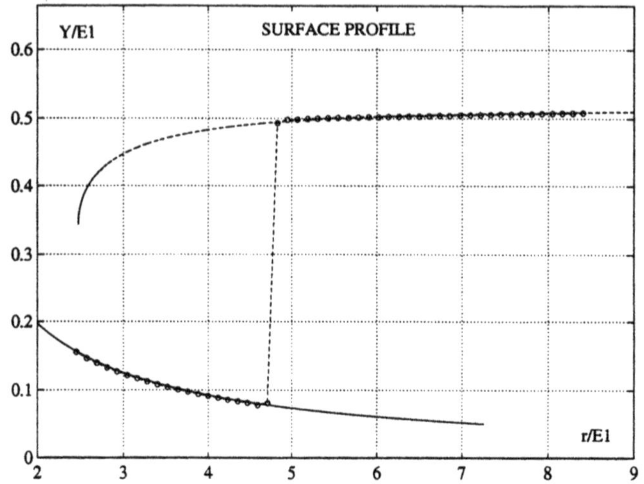

FIG. 7. Non dimensional free surface profile. Computed (dashed-circle line) and exact solutions.

Fig. 8 shows the radial velocity profile. The agreement of computed solution and exact solution is quite good. On the contrary, a typical error, that is present when artificial dissipation terms are used, is an underestimation of velocities in the subcritical region, due to a net loss of mass of the system. In fact, the mass error (according the present experience) is almost completely reflected on velocity with respect to depth.

Fig. 9 shows the energy profile. In this scheme, the hydraulic jump is a sort of black box whose main effect is a net loss of energy, imposed by boundary conditions. This plot is the most significant to show the efficiency of TVD second order corrections to upwind schemes, in cleaning out spurious oscillations around the shock. Without that corrections, the cross of the shock causes significant oscillations involving about 5 computing nodes.

Fig. 10 shows the ratio between the computed discharge and the discharge at the entrance of the channel. Clearly, this ratio must be exactly one everywhere in steady state condition, to satisfy the conservation of mass. This requirement is the most difficult to obtain crossing the shock, and is not obtained neither by artificial dissipation methods nor by simple upwind methods. Both give results that are always locally not conservative, but often are globally not conservative, that is they give an outlet discharge different (less) than inlet discharge, while across the shock a local increasing of mass is present. A small local increase of mass across the shock is still present even in the present upwind TVD method, but the relative amplitude of the phenomenon is reduced of about two order of magnitude. With any doubt, the mass conservation is the crucial point in free surface flows with transition of critical stage.

Fig. 11 shows the ratio between the specific force of the steady-state solution and the specific force at the entrance of the channel. The peculiar property of the flow in the divergent channel is the quite different growth rate of the specific force in the subcritical flow with respect to the supercritical flow, being the former much higher than the latter. The asymptotic trend (in space) of the specific force is a constant growth rate with respect to the radial coordinate, strictly related to the rapid asymptotic trend of the free surface to become horizontal.

Also in this case, the capability of the method to satisfy the conservation of linear momentum is significantly improved by TVD corrections, especially regarding oscillations across the jump, that are drastically cleaned with respect to the traditional upwind schemes.

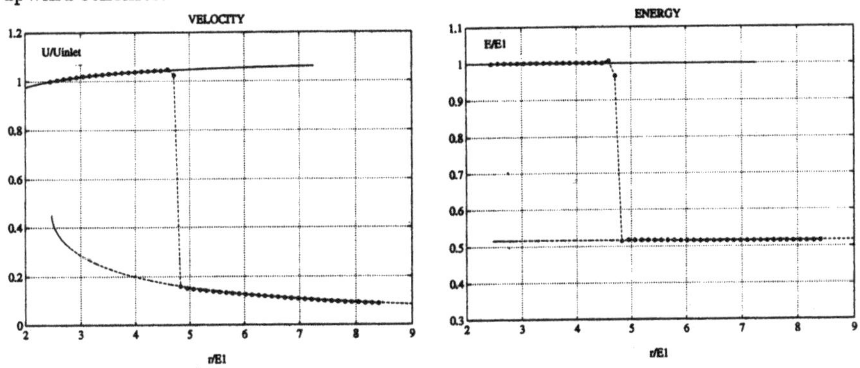

FIGG. 8, 9. Non dimensional velocity profile (left) and non dimensional energy profile (right). Computed (dashed-circle line) and exact solutions. Obviously, the ratio between upstream and downstream energy levels equals η.

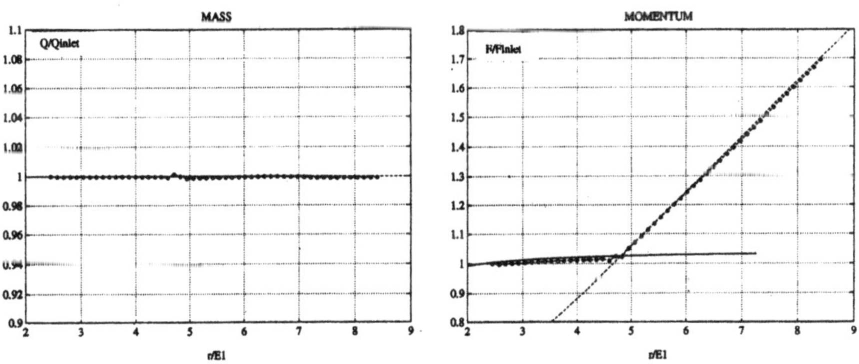

FIGG. 10, 11. Ratio between computed discharge and prescribed discharge at the entrance of the channel (left), and ratio between specific force and specific force at the entrance of the channel (right), versus non dimensional radial coordinate. Computed solution is the dashed-circle line.

6. Conclusions

The present paper deals with the conservation properties of the hydraulic jump in diverging channels. The physical scheme is very simple, supposing a complete radial symmetry of the problem. The numerical code is performed taking into account the recent advances in numerical analysis of shock phenomena in fluid dynamics.

In the first part of the paper, a brief review of the physics is given, essentially based on the work by Rubatta (1963, 1964).

It shown how commonly accepted solutions, such as that based on artificial dissipation terms, do not satisfy the basic conservation laws of the system.

Some essential element of upwind methods made second order accurate with TVD corrections is also given, also concerning the basic properties of total variation diminishing methods.

The second order TVD McCormack method is used, and a brief summary of the used algorithm is illustrated; the peculiar role of the source terms due to spatial coordinate curvature is put into evidence, showing the possibility to treat it in a mixed implicit-explicit way without any substantial additional complication of the numerical code.

Boundary and initial conditions are remarked, and the crucial importance of the formers is stressed; on the contrary, the relatively less important role of the latters is outlined.

Time histories of the state variables are shown, to show the stability of the steady state solution.

Results over the spatial distribution of physical quantities after reaching the steady state condition are given, and the present method is shown to be convergent to the exact solution.

The important role of the total variation diminishing method is put into evidence, especially regarding to the capacity of avoiding spurious oscillations near the shock; also regarding the capacity to preserve completely the basic conservation laws of the physical system.

Accuracy and stability of the method is encouraging for its application to geophysical flows in coastal and fluvial hydraulics, particularly in that cases in which transition from supercritical to subcritical flow takes place.

Extension to complex geometries surely requires extreme attention to boundary conditions, that are trivial only for uniform flow conditions at the boundaries.

7. Acknowledgements

This work was financially supported by the MAST G8 Coastal Morphodynamics research programme (Commission of the E. C. Directorate General for Science, Research and Development), and by CNR-GNFM, Italy. The author is particularly grateful to Dr. Marco Petti (University of Florence, Dept. Civ. Eng.) and to Prof. Vinicio Boffi (CNR-GNFM), for their interest in this support.

8. References

1. Alcrudo, F., Garcia-Navarro, P., Saviron, J. M., Flux difference splitting for 1D open channel flow equations, *International Journal for Numerical Methods in Fluids*, **14** (1992) 1009-1018.
2. Alcrudo, F., Garcia-Navarro, P., A high-resolution Godunov-type scheme in finite volumes for the 2D shallow-water equations, *International Journal for Numerical Methods in Fluids*, **16** (1993) 489-505.
3. Bellos, C. V., Soulis, J. V., Sakkas, J. G., Computation of two-dimensional unsteady open channel flow, *Advances of Water Resources* (1990).
4. Fennema, R. J., Chaudry, M. H., Implicit methods for two-dimensional unsteady free-surface flows, *IAHR Journal of Hydraulic Research*, **27** (3) (1989) 321-332.
5. Fennema, R. J., Chaudry, M. H., Explicit methods for 2-D transient free-surface

flows, *ASCE Journal of Hydraulic Engineering*, **116** (8) (1990) 1013-1034.
6. Fletcher, C. A. J., Computational Techniques for Fluid Dynamics, Springer Verlag, Berlin, 1988.
7. Garcia-Navarro, P., Alcrudo, F., Saviron, J. M., 1-D Open-channel flow simulation using TVD-McCormack scheme, *ASCE Journal of Hydraulic Engineering*, **118** (10) (1991) 1359-1372.
8. Glaister, P., An approximate linearised Riemann solver for the one-dimensional Euler equations with real gases, *Journ. of Comp. Physics*, **74** (1988) 382-408.
9. Glaister, P., Flux difference splitting for open-channel flows, *International Journal for Numerical Methods in Fluids*, **16** (1993) 629-654.
10. Hirsch, C., Numerical computation of internal and external flows, John Wiley & Sons, Chicester, 1990.
11. Jameson A., Schmidt W., Turkel, E., Numerical simulation of the Euler equations by finite volume methods using Runge-Kutta time stepping schemes, *AIAA Paper 81-1259*, AIAA 5th CFD Conference (1981).
12. Lax, P. D., Hyperbolic systems of conservation laws and the mathematical theory of shock waves, SIAM Publications, Philadelphia, 1979.
13. Montefusco, L., Valiani, A., Confronto tra modellazione fisica e matematica per la ricostruzione del campo di moto e della morfologia dell'alveo in ambiente fluviale, *Dipartimento di Ingegneria Civile dell'Università di Firenze*, 1993.
14. Roe, P. L., Generalized formulation of TVD Lax-Wendroff schemes, *ICASE Report 84-53*, NASA CR-172478, NASA Langley Research Center (1984).
15. Roe, P. L., Some contrbutions to the modelling of discontinuous flows, *SIAM Lectures of applied mathematics*, **22** (1985) 163-193.
16. Rubatta, A., Il risalto idraulico in canale divergente, *L'Energia Elettrica*, **XL** (10) (1963) 1-8.
17. Rubatta, A., Il risalto idraulico in canale convergente, *L'Energia Elettrica*, **XLI** (5) (1964) 1-6.
18. Valiani, A., Montefusco, L., Modello bidimensionale del deflusso per tracimazione di uno sbarramento in materiale sciolti, *XXII Convegno di Idraulica e Costruzioni Idrauliche*, Cosenza, (1990) 203-217.
19. Valiani, A., Rapid transients in free-surface flows with movable bed: a 2-D numerical solution, *HYDROCOMP '92, International Conference on Interaction of Computational Methods and Measurements in Hydraulics and Hydrology*, Budapest, Hungary, (1992).
20. Valiani, A., Un modello numerico semiimplicito per la trattazione di bruschi transitori in moti a superficie libera e fondo mobile, *XXIII Convegno di Idraulica e Costruzioni Idrauliche*, Firenze, (1992) E.261-276.
21. Yang, J. Y., Hsu, C. A., Chang, S. H., Computations of free surface flows, part 1: One dimensional dam break flow, *IAHR Journ. of Hydr. Res.*, **31** (1) (1993) 19-34.
22. Yang, J. Y., Hsu, C. A., Computations of free surface flows, part 2: Two dimensional unsteady bore diffraction, *IAHR J. Hydr. Res.*, **31** (3) (1993) 403-411.

Laboratory Studies

SALTATION HEIGHTS IN CHANNEL FLOWS

W. BRIAN DADE
Institute of Theoretical Geophysics, University of Cambridge
Cambridge CB2 3EQ, UK

and

ARTHUR R. M. NOWELL
School of Oceanography, University of Washington,
Seattle, WA 98195, USA

ABSTRACT

We report new observations for the average and maximal heights achieved by 0.1–cm diameter, smooth-glass spheroids in continuous, water-borne saltation over a bed of stationary grains which are otherwise identical to those in transport. These data provide a benchmark for trajectory models which describe the motion of individual bedload particles in channel flows of intermediate transport stage. One such model is presented. Agreement is good between predictions from this conceptual model and our observations of characteristic saltation heights in an idealized setting for which crucial model parameters are well-constrained.

1. Introduction

Particles in the bedload roll, slide, or hop along the bed. When individual particles move predominantly in discrete hops the particles are said to be in saltation. This transport mode has been known to be prevalent for coarse sand and larger material in flows of intermediate flow intensity since the seminal observations of Gilbert (1914). An analysis of this behavior also provides the foundation for one of the earliest, physically-based models for bedload transport (Einstein 1950), in which sediment movement is considered to occur during periods of continuous saltation punctuated by periods of rest or rolling on the bed.

In aqueous flows of high intensity two possible transformations of a saltation bedload take place. If near-bed concentrations of coarse material are very large due to the energetic capacity of the flow to carry grains, then the bedload can take on properties of a fluid-like grain flow. If particle trajectories are strongly correlated with turbulent fluctuations in the near-bed fluid flow, on the other hand, then particles become dispersed throughout the flow and are said to be in suspension.

There are large degrees of chance within and overlap between these modes of particle transport. A partitioning of the different modes has nevertheless remained useful for many approaches to the general problem of sediment transport. Individual particles saltating in water, for example, have been observed to attain maximal heights which are of the order of 2–5 particle diameters and to travel in discrete hops which are of the order of 10–1000 particle diameters in length (e.g., Abbot and Francis 1977). The maximal height attained by typical particles in saltation defines the thickness of the bedload transport layer and corresponds to an apparent hydrodynamic roughness for the sediment transporting flow.

The analysis of saltation trajectories is thus fundamental to studies of sediment transport in channel flows.

The usefulness of an analysis of saltation motion is demonstrated further by the advances made in the last decade or so in the ability to predict bedload transport in flows of intermediate transport stage, $1 < T_* < 10$, say, where $T_* = \tau_o/\tau_{cr}$, τ_o is the bed shear stress and τ_{cr} is the critical bed shear stress required to initiate particle motion. The advances have come in the form of mechanics-based models which describe the discrete trajectories of individual particles in continuous and intermittent saltation in water (e.g., Reizes 1978; Murphy and Hooshiari 1982; van Rijn 1985; Wiberg and Smith 1985, 1989; Wiberg 1987; Madsen 1991; Sekine and Kikkawa 1992). The models differ in their descriptions of bed microtopography, descriptions of the physics and longevity of grain interactions which occur at the bed, treatment of near-bed flow, and mathematical effort required to describe saltation trajectories. Each of the models, however, is deterministic in the calculation of the particle trajectories. Each of the models is also based on the assumption that saltation transport is driven primarily by the mean flow. Although at any instant in time the stochastic properties of the near-bed flow field will affect the motion of some bedload particles, individual particles in continuous saltation must nevertheless exhibit behavior centered about a time-averaged value. It is this characteristic behavior that is related to the mean flow and subsequently related to a time-averaged rate of sediment transport.

That this last notion is appropriate for a description of near-bed particle motion in flows of intermediate transport stage, $1 < T_* < 10$, can be demonstrated as follows. Consider that for the range of sediment grain sizes of interest here—coarse sand or larger—the individual particles are subject to nonlinear drag. Thus the still-water settling velocity w_s of a single grain is approximated by the expression $w_s = [4g(R-1)d/3C_D]^{1/2}$, where g is the acceleration due to gravity, R the ratio of the particle density ρ_p to that of the fluid ρ, d the nominal grain diameter, and C_D a drag coefficient with a value between approximately 0.4 and 1. The friction velocity $u_* = (\tau_o/\rho)^{1/2}$ serves as a velocity scale for the turbulent fluctuations in the transporting fluid flow. This velocity scale can be expressed as $u_* = [T_*\beta_{cr}\, g(R-1)d]^{1/2}$, where $\beta_{cr} = \tau_{cr}/g\rho(R-1)d$ has a typical value, say, between 0.02 and 0.1.

If we accept the conventional argument that turbulent fluctuations do not significantly affect particle trajectories until the turbulent velocity scale u_* takes on values near or in excess of the still-water settling velocity w_s, then the condition $u_*/w_s > 1$ corresponds to $T_* > (3\beta_{cr}C_D/4)^{-1}$. For reasonable estimates of β_{cr} and C_D from within the ranges of values of these parameters given above, a conservative estimate for the upper limit for the predominance of saltation motion of coarse material driven by the mean flow is then $T_* \approx 10$.

Similarly, a model for granular flow advanced by Hanes (1986) can be used to compute the volumetric concentration of solids at the top of a sheared grain layer comprising the bedload under intense flows. From this model we find that this concentration tends to zero as T_* approaches a lower bound of $\mu R\phi_b/(12C_D\beta_{cr})$, where $\phi_b \approx 0.6$ is the volumetric concentration of solids at the top of the underlying, stationary bed and $\mu \approx 0.5$ is the coefficient of dynamic friction for individual grains. This scaling indicates that, for values of C_D and β_{cr} as given above and $R = 2.6$ for quartz particles in water, densely-concentrated grain flows should not become important until T_* approaches and exceeds a value of about 10. For $T_* < 10$, we conclude that bedload transport is more appropriately considered in terms of the motion of individual grains.

An important parameter for any model for near-bed transport of particles in continuous saltation is the coefficient of restitution, e, which describes the degree to which momentum is retained by a moving particle in a momentary, semi-elastic collision with another particle. In the case of continuous saltation, the second particle is a stationary particle on the bed. Newton performed the first experiments to determine coefficients of restitution in low velocity impacts and found that for glass spheres $e = 15/16$. In engineering applications e is taken to be 0.95 for perfectly smooth glass spheres in low velocity, direct central impacts (Weaver 1930). Recent work has indicated that e is a weakly decreasing function of impact velocity due to inelastic deformation at colliding surfaces. Even for direct central impacts between glass spheres at velocities approaching 3 m s^{-1} in air, however, values of e are experimentally observed to be no less than 0.8 (Lun and Savage 1986).

Wiberg and Smith (1985, 1989), on the other hand, calibrated their saltation model with the data of Abbot and Francis (1977) for composite-material casts of irregularly-shaped, natural grains saltating over a stationary bed of similar particles. Wiberg and Smith found that a value of $e = 0.4$ was appropriate to match the trajectories reported by Abbot and Francis, and subsequently achieved considerable success in predicting rates of bedload transport in channel flows. One useful result of the Wiberg model is the relationship between bed shear stress and the average maximal height z_{max} achieved by saltating particles of diameter d. Wiberg and Rubin (1989) re-expressed the original results in terms of the transport stage of channel flows. This relationship is given by

$$Z_{max} = \frac{z_{max}}{d} = \frac{a_1 T_*}{1 + a_2 T_*}, \tag{1}$$

where $a_1 = 0.68$ and $a_2 = 0.0204\,(\ln[d])^2 + 0.0220\,\ln[d] + 0.0709$ for particles with density equal to that of quartz and for which the diameter is expressed in centimeters. Alternative relationships have been summarized by Bridge and Bennet (1992).

Sekine and Kikkawa (1992) calibrated their mechanics-based model with newly acquired data for solid glass spheres with diameters of 0.5 cm and which were saltating over a loose bed of similar particles. They found that $e = 0.65$. Madsen (1991) implicitly used a value of $e = 1$ in his conceptual model for continuous saltation.

There are sound reasons for accepting a value of e significantly different from the expected value of just less than unity in calculations of saltation trajectories in natural settings. Some observations of individual particles in bedload transport suggest, for example, that a continuous saltation model may be inappropriate (e.g., Gordon et al. 1972; Abbot and Francis 1977). Many saltation trajectories do not commence with a momentary, semi-elastic collision at the bed immediately following the previous flight. Some incoming grains have been observed instead to roll briefly over stationary particles before lifting off from the bed to make another saltation. A reduced value of e may thus include the effects of momentum loss due to friction at the colliding surfaces of irregularly-shaped or rough-surfaced grains (as in the experiments of Abbot and Francis 1977). Additionally, a reduced value of e can incorporate the redistribution of momentum due to the displacement of particles comprising a loose bed (as in the experiments of Sekine and Kikkawa 1992, and certainly in experiments with wind-blown sand). Without *a priori* knowledge of the nature of particle interactions and the value of e, however, there is no unambiguous test of models for trajectories in continuous, water-borne saltation.

We report here new observations that provide such a test for mechanics-based models which describe the continuous saltation of spheroids with well-constrained, low-velocity impact response in channel flows. We also present a theoretical analysis of particle saltation which is distinguished from other models by its explicit incorporation of the effects on nonlinear drag force and a simplification of the equations which describe the motion of individual particles. Our approach is especially appealing in that the results include analytical descriptions of particle trajectories. Predictions from this model agree with new observations of saltation heights in an idealized experimental setting.

Our conceptual model for continuous saltation in two dimensions is described next. The starting point for this approach is similar to that posed by Owen (1964) for saltation of particles in air and later used in various forms by many of the workers cited above in their studies of continuous saltation of particles in water. Descriptions of experimental methods, materials and results then follow.

2. A model for particle motion during continuous saltation

2.1. Near-bed flow

Calculations from the model of Dade (1993) suggest that the representative mean fluid velocity u in the vicinity of coarse grains of the bed underlying channel flows and in the absence of sediment transport is $8.5u_* \pm 5\%$. This estimate corresponds approximately to the fluid velocity at the tops of bed grains in a hydrodynamically rough flow, given by

$$u/u_* = 2.5\ \ln[d/z_o] \tag{2}$$

where $z_o = d/30$ (Schlichting 1979). We assume that $u = 8.5u_*$ in our model calculations of continuous saltation trajectories for 0.1–cm diameter particles. We also assume that there is no mean flow in the vertical and cross-stream directions.

2.2. Saltation trajectories

A description of the characteristic trajectories of particles in dilute, continuous saltation can proceed from a consideration of the relevant physical forces which affect the motion of individual particles. These include the forces required to accelerate both the particle and the fluid displaced by the particle, the drag arising from relative velocities between the particle and the ambient fluid, and body forces. Following the arguments of Reizes (1978) and

Murphy and Hooshiari (1982), we neglect the effects of lift forces and unsteady drag due to the history of particle motion. We also neglect the effects of turbulent fluid velocities, and assume that the characteristic motion of a particle is driven by the mean flow. We consider the near-bed fluid flow to be steady, uniform and unaffected by the presence of the grains in transport. Interparticle forces are ignored because the mobile particles are considered to be noncohesive and in dilute, continuous saltation with negligible bed contact other than during momentary, intersaltational collisions.

The force balance that exists in the downstream or x direction is then given by

$$\frac{\pi}{12} d^3 \rho_f (2R + 1) \frac{du_r}{dt} = -F_{Dx} \tag{3a}$$

where $u_r = u_r(0)$ at $t = 0$. The balance in the vertical or z direction is given by

$$\frac{\pi}{12} d^3 \rho_f (2R + 1) \frac{dw_r}{dt} = -F_{Dz} + F_g \tag{3b}$$

where $w_r = w_r(0)$ at $t = 0$. In Eq. 3, $u_r = (u_P - u)$ and $w_r = (w_P - w)$ are the relative velocities between particle and fluid, F_{Dx} and F_{Dz} are the drag forces in the respective directions, F_g is the weight of the submerged particle, and $R = \rho_p/\rho$. The function of R in parentheses on the left side of Eq. 3 incorporates the effects of added mass owing to the acceleration of fluid immediately surrounding a spheroidal grain. Because there is no mean flow in the cross-stream direction and no body forces acting in that direction, cross-stream motion is ignored.

Drag forces in the horizontal and vertical directions are defined as

$$F_{Dx} = \frac{\pi}{8} d^2 \rho\, C_D u_r |\mathbf{u}_r| \quad \text{and} \quad F_{Dz} = \frac{\pi}{8} d^2 \rho\, C_D w_r |\mathbf{u}_r|, \tag{4}$$

where $|\mathbf{u}_r| = (u_r^2 + w_r^2)^{1/2}$ is the magnitude of the relative velocity vector between particle. Following Clift et al. (1978), we use an empirical correlation between drag force and the spherical particle Reynolds number $Re_p = |\mathbf{u}_r| d/\nu$ given by

$$C_D = 24\, A/Re_p, \tag{5a}$$

where $$A = (1 + 0.15 Re_p^{0.687}) \qquad \text{for } Re_p \le 1000, \tag{5b}$$

and $$A = 0.018\, Re_p \qquad \text{for } Re_p > 1000. \tag{5c}$$

Eq. 5 recreates the standard drag curve for spheres in steady motion to within ± 10 percent. The coefficients in Eq. 5 will obviously vary for extremely non-spherical particle shapes. In practice other expressions for nonlinear drag, embodied in the term A, can be applied (e.g., Soulsby 1993). Use of this type of expression facilitates the analysis of near-bed motion of fine particles and as mentioned in §1 represents one aspect of our model that distinguishes it from other approaches.

The submerged weight of a spherical particle is given by

$$F_g = -g\,\rho_f(R-1)\frac{\pi}{6}d^3 . \tag{6}$$

Substitution of Eqs. 4–6 into Eq. 3, rendering the results dimensionless in terms of the length scale d, velocity scale ν/d, time scale d^2/ν, and subsequent solution yield expressions for the relative velocity and displacement of a particle in the horizontal and vertical directions during a single trajectory. These are given in terms of A as defined in Eq. 5, $B = 36/(2R+1)$, $\Xi = g(R-1)d^3/\nu$, and $Re_x = ud/\nu$ with u given in Eq. 2. The upper case symbols U_r, X, W_r, and Z represent the dimensionless quantities of their lower case equivalents.

Dimensionless velocity and displacement in the downstream direction are given by

$$U_r(T) = U_r(0)\exp[-A\,B\,T] \tag{7a}$$

and

$$X(T) = Re_x\,T + \frac{U_r(0)}{A\,B}\left(1-\exp[-A\,B\,T]\right) + X(0) \tag{7b}$$

where $X(0)$ is the downstream position of the particle at the beginning of a saltation. The analogous expressions for the velocity and displacement of a particle in the vertical direction are given by

$$W_r(T) = \left(W_r(0) + \frac{\Xi}{18A}\right)\exp[-A\,B\,T] - \frac{\Xi}{18A} \tag{8a}$$

and

$$Z(T) = -\frac{\Xi\,T}{18A} + \frac{1}{A\,B}\left(W_r(0) + \frac{\Xi}{18A}\right)\left(1-\exp[-A\,B\,T]\right) + Z(0), \tag{8b}$$

where $Z(0)$ is the vertical position of the particle at the beginning of a saltation.

Dimensionless time of flight T_j for a single saltation is obtained from Eq. 8b as the time required by a particle to return to its original elevation $Z(0)$. This is given by

$$T_j = \frac{18}{B\,\Xi}\left(W_r(0) + \frac{\Xi}{18A}\right)\left(1-\exp[-A\,B\,T_j]\right). \tag{9}$$

Estimates of T_j from Eq. 9 are found through iteration. Estimates of T_j facilitate calculations of the downstream displacement, velocity, and average acceleration of particles during a typical trajectory. Average particle velocity, for example, is estimated in dimensionless terms as $[X(T_j) - X(0)]/T_j$. Trajectory duration can also be compared to the effective particle response time $d^2/(ABv)$. This formulation of the response time implies an asymmetry in the rapidity with which a particle accelerates under fluid forcing that is dependent on the nonlinear drag (embodied in the term A).

In obtaining the solutions given in Eqs. 7–9, Re_p is calculated at the start of each trajectory and held constant thereafter. The implication is that the maximal drag experienced by a particle at the onset of a trajectory characterizes the subsequent acceleration. This simplification is based on the observation that for coarse material in subaqueous transport, jump times are of the same order or less than the particle response time. Thus discrete saltations are dominated to a large degree by their initial conditions. Note, too, that in Eqs. 7 and 8 and the expressions that follow, there is no explicit dependence of discrete particle motion on a critical bed shear stress (cf. Madsen 1991).

The time-averaged elevation of a particle during continuous saltation is given by

$$Z_{av} = -\frac{\Xi\, T_j}{36A} + \frac{1}{A\,B}\left(W_r(0) + \frac{\Xi}{18A}\right)\left(1 - \frac{1-\exp\left[-A\,B\,T_j\right]}{A\,B\,T_j}\right) + Z(0)\ . \quad (10)$$

Maximal trajectory height above the bed, on the other hand, is achieved when $W_p(T) = 0$. Applying this condition to Eqs. 8a and 8b we find that

$$Z_{max} = \frac{1}{A\,B}\left(W_r(0) + \frac{\Xi}{18A}\ln\left[\frac{\frac{\Xi}{18A}}{\frac{\Xi}{18A} + W_r(0)}\right]\right) + Z(0)\ . \quad (11)$$

As mentioned above, these equations suggest that aspects of saltation trajectories in equilibrium bedload layers need not be dependent on a knowledge of the critical bed shear stress required for the initiation of transport. One can instead consider the inverse problem, however, and note that non-momentary contact between moving and bed grains becomes important and interparticle friction no longer negligible in flows with bed shear stresses for which Z_{av} fails to exceed, say, 1.5. The cessation of particle motion must then occur at slightly lower bed shear stresses with actual values of the critical stress for cessation of motion depending on, among other factors, the value of the coefficient of dynamic friction for intermittently rolling and sliding grains. This aspect of low-stage, bedload transport is not addressed here.

2.3. Particle-bed collisions

At the end of each flight in continuous saltation, moving particles momentarily collide with the stationary bed. Averaging all possible contact geometries leads to an idealized treatment of the bed as a smooth plane inclined 45° from the horizontal, as shown in Figure 1. In low angle impacts, downstream particle momentum is instantaneously arrested at $Z(0) = 0.5$ and translated into upward momentum for the subsequent trajectory. Kinetic energy is potentially lost through semi-elastic interaction and friction at the colliding surfaces.

These notions lead to a straightforward formulation for the initial velocities for a typical particle trajectory in a dilute layer of continuously saltating grains. The relevant conditions, taken at $Z(0) = 0.5$, are

$$U_p(0) = 0 \qquad W_p(0) = e\,[(U_p(T_j))^2 + (W_p(T_j))^2]^{1/2}, \tag{12a}$$

where e is the coefficient of restitution. In terms of Eqs. 7–11, Eq. 12a becomes

$$U_r(0) = -Re_x \qquad W_r(0) = e\,[(Re_x + U_r(T_j))^2 + (W_r(T_j))^2]^{1/2}. \tag{12b}$$

The initial conditions for discrete trajectories defined in Eq. 12 with $e = 1$ and $W_p(T_j) = 0$ correspond to those used by Madsen (1991).

Note that for the idealized impact geometry adopted in our model, the conservation of particle momentum requires that the impact and ejection angles α_i and α_e, respectively, have values which satisfy the relationship

$$e = \frac{\tan(\alpha_e - 45^\circ)}{\tan(\alpha_i + 45^\circ)}\cdot \tag{13}$$

For idealized ejection angles oriented near 90° from the horizontal datum and e near unity, impact angles must be around 10° or less. Such low incident angles are expected from the arctangent of the ratio $W_p(T_j)/U_p(T_j)$ calculated from Eqs. 7 and 8 subject to the initial conditions defined in Eq. 12. The present model, including the idealized geometry of particle-bed contacts shown in Figure 1, is therefore internally consistent.

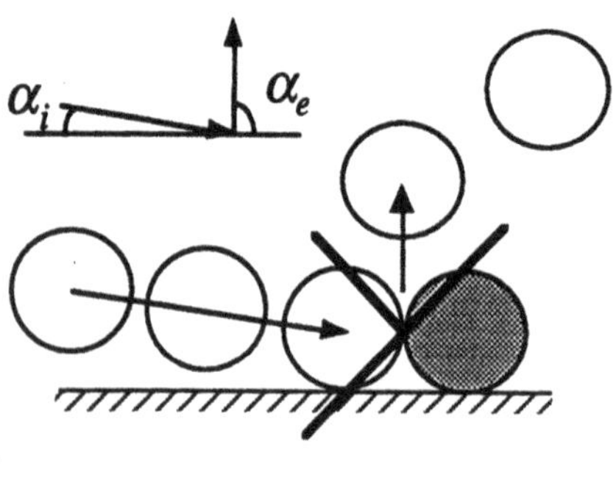

FIGURE 1. Idealized geometry of impacts between saltating and stationary-bed particles. Channel flow is from left to right. The shaded grain represents a bed particle.

An important aspect of water-borne saltation embodied in the conditions of Eqs. 12 and 13 is that continuous flights are driven primarily by particle momentum derived from the flow, and not from inertial impact. The role of granular roughness in the translation of horizontal momentum into vertical momentum is thus important in water-borne saltation. This distinguishes the process of subaqueous saltation from aeolian saltation.

This difference can be explored further with a calculation of the heights attained by particles settling in still fluids to and rebounding from a perfectly smooth bed for which $e = 0.9$. Consider a glass sphere with a diameter of 0.1 cm and with a density of 2600 kg m^{-3}. In still air with a nominal density of 1 kg m^{-3} and with a nominal kinematic viscosity of

1.6×10^{-5} m^2 s^{-1}, the sphere would impact the the bed with a terminal settling speed of about 7.4 m s^{-1}. From Eqs. 9 and 11 we find that the particle would rebound to a height of approximately 1.5 m, or about 1500 particle diameters, and return to the bed over the span of about 1 s.

In still water with a nominal density of 1000 kg m^{-3} and a kinematic viscosity of 10^{-6} m^2 s^{-1}, the same glass sphere would impact the bed with a terminal settling speed of about 15 cm s^{-1} and the particle center would rebound to a height of only 0.16 cm, or about 1.6 particle diameters, and resettle to the bed during a span of about 1/24 s.

These calculations reinforce the notion that aeolian saltation dynamics are driven to a large degree by the inertia of the individual particles. In water, however, saltation trajectories reflect to a much greater extent the momentum imparted to individual particles by the ambient fluid flow. The conditions given in Eq. 12 accommodate these differences.

Our conceptual model outlined in Eqs. 2–12 is most readily applied to water-borne saltation. It can be nominally applied to the case of continuous saltation in air, however, with the modification $u/u_* = \Gamma(Z_{av})$. This accommodates the fact that the large rebound height of a grain in air results in exposure of the wind-blown grain to fluid velocities which are relatively much greater than are seen by a water-borne particle travelling near the tops of bed roughness. In cases of wind-blown transport over a loose bed, moreover, values of e need to be chosen to accommodate significant redistribution of momentum during the scatter of bed particles involved in high-energy, intersaltational impacts (e.g., Hunt and Nalpanis 1986; Mitha et al. 1986; Haff and Anderson 1993).

In closing this presentation of our conceptual model for subaqueous saltation, we point out that many of the Eqs. 2–12 are coupled and must be solved iteratively. For example, $W_r(T_j)$ depends on T_j which depends on $W_r(0)$, which in turn is a function of $W_r(T_j)$. Solutions to this system of equations, however, are readily obtained with very few iterations.

3. Methods and Materials

A laboratory experiment was conducted to obtain average and maximal trajectory heights for smooth, glass spheres in dilute, continuous saltation transport driven by channel flows over planar beds of fixed roughness.

The experiment was performed in the racetrack flume at the Friday Harbor Laboratories of the University of Washington. The test section of the flume is an 8–m long and 0.7–m wide channel with Plexiglass walls and floor. Flow depth was maintained at 0.12 m. Flow is driven by a series of paddles in the return section of the flume, and entry into the test section is constrained by several narrow sub-channels which straighten the flow.

Before use all flume surfaces were soaked for 48 h in a slow flow of aqueous 3% hydrogen peroxide solution and then cleaned thoroughly with a weak detergent. All runs were made with filtered tap water maintained at 16.5 ± 0.1 °C throughout the water column. Nominal fluid density and viscosity are taken to be 1000 kg m^{-3} and 0.011 Poise, respectively.

The experiment consisted of tracking in two dimensions the motion of saltating particles in five flume flows over a bottom-flush plate of immobile roughness consisting of grains identical to those in motion. In each run, near-bed particle trajectories were viewed from the sidewall of the flume 6 m downstream from the point of flow entry into the straight test section and 0.90–0.95 m downstream from the upstream edge of the roughness plate. Particles were haphazardly introduced into the steady flow at the point of entry into the test section. Several minutes (and thus several "laps" in the recirculating flume) passed between the time of particle introduction and the beginning of video recording of near-bed particle motion.

As particles passed through the view point, trajectories were recorded on videotape at image acquisition rates of 30 frames s^{-1} with a NEC TI 22PII CCD video camera equipped with a Tamron 1:2.5, 90 mm telemacro lens and positioned at the sidewall of the flume. Camera optics were focussed for an approximately 3–cm high by 3–cm wide by 0.5–cm deep field of view positioned along the channel centerline and with the channel bottom located in the lower third of the field. In each run a total of 500 to 1000 trajectories was recorded during a period of approximately 10 minutes. An image of a scale precisely marked in 1–mm increments was recorded prior to each run. This scale provided the basis for absolute estimates of particle displacement made in subsequent image analyses.

High-definition contrast between the glass spheres in transport (described below) and the ambient environment, necessary for image analyses, was achieved by making the video recordings at night in a completely dark room. The view point was illuminated with an approximately 0.2–cm thick sheet of light oriented longitudinally in the flow and created by passing the beam of a 10 mW He-Ne laser through a cylindrical lens. The stationary bed was spray-painted flat black and so bed particles did not saturate the recording optics with scattered light. Scattering of the laser light by the mobile glass particles, however, made them appear as points of light against the otherwise dark background. An important point is that not just the upper hemispherical surfaces of saltating particles were illuminated by the overhead light source, but rather the entire spheres. Because recording optics were nominally at infinite focal distance from the plane of light, incident light on camera lenses was paraxial. Accordingly, no corrections for the refraction at water-Plexiglass and Plexiglass-air interfaces were necessary.

Trajectories recorded at elevations between 0.1 cm and 2 cm above the plane of the smooth floor of the flume were later analyzed on an automated system that consists of a Motion Analysis Corporation VP–310 video processor and Expertvision software supported by a Sun Microsystems 3/160 workstation. Given the scale of the optical configuration, a single, illuminated sphere easily saturated one or more pixels in the field of view. Software program parameters were set to optimize data collection rate and accuracy. A general review of particle-imaging techniques is given in Adrian (1991).

Particle concentrations were sufficiently dilute and particle-background contrast sufficiently high that the portions of trajectories recognized by the system were of very high quality. Digitized trajectories for each run produced by the automated Motion Analysis system were edited to eliminate trajectories exhibiting any upward-directed acceleration (indicating incipient suspension). Particle trajectories which appeared stationary or exhibited prolonged contact with the bed were also removed from the image records. Thus resting and rolling particles were edited from the recorded images as well. The total number of trajectories eliminated was negligible. The overall impression during the experiment and from the video tapes is one of mobile particles in continuous saltation.

Recorded trajectories do not include complete saltations. Thus the experimental configuration precluded any direct measurements of trajectory length. Our estimates of z_{max}, moreover, represent lower statistical bounds for the true values. Because the estimated total flight duration t_j was typically of order 0.1 s (calculated from Eq. 9) and the duration of recorded portions of trajectories typically only a fraction of t_j, there were insufficient numbers of video frames per trajectory to obtain accurate measures of particle velocities. No attempts were made to gather rolling-grain velocities or collision statistics because fixed-roughness occluded the view right at the bed and also because of the velocity-measurement bias. To reiterate, however, data reported here pertain only to saltating grains continuously moving at or above the tops of roughness grains and exhibiting little or no rolling contact with the bed. This restriction is consistent with the assumptions of the model for heights of particles in continuous saltation described by Eqs. 10 and 11.

Flow characterization consisted of velocity measurements made at the view point just prior to each recording run with a TSI model 9100–8, two-component, three-beam, laser Doppler velocimeter. The optics of the velocimetric system were configured to measure backscattered light from a 1–W argon-ion laser and the signal-processing parameters were set to optimize rates of data acquisition. Nominal friction velocities at the view point were estimated from measured free-stream velocities of flows over rough plates according to Schlichting (1979). These values were checked against i) measurements of the near-bed Reynolds stress and ii) the slope of a semi-log plot (cf. Eq. 2) of the mean flow velocity and elevation above the bed (Dade 1992). Particle transport in our experimental setting occurred in well-developed, turbulent boundary layers of steady channel flows.

Particles used in this study were technical-grade glass spheres with polished-smooth surfaces. Nominal diameter of these spheres is 0.1 ± 0.02 cm and nominal density 2600 kg m^{-3} (Cataphote Microbeads Class-X Size-TG–1, Batch # 82110). As reported by the supplier, these particles are spheroids with no more than 15 % of the particles exhibiting irregular shapes. The average settling speed of these particles in still water is about 15 cm s^{-1}.

This particle size was chosen for ease of visualization. Additionally, this particle size is in the middle of the range of nonlinear drag effects described by Eq. 5b. In order to compare predictions from our conceptual model to those of the Wiberg model (Eq. 1), the nominal critical bed shear stress required for the onset of significant transport of these particles is taken to be 0.612 Pa, corresponding to a dimensionless grain volume Ξ equal to 1.57×10^4 and a critical Shields parameter β_{cr} equal to 0.039 (cf. Smith 1977).

The roughness plate consisted of glass particles identical to those in motion but adhesively fixed to a Plexiglass panel in haphazard arrangement. As was mentioned in the introduction e is typically taken to be 0.95 for perfectly smooth glass in low-velocity, direct-central impacts (Weaver 1930). In glancing impacts e is taken to be exactly unity (Engel 1976). The coefficient of restitution for Plexiglass in low-velocity, room-temperature, direct-central impacts is 0.9 ± 0.05 (Engel 1976). For our highly idealized, experimental settings in which smooth, glass spheres are in continuous saltation over beds of stationary but otherwise identical particles mounted on a Plexiglass plate, the coefficient of restitution is thus expected to take on a value in the range $1 < e < 0.8$.

4. Results

Vertical distributions of the measurements of average relative saltation trajectory height Z_{av} and maximal relative saltation trajectory height Z_{max} within dilute saltation layers are summarized in Figure 2. The histograms for Z_{av} effectively depict the time-averaged distribution of particulate mass in dilute saltation layers. The dashed lines in the Z_{av} panels represent median values of observed Z_{max} for each flow condition, indicating the degree to which the mass distribution is (or is not) vertically uniform. The vertical distributions of both Z_{av} and Z_{max} are necessarily truncated at the tops of the roughness grains but are otherwise generally skewed toward large elevations. Accordingly they can be described in terms of log-normal or gamma distributions. If gamma distributions are applied to the histograms for observed Z_{av}, all distributions interestingly exhibit a coefficient of variation of 45% ± 3%.

First-moment estimates of Z_{av} and Z_{max} as a function of bed shear stress are summarized in Table 1. Because the vertical distributions of mass are not normally distributed, median values are more meaningful measures of typical trajectory dimensions than are means. Plots of median values of measured Z_{av} and Z_{max} as functions of bed shear stress appear in Figure 3. These are compared with the corresponding predictions of our conceptual model for representative values of e, and with predictions calculated from Eq. 1.

TABLE 1: Statistics for measured heights of particles in dilute saltation layers; $d = 0.1$ cm, $R = 2.6$

u_∞ (m s^{-1})	0.56		0.67		0.72		0.86		0.89	
τ_o (Pa)	1.02		1.52		1.76		2.50		2.60	
$u_* d/\nu$	29		35		38		45		46	
u/u_*	8.8		8.7		8.6		8.4		8.3	
no. obs.	621		845		740		741		457	
	mean	median	mean	median	mean	median	mean	median	mean	median
Z_{av}	3.1 (± 0.4)	1.8 (± 0.1)	4.1 (±0.4)	2.3 (± 0.2)	3.7 (± 0.4)	2.2 (± 0.1)	5.0 (± 0.5)	2.5 (± 0.1)	3.6 (± 0.3)	2.9 (± 0.2)
Z_{max}	3.8 (± 0.4)	2.4 (± 0.1)	4.9 (± 0.4)	3.1 (± 0.2)	4.6 (± 0.4)	3.1 (± 0.2)	5.7 (± 0.5)	3.4 (± 0.2)	4.5 (± 0.3)	3.8 (± 0.4)

Values of u/u_* are estimates of the flow velocity at the tops of grain roughness from a model for near-bed channel flow outlined in Dade (1993). Approximate 95 percent confidence intervals on mean and median statistics of Z_{av} and Z_{max} are given in parentheses.

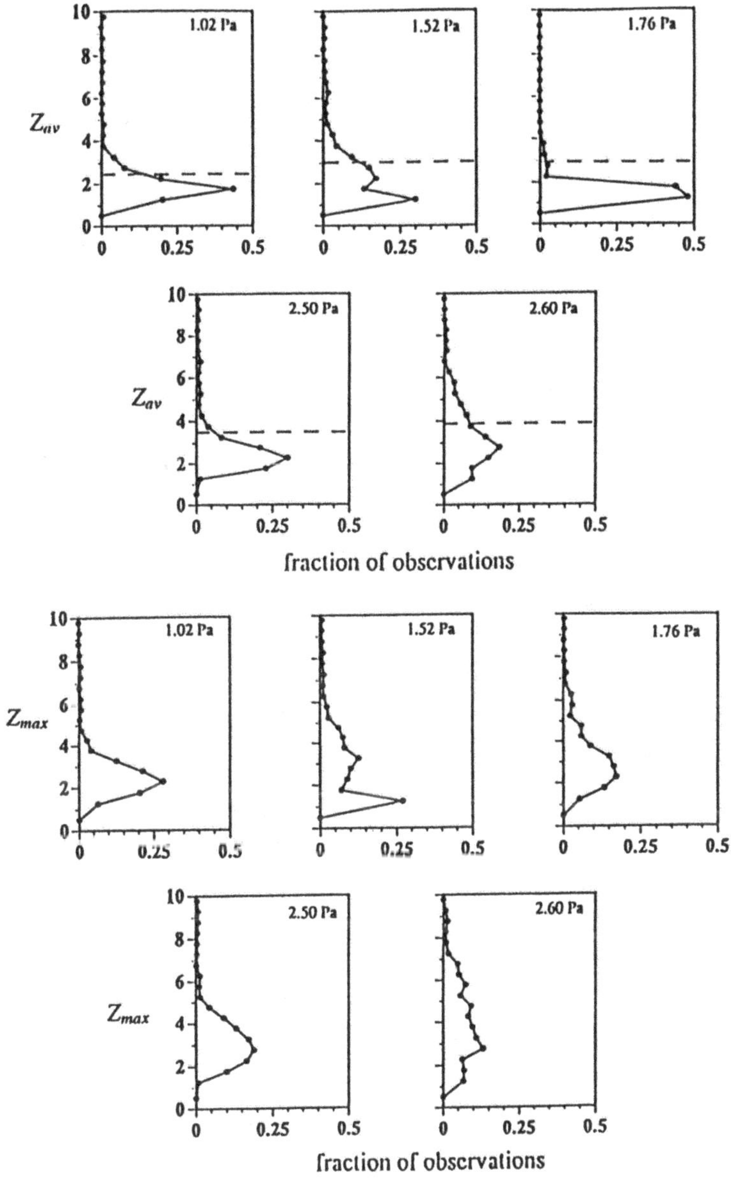

FIGURE 2. Vertical distributions of measured Z_{av} and Z_{max} of 0.1–cm diameter glass spheroids in dilute saltation layers driven by steady channel flows. The bed shear stress is shown in the upper right corner of each panel. The dashed lines indicate the median value of Z_{max} for each run.

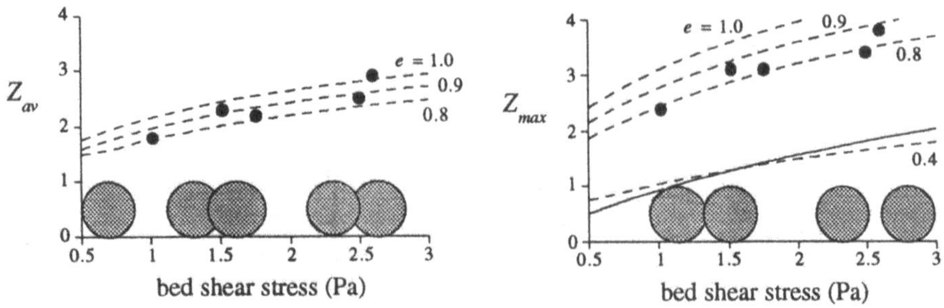

FIGURE 3. Average trajectory height Z_{av} and maximal trajectory height Z_{max} as functions of bed shear stress. The solid symbols represent median measured values The dashed lines show model predictions from Eqs. 7 and 8 for the indicated values of e, $u/u_* = 8.5$ and $Z(0) = 0.5$. The solid line shows calculations from Eq. 1 for 0.1–cm diameter grains with a critical bed shear stress of 0.61 Pa. The shaded grains represent stationary bed roughness.

5. Discussion and conclusions

The observations summarized in Table 1 and Figures 2 and 3 represent the behavior of individual spheroids in continuous, water-borne saltation over a bed of grains which are stationary but otherwise identical to those in transport. Intensities of the sediment-transporting channel flows are in the range $1 < T_* < 5$. Ratios of the flow friction velocity and particle settling velocity are in the range $0.2 < u_*/w_s < 0.5$.

The distributions of saltation heights shown in Figures 2 and 3 undoubtedly represent the effects of stochastic properties of bed microtopography, intersaltational collisions, and the near-bed flow field. Parameters that describe in average terms the near-bed velocity of the transporting flows as well as the efficiency and geometry of particle impacts in our idealized experimental system, however, are well constrained by *a priori* considerations. These data thus constitute a useful benchmark for mechanics-based models which describe water-borne saltation.

The conceptual model for continuous saltation described here is not meant to supplant the more comprehensive approaches taken by earlier workers. Our model, for example, does not consider the effects on particle saltation of prolonged frictional contact with the bed, the stochastic properties of bed microtopography and geometry of intersaltational collisions, unsteady drag forces during acceleration, lift forces during flight, or turbulence. The model includes the essential physics of continuous saltation, however, and predictions agree reasonably well with predictions for typical saltation heights obtained from the more complete numerical calculations of Wiberg and Smith (1985, 1989) or Sekine and Kikkawa (1992) when similar values of the coefficient of restitution are used.

Predictions from our conceptual model also agree with median estimates of the experimentally observed average and maximal heights attained by glass spheres in continuous saltation. Indeed, approximately 90% of the overall variance in the median estimates of the observed average and maximal heights of spheres in saltation is explained by our model and $Z(0) = 0.5$, $u/u_* = 8.5$, and $e = 0.85$. This agreement between theory and well-constrained experiments provides confirmation in general for mechanics-based descriptions of continuous saltation in channel flows.

6. Acknowledgements

P. Hill and P. Jumars made helpful comments on an earlier version of the text. Support for this study was provided by the U.S. Office of Naval Research.

7. References

Abbot, J.E., and J.R.D. Francis, *Phil. Trans. Roy. Soc. Lond.* **A284** (1977) 225–254.

Adrian, R.J., *Annu. Rev. Fluid Mech.* **23** (1991) 261–304.

Bridge, J.S., and S.J. Bennet, *Water Resources Res.* **28** (1992) 337–363.

Clift, R., J.R. Grace, and M.E. Weber, *Bubbles, Drops, and Particles* (Academic Press, 1978) 380 p.

Dade, W.B., unpublished Ph.D. dissertation, University of Washington (1992) 270 p.

Dade, W.B., *Limnology and Oceanography* **38** (1993) 52–69.

Einstein, H.A., *Tech. Bull. No. 1026* (Soil Conservation Service, U.S. Department of Agriculture, Washington, D.C., 1950) p. 1–71.

Engel, P.A., *Impact Wear of Materials* (Elsevier, 1976) 339 p.

Gilbert, G.K., *Prof. Paper No. 86*, (U.S. Geological Survey, Washington, D.C., 1914) 263 p.

Gordon, R., J.B. Carmichael, and F.J. Isackson, *Water Resources Res.* **8** (1972) 444–459.

Haff, P.K. , and R.A. Anderson, *Sedimentology* **40** (1993) 175–198.

Hanes, D.M., *Acta Mechanica*, **63** (1986) 131–142.

Hunt, J.C.R., and P. Nalpanis, in *Proceedings of the International Workshop on the Physics of Blown Sand, Aarhus, Memoirs no.* 8, eds. O. E. Barndorff-Nielsen, et al., vol. 1 (1985) 9–36.

Lun, C.K.K., and S.B. Savage, *Acta Mechanica* **63** (1986) 15–44.

Madsen, O. S., *Proceedings of the ASCE, Coastal Sediments '91*, (Seattle, WA, 1991) p. 15–27 + *corrigendum*.

Mitha, S., M.Q. Tran, B.T. Werner, and P.K. Haff, *Acta Mechanica* **63** (1986) 267–278.

Murphy, P.J., and H. Hooshiari, *ASCE J. Hydraul. Div.* **108** (1982) 1251–1267.

Owen, P.R., *J. Fluid Mech.* **20** (1964) 225–242.

Reizes, J.A., *ASCE J. Hydraul. Div.* **104** (1978) 1305–1321.

Sekine, M., and H. Kikkawa, *J. Hydraul. Eng.* **118** (1992) 536–558.

Schlichting, H., *Boundary-Layer Theory* (7th ed., McGraw-Hill, 1979) 817 p.

Smith, J.D., in *The Sea* vol. 6, eds. E. D. Goldberg, et al. (John Wiley, 1977) p. 539–577.

Soulsby, R.L., *Report SR 351* (HR Wallingford, 1993).

van Rijn, L.C., *J. Hydraul. Eng.*, **110** (1985) 1431–1456.

Weaver, F.N., *Applied Mechanics* (The Ronald Press Company, 1930) 322 p.

Wiberg, P.L., unpublished Ph.D. dissertation, University of Washington. (1987) 132 p.

Wiberg, P.L., and D.M. Rubin, *J. Geophys. Res.* **94(C4)** (1989) 5011–5016.

Wiberg, P.L., and J.D. Smith, *J. Geophys. Res.* **90(C4)** (1985) 7341–7354.

Wiberg, P.L., and J.D. Smith, *J. Hydraul. Eng.* **115** (1989) 101–123.

MEASUREMENT OF THE VELOCITY OF SEDIMENTATION BY AN X-RAY ATTENUATION TECHNIQUE

R. ANTHORE ‡, F. BOSTEL ‡, M. MASMOUDI ‡,

UFR des Sciences, URA 808

76821 Mont Saint Aignan, France

and

F. FEUILLEBOIS ‡,

Laboratoire d'Aérothermique du CNRS, 4 ter, route des Gardes,

92190 Meudon, France

‡*Groupement de Recherche "Physique des Milieux Hétérogènes Complexes", CNRS*

ABSTRACT

The concentration in a dispersion of particles is measured with an experimental technique based on the attenuation of an X-ray beam. For very dilute polydisperse suspensions of nearly spherical particles in sedimentation, the histogram is obtained. For quasi-monodisperse suspensions of spherical particles, the average velocity of sedimentation is derived from the time evolution of the concentration profile; in the case of dilute suspensions of interacting particles the velocity of sedimentation obtained in this way was found to be in agreement with the theoretical prediction by Batchelor & Wen. For concentrated quasi-monodisperse suspensions of spherical particles, the average velocity of sedimentation is well fitted by an extended form of Richardson & Zaki correlation.

1. Introduction

The concentration in a dispersion of particles is measured with an experimental technique based on the attenuation of an X-ray beam. This article gives a survey of some applications to sedimentation.

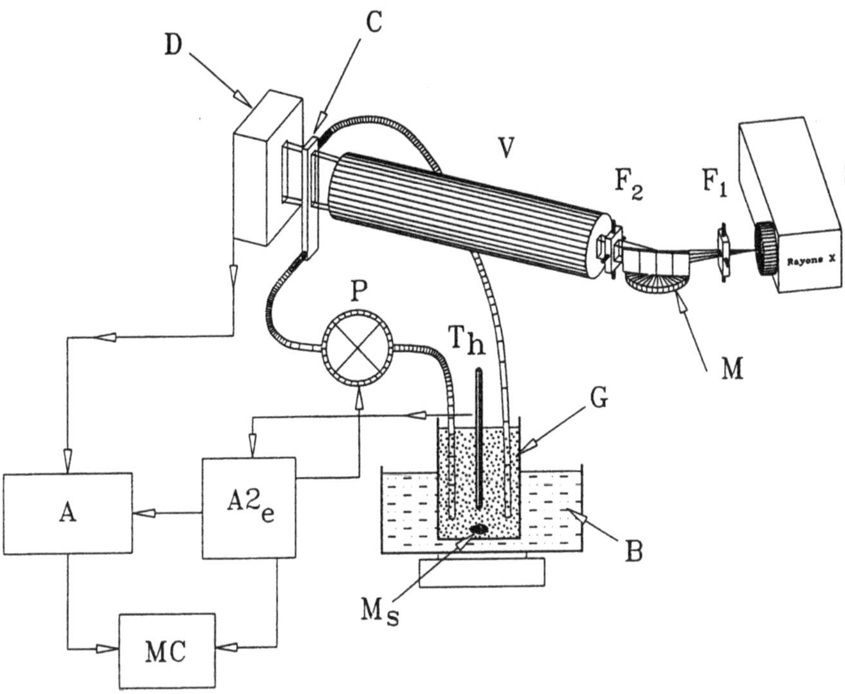

Figure 1: Experimental setup.

The experimental setup and procedure are described first. The measurement of the average velocity of sedimentation and the comparison with theory are then discussed. The spreading of the top front and the evolution of the bottom front are considered in more detail. For very dilute suspensions it is explained how the measurement of the histogram is obtained from the concentration profiles; typical results are then presented.

2. Experimental setup and procedure

The experimental setup used in the laboratory is represented in fig. 1. An X-ray beam emerging from the source (S) is reflected by a quartz cylindrical monochromator (M) so as to become purely monochromatic and free of harmonics. Slots (F_1,F_2), and a void enclosure (V) allow to eliminate parasitic rays and to obtain a homogeneous beam in the entrance plane of the cell (C). The dimensions of the cross section of the beam by this plane are 50 mm × 10 mm. The suspension of particles is kept in a closed tank (G). It is kept homogeneous by a mechanical agitator (Ms) and stabilized in temperature within ± 0.2°C by a thermostated bath (B). The temperature is controlled by the thermometer (Th). At the beginning of each experiment, the suspension is pumped into the cell (C) by a system of peristaltic pumps (P). The fraction of the beam which is not absorbed by the suspension in the cell is measured on a linear detector (D). The cell is located as close as possible to the detector so as to eliminate any parallax error. The detector (D) and its associated electronics (A) provide 200 measurement points along the vertical, each of which with a spatial resolution of 200 μm. That is, the measurement length along the vertical is 4 cm. All measurement points are explored simulteaneously and there is no need to move the detector. The data are recorded in a multispectral analyzer (Mc). The whole experimental process is driven by a microcomputer (A2e).

The part of the X-ray beam diffused by the particles is negligible. The transmitted intensity follows a classical exponential decay law [1]:

$$I = I_0 \exp[-\alpha_p l_p - \alpha_f (l - l_p)] \tag{1}$$

where I_0 is the intensity of the incident beam, α_p and α_f are the linear absorption coefficients of the particles and fluid, respectively, l is the thickness of the cell and l_p is the total equivalent thickness of particles crossed by the X-ray beam. Then

$$I_f = I_0 \exp[-\alpha_f l] \tag{2}$$

is the X-ray intensity transmitted through the cell when filled up with fluid. The transmission factor is defined as

$$T = \frac{I}{I_f}.$$

From (1) and (2), the transmission factor becomes

$$T = \exp[-l(\alpha_p - \alpha_f)\Phi].$$

The system is calibrated by first filling the cell with pure fluid then with homogeneous mixtures of fluid and particles in various concentrations.

By then measuring the transmitted intensity $I(z, t_s)$ at each vertical location z and time t_s (t_s denoting the time from the beginning of the sedimentation experiment), the transmission factor is obtained as $T(z, t_s) = I(z, t_s)/I_0$. The average volume concentration $\Phi(z, t_s)$ along the X-ray beam is given by

$$\Phi(z, t_s) = \Phi_0 \frac{\log(T(z, t_s))}{\log(T_0)}, \tag{3}$$

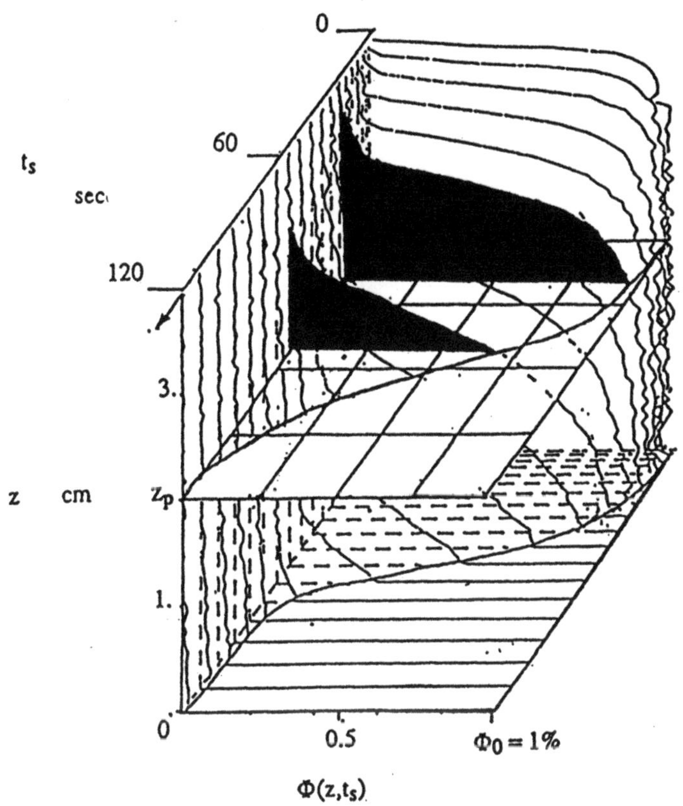

Figure 2: Evolution with time t_s of the profile in volume concentration Φ vs. height z for a suspension of glass beads with diameter 150 μm in a fluid with viscosity 1 Pa.s. The initial volume concentration is $\Phi_0 = 1\%$.

where Φ_0, the initial total volume concentration, and T_0, the initial transmission factor, are independent of z.

A typical sedimentation experiment may last from 2.5 sec to several hours. In any case, 60 concentrations profiles are recorded at predetermined times.

Typical concentration profiles from the X-ray absorption experiment are shown in fig. 2.

3. Average velocity of sedimentation

The evolution of the concentration profiles presented in fig. 2 is used to obtained an average velocity of sedimentation which can be compared to theory. Some theoretical background on sedimentation will be presented first.

3.1 Theoretical background

A characteristic number useful in sedimentation is the Péclet number describing the relative effect of gravity and Brownian diffusion. For interacting spherical particles, the relevant Péclet number is written as

$$Pe = \frac{V_r a}{D} \tag{4}$$

where V_r is the relative velocity of two particles, a is a radius and D the coefficient of Brownian diffusion. a and D are averages between the values of both particles. The Péclet number defined in (4) is small for Brownian particles (typically smaller than 1μm) and large for non-Brownian particles (typically larger than 1μm). For dilute suspensions for which only pair interactions are important, two main results have been obtained for the average velocity of sedimentation using the low Reynolds approximation for the fluid motion around the particles.

For a monodisperse suspension of Brownian spheres, $V_r \to 0$ and D is large, thus $Pe \to 0$; the result obtained for a homogeneous suspension [2] is:

$$V = V_0(1 - 6.55\Phi_0)$$

where V_0 is the velocity of sedimentation of a single sphere and Φ_0 is the volume concentration.

For a nearly monodisperse suspension of non-Brownian spheres, $V_r \to 0$ but $D \to 0$ faster, thus $Pe \to \infty$; the result obtained for this case [3] for a homogeneous suspension is:

$$V = V_0(1 - 5.6\Phi_0) \tag{5}$$

The two preceding results are exact in the sense that they were obtained with the limited set of assumptions quoted above. We are concerned here with non-Brownian suspensions so that the second result is the relevant one.

For concentrated suspensions, the problem is complicated since the hydrodynamic interactions between more than two particles become important. Moreover for non Brownian particles the structure of the suspension also depends on these hydrodynamic interactions. There is no exact result for this case, so that empirical relationships have to be used. Such a relationship will be proposed later.

3.2 Experimental determination of the average velocity of sedimentation

From the time evolution of the concentration profile, fig. 2, the experimental average velocity of sedimentation is obtained as a flux across a horizontal reference plane. One such reference plane is represented at $z = z_p$ in fig. 2.

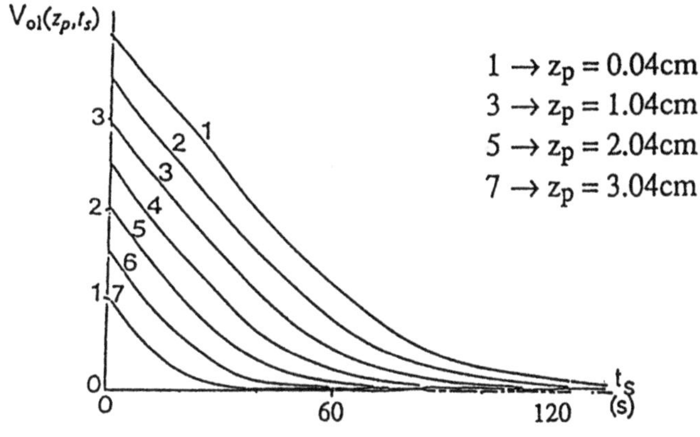

Figure 3: Evolution with time t_s of the total volume of particles located above a reference plane $z = z_p$, for various choices of this reference plane. The top of the cell is located at $z = z_0 = 4$ cm.

The total volume of particles located above this plane at a given time t_s is practically obtained from a discrete sum over the volume concentration from the plane $z = z_p$ up to the top plane $z = z_0$:

$$\mathrm{Vol}(z_p, t_s) = A \sum_{z=z_p}^{z=z_0} \Phi(z, t_s)\Delta z. \tag{6}$$

Here A represents the area of interior cross section of the cell. A typical variation of this volume Vol with time t_s for various reference planes z_p is represented in fig. 3. When a curve is straight, the homogeneous part of the suspension is passing through the reference plane. The trailing part of each curve is characteristic of the reference plane then being in the top inhomogeneous front of the suspension. From the straight parts of the curves, the average velocity of sedimentation in the homogeneous part of the suspension, defined as the flux of particles across a given reference plane, is obtained as

$$V = \frac{1}{A\Phi_0} \frac{d\,\mathrm{Vol}(z_p, t_s)}{dt_s}. \tag{7}$$

3.3 Experimental results

The experimental results for the average velocity of sedimentation in a dilute suspension of glass beads with diameter 150 μm are shown in figure 4. Taking into account the error bars which were estimated from the experiment to be ± 0.025, and applying a linear regression analysis [4] to the points in this figure, we obtained the constants B, S of the linear regression line

$$V = B + S\Phi_0$$

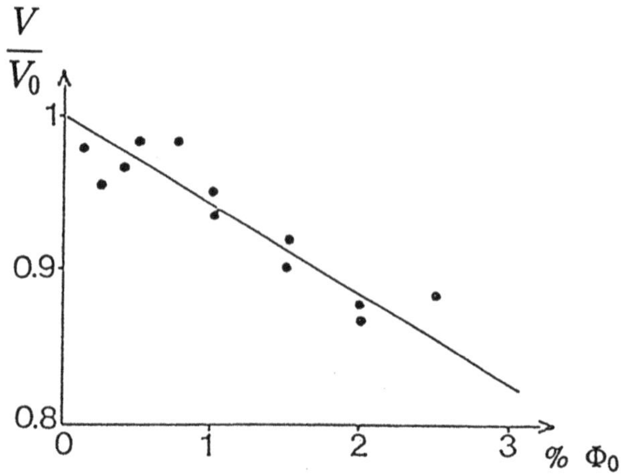

Figure 4: Average velocity of sedimentation normalized by the velocity of sedimentation of a single sphere vs. the volume concentration in a homogeneous suspension, for low concentrations. Dots represent the experiment and the solid line is the least squares fitting.

together with their 90% confidence limits:

$$B = \frac{V(0)}{V_0} = 1.00 \pm 0.05 \tag{8}$$

$$S = -5.3 \pm 1.1 \tag{9}$$

The result (8) shows that Stokes' law is verified with a 5% precision. The result (9) for the slope is in good agreement with Batchelor & Wen's theoretical result $S = -5.6$, equ.(5). The low precision on S obtained from the regression analysis can also be understood as follows. The absolute precision on $S\Phi_0$ is

$$\Delta(S\Phi_0) = \Delta V + \Delta B \simeq 0.075.$$

$S\Phi_0$ being a correction term for V, it is generally much smaller than V. Then

$$\frac{\Delta(S\Phi_0)}{S\Phi_0} \gg \frac{\Delta V}{V}.$$

The relative error on $\Delta(S\Phi_0)$ may thus be important. The error $\frac{\Delta S}{S}$ may then also be important. More details and a discussion may be found in [5].

Experimental results valid for the whole range of concentrations from dilute to concentrated were obtained by following the middle of the top front of the suspension. They are shown in fig. 5 under the label "top front". As no theoretical results are presently available, as said earlier, a correlation has to be used to fit these results.

Richardson & Zaki's [6] correlation for monodisperse suspensions of spheres is probably the best known one:

$$V = V_0(1 - \Phi_0)^{4.65}.$$

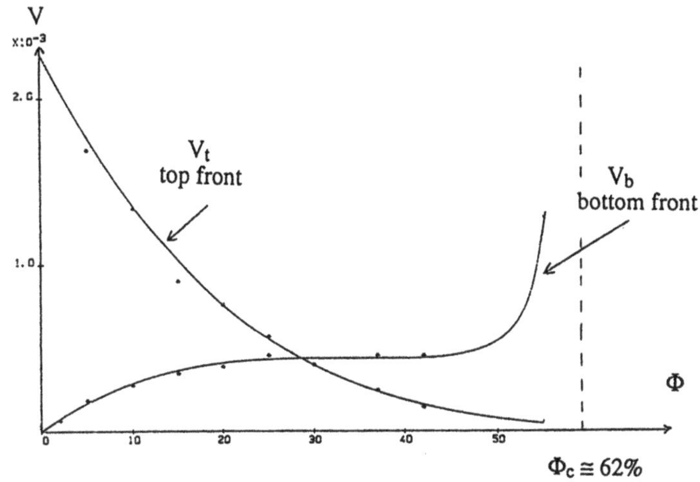

Figure 5: Variation of the velocities of the top and bottom fronts with concentration. The experimental results for the top front are well fitted by the modified Richardson and Zaki correlation (10).

An improved version is proposed here:

$$V = V_0(1 - \Phi_0/\Phi_c)^{5.6\Phi_c}. \tag{10}$$

This expression is consistent with the Batchelor and Wen result (5) in the limit $\Phi_0 \to 0$ and gives a zero velocity in the limit of dense packing $\Phi_0 \to \Phi_c$ which is consistent with experiment.

The experimental results in fig. 5 are shown to be well correlated by (10).

4. The spreading of the top front

The spreading of the top front of the suspension during sedimentation may be due to several causes.

Some spreading is due to polydispersity, the larger and more dense particles falling faster.

On the other hand a sharpening of the front is to be expected on the basis of the expressions for the average velocity of sedimentation (5)(10) which decrease with increasing concentration. This is because isolated particles which are left behind the cloud will then catch up with it. This effect is actually observed in monodisperse suspensions for $\Phi_0 > 10\%$.

Some spreading in a monodisperse suspension is to be expected from the dynamics of doublets. Doublets of particles fall faster than isolated particles. Then the top part of the cloud is poorer in doublets after some time. The concentration decreases there, thus the top front spreads. Figure 6 shows that spreading increases with concentration when $\Phi_0 < 10\%$; this behaviour may be due to this dynamics of doublets.

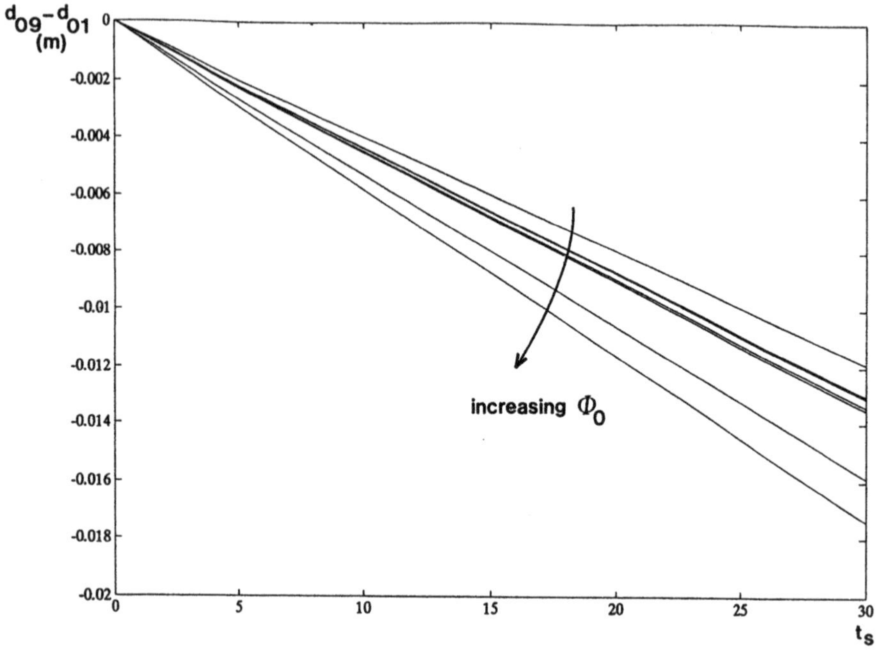

Figure 6: Spreading of the top front increases with concentration in a semi-dilute suspension ($\Phi_0 <$ 10%). Spreading is measured by time evolution of $d_{0.1}^{0.9}$, distance between the planes where the local concentration is $0.1\Phi_0$ and $0.9\Phi_0$, respectively. The values of Φ_0 on the lines is, from top to bottom, 0.0039, 0.005, 0.0075, 0.01, 0.015, 0.02, 0.025, respectively.

5. The evolution of the bottom front

Measurement of the time evolution of the concentration profile also provides the upward velocity and spreading rate of the bottom front.

We consider here that the suspension is concentrated enough so that the velocity of the top front V_t is practically equal to the velocity of the homogeneous part of the suspension. To simplify the notation, let us change Φ_0 to Φ. The velocity of the bottom front is calculated from a mass balance to be

$$V_b = \frac{V_t \Phi}{\Phi_c - \Phi}.$$

It is plotted vs. the concentration Φ in figure 5. The bottom front is generally sharp, thus direct measurement is easy; However, when $\Phi \sim \Phi_c$, V_b varies widely and the front spreads, thus the contrast is lower.

The velocity of the isoconcentration planes can be obtained from Kynch's theory [7]. Let $j = \Phi V_t$ be the flux of particles. The equation of continuity is

$$\frac{\partial \Phi}{\partial t} + \frac{\partial j}{\partial z} = 0$$

which may be rewritten as

$$\frac{\partial \Phi}{\partial t} + \frac{\partial j}{\partial \Phi}\frac{\partial \Phi}{\partial z} = 0.$$

Thus $\frac{\partial j}{\partial \Phi}$ is the velocity of the isoconcentration planes. It is measured as the slope in a plot of j vs Φ.

For high concentrations, the X-ray absorption technique appcars as an alternative to the acoustic technique used by [8].

6. Very dilute suspensions; measurement of the histogram

If the suspension is very dilute, hydrodynamic interactions between particles are negligible. The histogram can then be obtained from the time evolution of the concentration profile.

Particles with a diameter say D_s fall during a time t_s a distance $z_0 - z$ from top (fig. 7). Their velocity is

$$\frac{z_0 - z}{t_s} = K D_s^2 \tag{11}$$

where from Stokes' law

$$K = \frac{(\rho_p - \rho) g}{18\mu},$$

in which ρ_p is the density of particles and ρ, μ are respectively the fluid density and viscosity. Thus D_s is obtained from fig. 7 and equ. (11). The same result for D_s could be obtained using another time and another falling distance, e.g. t'_s and $z_0 - z'$,

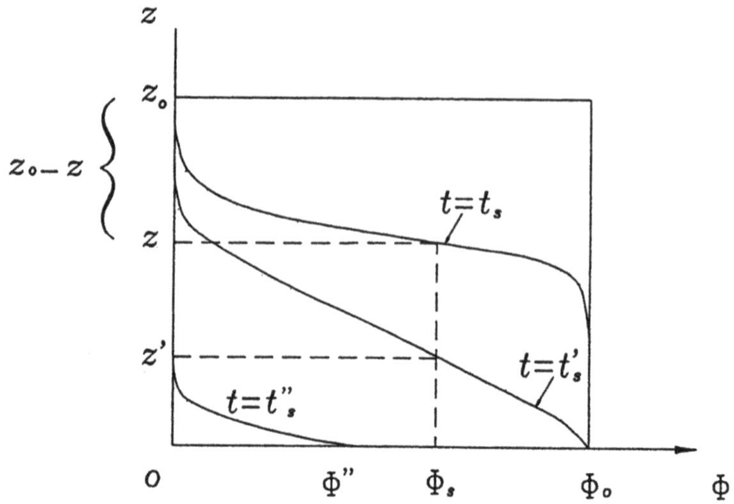

Figure 7: Time evolution of the concentration profile

fig. 7. All particles with diameter less than D_s fall slower. Thus Φ on the plot is also the concentration of cumulated sizes less than D_s. For each time t_s, a plot of Φ/Φ_0 vs. D_s can thus be constructed and we obtain the cumulated size distribution shown in fig. 8. For some late times for which the particle cloud has nearly left the measurement window, like t''_s in fig. 7, the whole range of Φ is not covered that is Φ is less than some Φ''. The corresponding curve in the Φ/Φ_0 vs. D plot, fig. 8, then has a restricted range. All the curves obtained from the various times t_s should be anyway superimposed in fig. 8.

A typical result of a cumulated size distribution obtained with this technique is presented in fig. 9. It concerns silica beads with a 1% volume concentration. The cumulated size distribution obtained from direct microscopic observation, which should be considered as a reference method, is also shown for comparison in the same figure. The results for the median diameter obtained from both methods differ by about 10% and there is some discrepancy in the spreading. This is mainly due to the fact that the hydrodynamic interactions were ignored when using the results from the X-ray absorption. Using an average correction factor from eq. (5) provides an improved cumulated size distribution, fig. 10, which fits better with the result from microscopic observation. However, there is still some discrepancy. Since the suspension is polydisperse, different correction factors for each particle size should be used to account for the hydrodynamic interactions [3]. In other words, using an average correction factor is not sufficient. A work on this topic is in progress. It is anyway recommended to use as low a concentration as possible, since the effects of hydrodynamic interactions increase with concentration.

The technique provides an equivalent diameter and a cumulated size distribution

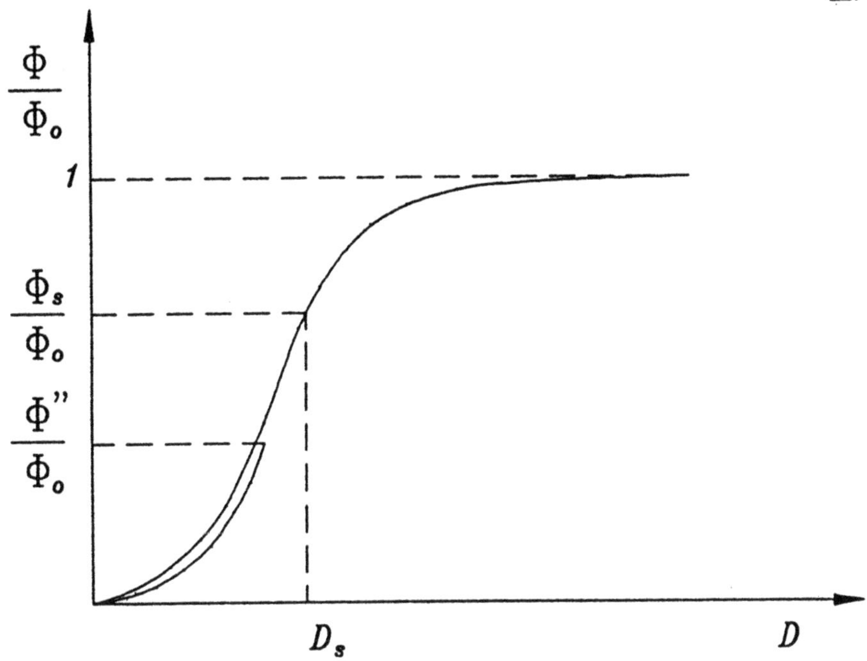

Figure 8: Cumulated size distribution; principle of construction.

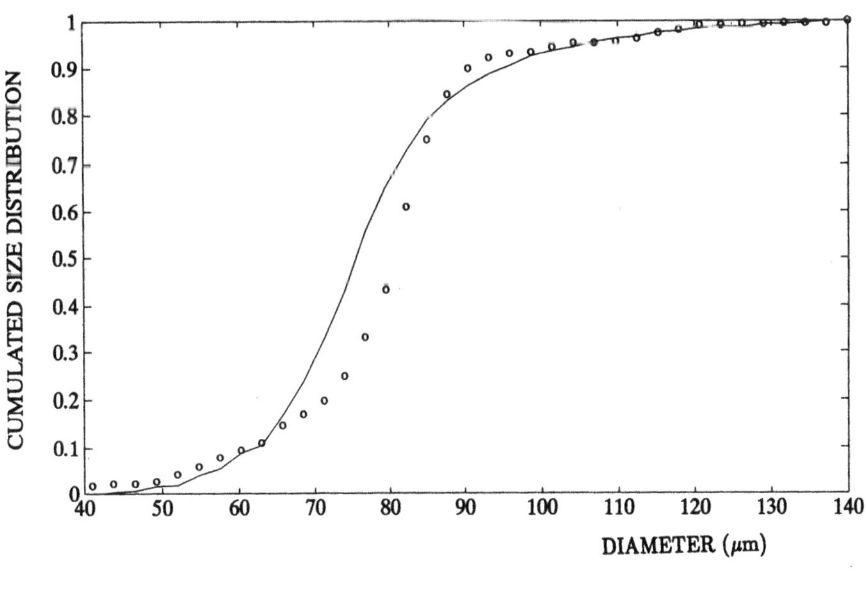

Figure 9: Cumulated size distribution; example of application. The solid line represents the result from the X-ray absorption technique and the circles the result from direct microscopic observation.

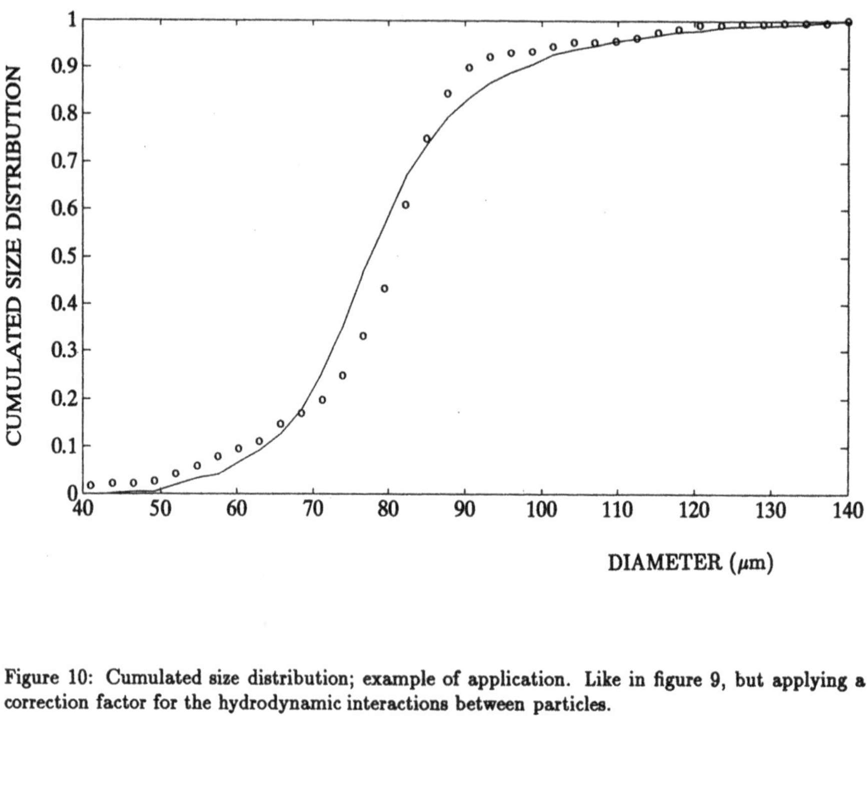

Figure 10: Cumulated size distribution; example of application. Like in figure 9, but applying a correction factor for the hydrodynamic interactions between particles.

for nearly spherical particles. Now for non-spherical particles, the interpretation of the results obtained from X-ray absorption is still difficult.

The results provided by this direct measurement technique are presently similar to the ones obtained by taking samples of a suspension with a pipette. But the investment in time for measurement and analysis is reduced at least 100 times. It is moreover possible by using various supporting fluids to have access to equivalent particle diameters in the range from 0.1 to 300 μm, which is difficult with other techniques like sieving.

7. Conclusion

X-ray absorption is a very versatile way of studying the suspensions of solid particles.

• For very dilute polydisperse suspension of nearly spherical particles in sedimentation, the histogram is obtained.

For non-spherical particles, further theory is needed to interpret the experiments.

• For dilute quasi-monodisperse suspensions of spherical particles, the average velocity of sedimentation is in agreement with the theoretical prediction by Batchelor & Wen.

A work is in progress to confirm Batchelor & Wen theoretical predictions for polydisperse suspensions. Further theory is needed to interpret the spreading of the top front when hydrodynamic interactions between particles are important.

• For concentrated quasi-monodisperse suspensions of spherical particles, the results for the average velocity of sedimentation are well fitted by an improved Richardson & Zaki correlation.

Further theory is needed to model concentrated suspensions.

8. References

[1] H.P. Klug and L.E. Alexander. *X-ray diffraction procedure* (Wiley Interscience, 1954).

[2] G. K. Batchelor. Sedimentation in a dilute dispersion of spheres. *J. Fluid Mech.* **52** (1972) 245.

[3] G. K. Batchelor and C. S. Wen. Sedimentation in a dilute polydisperse system of interacting spheres. part 2: Numerical results. *J. Fluid Mech.* **124** (1982) 495.

[4] J.S. Bendat and A.G. Piersol. *Random data: analysis and measurement procedures.* (Wiley Interscience, 1971).

$R = \frac{R_\delta^2}{2}$.

The estimation of the turbulent character of the flow above ripples is very delicate : our experimental conditions would be supposed to lead to a turbulent flow according to Sleath (1984) and should produce a laminar one according to Sato & *al.* (1984) visualisations.

The objective of the present work was to analyse the flow regime above a vortex rippled bed for a Reynolds number R of 1700 $\left(R = \frac{\overline{u_0}\, a_0}{\upsilon}\right)$ and $\frac{h}{L}$ = 0.17 (where : h is the rippled height and L the wave lenght of the ripples).

2. Basic equations

The fluid motion is governed by the Navier-Stokes and continuity equations. Using the Einstein summation convention, the differential problem reads (incompressible flow) :

$$\frac{\partial u_i}{\partial t} + u_j \frac{\partial u_i}{\partial x_j} = -\frac{1}{\rho}\frac{\partial p}{\partial x_i} + \upsilon \frac{\partial^2 u_i}{\partial x_j^2} \tag{1}$$

$$\frac{\partial u_i}{\partial x_i} = 0 \tag{2}$$

In order to describe a turbulent flow, it is convenient to use the following decomposition of the velocity :

$$u_i(t) = \overline{u}_i(t) + u'_i(t) \tag{3}$$

where $\overline{u}_i(t)$ is the time-mean value of $u_i(t)$ during a fraction of the wave cycle and $u'_i(t)$ is the fluctuating component of $u_i(t)$. The flow has been split into fifty phases for each wave cycle so that $\overline{u}_i(t)$ is the mean value of $u_i(t)$ during a time window of T/50.

As we are interested in turbulence production, we need the turbulent kinetic energy transport equation :

$$\underbrace{\frac{\partial \overline{q'^2}}{\partial t} + \overline{u}_j \frac{\partial \overline{q'^2}}{\partial x_j}}_{(a)} = \underbrace{-\overline{u'_k u'_i}\frac{\partial \overline{u_i}}{\partial x_k}}_{(b)} \underbrace{- \upsilon \overline{\frac{\partial u'_i}{\partial x_j}\left(\frac{\partial u'_i}{\partial x_j} + \frac{\partial u'_j}{\partial x_i}\right)}}_{(c)}$$

$$\underbrace{-\frac{\partial}{\partial x_k}\left[\overline{u'_k\left(q'^2 + \frac{p'}{\rho}\right)} - \upsilon \frac{\partial \overline{q'^2}}{\partial x_k} - \upsilon \frac{\partial \overline{u'_i u'_k}}{\partial x_i}\right]}_{(d)} \tag{4}$$

$\overline{q'^2}$ being the turbulent kinetic energy,

where : (a) represents the total variation of turbulent kinetic energy

(b) is the turbulent energy production term ; it can be positive or negative

(c) represents the viscous dissipation of turbulent energy. Having no access to the velocity instantaneous gradients, we cannot calculate this term

(d) represents the turbulent diffusion. The experimental determination of this term is only possible when we can know the instantaneous turbulent kinetic energy and pressure fluctuations, which is not our case.

Assuming the flow to be two-dimensional, the production term $-\overline{u'_k u'_i}\frac{\partial \overline{u_i}}{\partial x_k}$ reduces to:

$$-\overline{u'_k u'_i}\frac{\partial \overline{u_i}}{\partial x_k} = -\overline{u'^2}\frac{\partial \bar{u}}{\partial x} - \overline{v'^2}\frac{\partial \bar{v}}{\partial y} - \overline{u'v'}\left(\frac{\partial \bar{u}}{\partial y} + \frac{\partial \bar{v}}{\partial x}\right) \tag{5}$$

The global analysis of the vortices evolution is defined by the vorticity transport equation :

$$\frac{\partial \Omega_i}{\partial t} + u_j \frac{\partial \Omega_i}{\partial x_j^2} = \upsilon \frac{\partial^2 \Omega_i}{\partial x_j^2} + \Omega_j \frac{\partial u_i}{\partial x_j} \tag{6}$$

where Ω represents the vorticity rate.

3. Experimental set up

The experiments are carried out by using a wave channel, 9 m long, 0.8 m wide and 1 m height. Surface gravity waves are generated by a piston driven by a direct current motor connected to an eccentric. At the other end of the wave channel an absorbing beach is displayed. It is composed of filtering elements and is characterized by a reflection coefficient which is less than 5 % in our conditions. In the test section, approximately 5 m far from the wave generator, artificial ripples are set. The ripples are artificial because our wave channel is not equiped to introduce sediments. Thus we do not take into account the presence of sediments in suspension and the porous feature of the bottom which are important parameters of the flow, notably for the mass transport. The ripples which are symmetrical with rounded crests, have dimensions which correspond to those observed in movable bed channel under the same wave conditions (Sato *et al.* (1984), Sleath (1984)). The height h and the wavelength L of the ripples are 3.10^{-3} m and $1.8.10^{-2}$ m respectively. The bottom profile is obtained by means of a form milling-cutter from an alluminium plate in such a way that the wall profile is described by the parametric relationships :

$$x = X - \frac{h}{2}\sin\left(\frac{2\pi X}{L}\right) \tag{7}$$

$$y = \frac{h}{2}\left(\cos\left(\frac{2\pi X}{L}\right) + 1\right) \tag{8}$$

where X is a dummy variable and (x,y) is an orthogonal coordinate system ; the x-axis lies along the direction of wave propagation and the y-axis is directed upwards. A ripple sketch and wave conditions are given on figure 1.

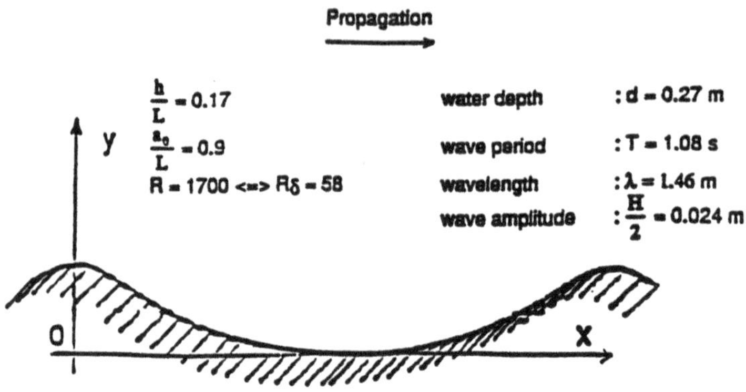

Figure 1 Bottom profile and wave characteristics

4. Data processing method

A temporal reference is required to obtain the coherent velocity fields at any time for non-permanent flow by means of one point measurements. This reference is provided by the free surface elevation. It is measured by means of a resistive probe located at a fixed position and the sampling of the output signal from laser anemometer is synchronized with the evolution of the free surface.

On the assumption that the temporal evolution of the flow is periodic, we make a dual partition of the data samples :

- partition into as many parts as synchronisms ; these parts are called waves. Let w be any wave index, W the total number of waves and T(w) the time period.

- partition of each wave into K intervals called phases. Let k be the phase index.

Thus, for example, the instantaneous value of the horizontal velocity component coming out from the velocimeter can be written as u(i,k,w) with $1<i<N(k,w)$ where N(k,w) is the number of data for the interval k and wave w. Figure 2 shows an example of sample and figure 3 the dual partition. In figure 2 each point represents a validated measurement and vertical dashed lines the synchronisms. We chose K equal to 50 and W is typically about 10.

Velocity measurements are performed in coincidence mode with a coincidence time window of 100 μs.

Thus, we can get the cross correlation of fluctuations and use the Reynolds decomposition to exhibit the turbulence production term.

For each of the K.W intervals we compute the following quantities in this way :

- the mean values of velocity components :

$$\bar{u}(k,w) = \frac{1}{\mathrm{N(k,w)}} \sum_{i=1}^{N(k,w)} u(i,k,w) \tag{9}$$

$$\bar{v}(k,w) = \frac{1}{\mathrm{N(k,w)}} \sum_{i=1}^{N(k,w)} v(i,k,w) \tag{10}$$

- the variances of velocity fluctuations from linear regression :

$$\overline{u'^2}(k,w) = \frac{1}{\mathrm{N(k,w)}} \sum_{i=1}^{N(k,w)} [u(i,k,w) - (A(k,w).t(i,k,w) + B(k,w))]^2 \tag{11}$$

$$\overline{v'^2}(k,w) = \frac{1}{\mathrm{N(k,w)}} \sum_{i=1}^{N(k,w)} [v(i,k,w) - (C(k,w).t(i,k,w) + D(k,w))]^2 \tag{12}$$

where A, B, C, D are the linear regression coefficients.

- The cross correlation of velocity fluctuations :

$$\overline{u'v'}(k,w) = \frac{1}{\mathrm{N(k,w)}} \sum_{i=1}^{N(k,w)} [u(i,k,w) - (A(k,w).t(i,k,w) + B(k,w))][v(i,k,w) - (C(k,w).t(i,k,w) + D(k,w))] \tag{13}$$

The linear regression is used to take into account the non-stationary feature of the flow. The flow being assumed periodic, we can built a "mean wave" in order to eliminate the small variations which can be observed from one cycle to the other. This "instantaneous mean field" is obtained by performing an ensemble average for each of K phases as follows:

$$Q(k) = \frac{1}{\sum_{w=1}^{W} \mathrm{N(k,w)}} \sum_{W=1}^{W} N(k,w).Q(k,w) \tag{14}$$

where Q(k) represents the horizontal and vertical mean and fluctuant velocity components : $\bar{u}$(k), $\bar{v}$(k), $\overline{u'^2}(k)$, $\overline{v'^2}(k)$ and the cross correlation $\overline{u'v'}(k)$. The mean period of this mean wave is $T = \frac{1}{W}\sum_{w=1}^{W} T(w)$ where T (w) is the period of the w wave.

The calculation of velocity fluctuations for each of the K.W. intervals before the reconstitution of the mean wave avoids to integrate the variations of the velocity evolution for different waves.

Then we can calculate :

- the spanwise component of vorticity : $\Omega(k) = \dfrac{\partial \bar{v}(k)}{\partial x} - \dfrac{\partial \bar{u}(k)}{\partial y}$ (15)

- the 2D turbulent kinetic energy : $\overline{q'^2}(k) = \dfrac{1}{2}(\overline{u'^2}(k) + \overline{v'^2}(k))$ (16)

- the turbulent energy 2D production term :

$$- \overline{u'v'}(k)\left[\frac{\partial \bar{u}(k)}{\partial y} + \frac{\partial \bar{v}(k)}{\partial x}\right] - \overline{u'^2}(k)\frac{\partial \bar{u}(k)}{\partial x} - \overline{v'^2}(k)\frac{\partial \bar{v}(k)}{\partial y} \quad (17)$$

We have chosen a very thin grid for the measurements (vertical step of 0.25 mm close to the bed) in order to reduce the error on the calculated terms involving derivatives.

5. Vorticity contours

Figure 2 shows the time evolution of the horizontal component of the mean velocity above the ripple crest at y = 12.2 mm. This figure allows an easy identification of the different phases of the flow.

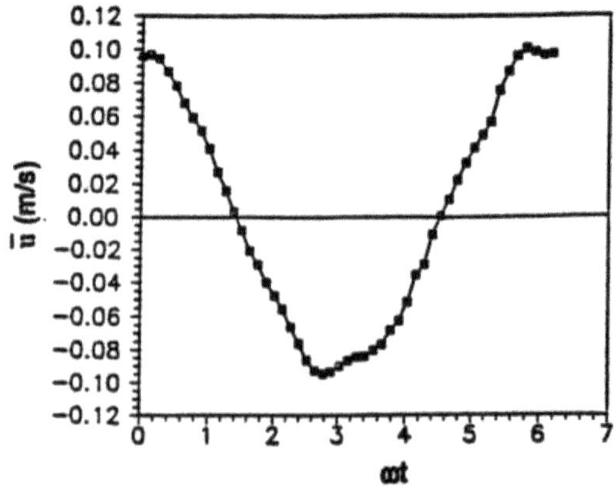

Figure 2 Temporal evolution of horizontal component of mean velocity at x = 0, y = 12.2

Vortices are characterized by high vorticity areas. Figure 3 presents isovalue curves of the spanwise component of vorticity. Eddies travel close to the bed and are convected along about two and an half times the ripple wavelength.

6. Trajectories of fluid particles

In order to have a Lagrangian approach of the flow near the ripples for our experimental velocity measurements, we have reconstructed the trajectories of the fluid particles by means of a formula taking into account the acceleration (second order approximation) :

$$\begin{aligned} x(t+dt) &= x(t) + dx = x(t) + U(x(t),t)dt + 0.5\ U(x(t),t)dt - 0.5\ U(x(t-dt),\ t-dt)dt \\ &= x(t) + [1.5\ U(x(t),t) - 0.5\ U\ (x(t-dt),\ t-dt)]dt \qquad (18) \end{aligned}$$

This formula assumes that the variations of velocity are equal (and consequently the acceleration) between t-dt and t and between t and t+dt. The positions of the fluid particles are calculated phases after phases and the velocities are locally obtained by interpolation (in $1/r^2$) from adjoining points.

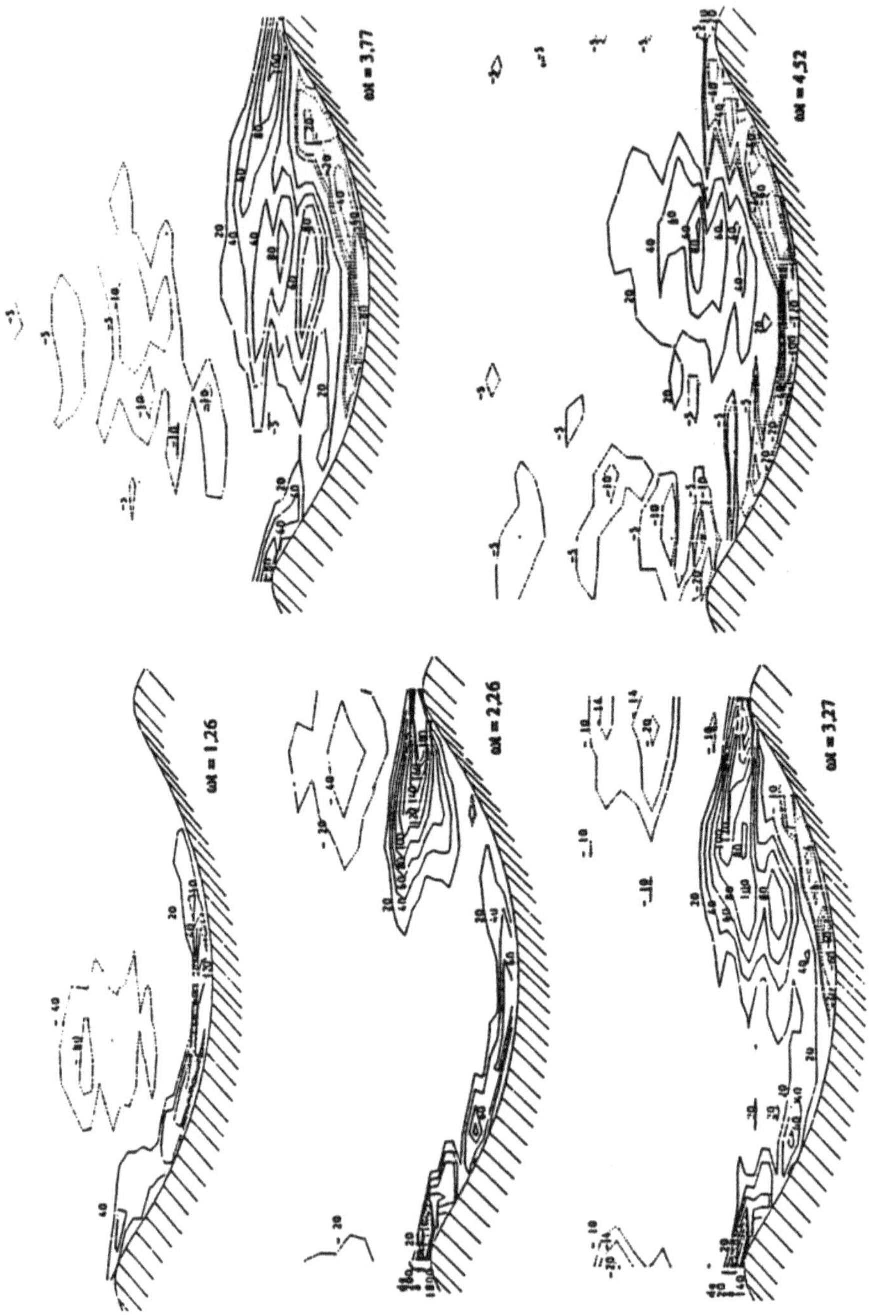

Figure 3 Controus of spanwise component of vorticity (rd/s)
(—) positive vorticity, (---) negative vorticity

Figures 4 and 5 show two examples of results obtained for R = 1700. The starting points of the trajectories approximatively correspond to the center of the vortices just before their convection, vortex with a positive vorticity for figure 4 and negative for figure 5.

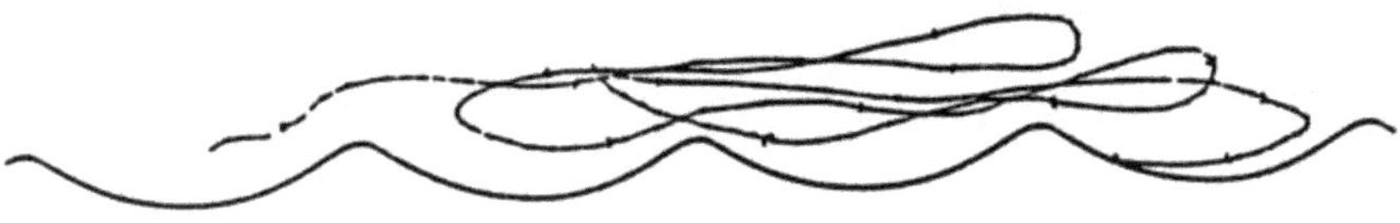

Figure 4 Trajectory of fluid particles ; start phase : $\omega t = 4.27$

Figure 5 Trajectory of fluid particles ; start phase : $\omega t = 1.38$

The particles travelling is about 2.5 ripples downstream and 2 ripples upstream. It corresponds to the motion of the vortices obtained from the isovalue lines of the vorticity. Nevertheless we have not to forget that our ripples are artificial ones and that we consequently do not take into account the interaction of the sediments with the flow.

7. Turbulence analysis

The time evolution of the velocity fluctuations exhibits peaks in some well defined areas of the flow field. Figure 6 shows such peaks above the ripple trough. These peaks rise up to 15 % of the amplitude of the horizontal velocities just outside the disturbed zone. Thus, according to Sleath (1984) the flow appears to be locally turbulent for some fractions of the wave cycle.

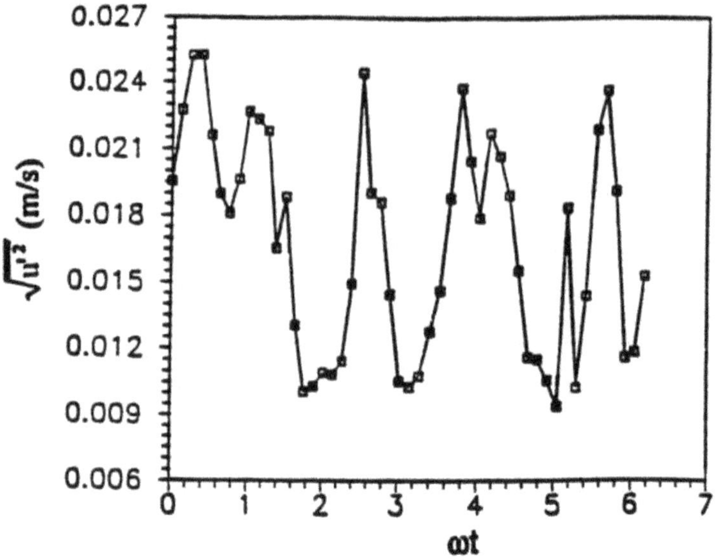

Figure 6 Temporal evolution of u r.m.s. at x = 9, y = 3.7

On figure 7 are drawn isovalue curves of the 2D turbulent kinetic energy $\overline{q'^2} = \frac{1}{2}(\overline{u'^2} + \overline{v'^2})$. The turbulence is localised and is associated with the vortices in the vicinity of the ripples.

In order to study the origin of turbulence, figure 8 shows contours of turbulent energy production. A strong positive production is associated to the fronts (relatively to the convection or extension direction) of the vorticity areas. Very close to the bottom, the turbulence production is always negative as can be seen on figure 9 which shows the temporal evolution of the turbulent energy production above the ripple crest. Consequently, the turbulence seems to be produced by vortices dynamics and a very active dissipation of this turbulent energy appears to take place within the boundary layer.

8. Concluding remarks and future prospects

The realised measurements using a two-component laser-Doppler velocimeter allowed us to show that for a Reynolds number R of 1700, the flow in the vicinity of the rippled bed is locally turbulent for some fractions of the wave cycle. Previous works (Lhuissier, Marin & Vittori, 1991) indicate that the flow is chaotic for this Reynolds number. Figure 10 gives an example of strange attractor (typical of a chaotic flow) which has been obtained. These results show that for our experimental conditions the turbulence is not fully developped and that a more determinist approach may be done in the analysis of the phenomena. The turbulence appears to result from the interaction between vorticity areas and the surrounding flow.

In another way, the exploitation of the evolution of the velocity field with the wave phasis allows to determine the mean trajectory of a fluid particle according to an initial spatio-temporal position of it (initial position for a given phasis of the wave). Such an approach, considering two neighbouring particles, must allow to obtain the coefficients of diffusion for this type of flow as for a fully developped turbulent flow.

This work has been supported by D.R.E.T. ; contract N° 88 - 171.

9. Acknowledgements

We thank Mr BROSSARD for his participation at this work.

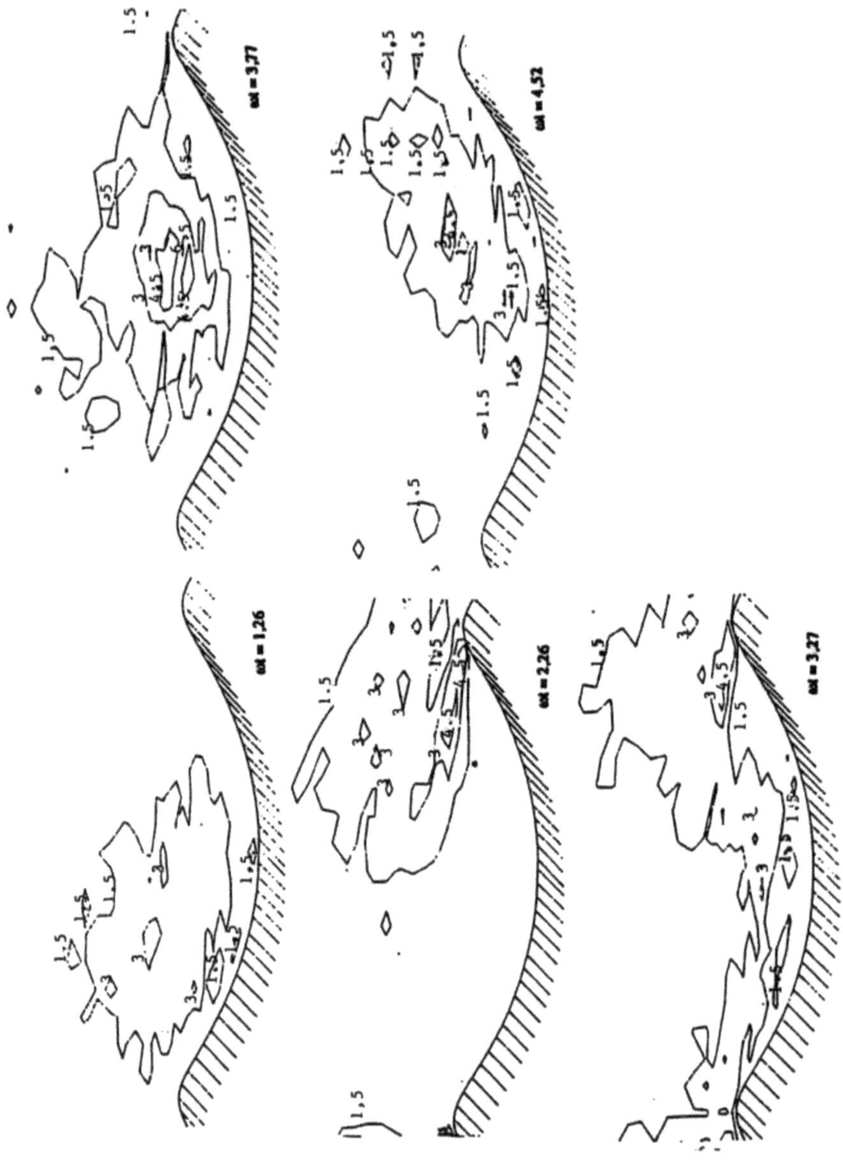

Figure 7 Contours of turbulent kinetic energy $\overline{q'^2} = \frac{1}{2}(\overline{u'^2} + \overline{v'^2}).10^{-4}\ m^2/s^2$

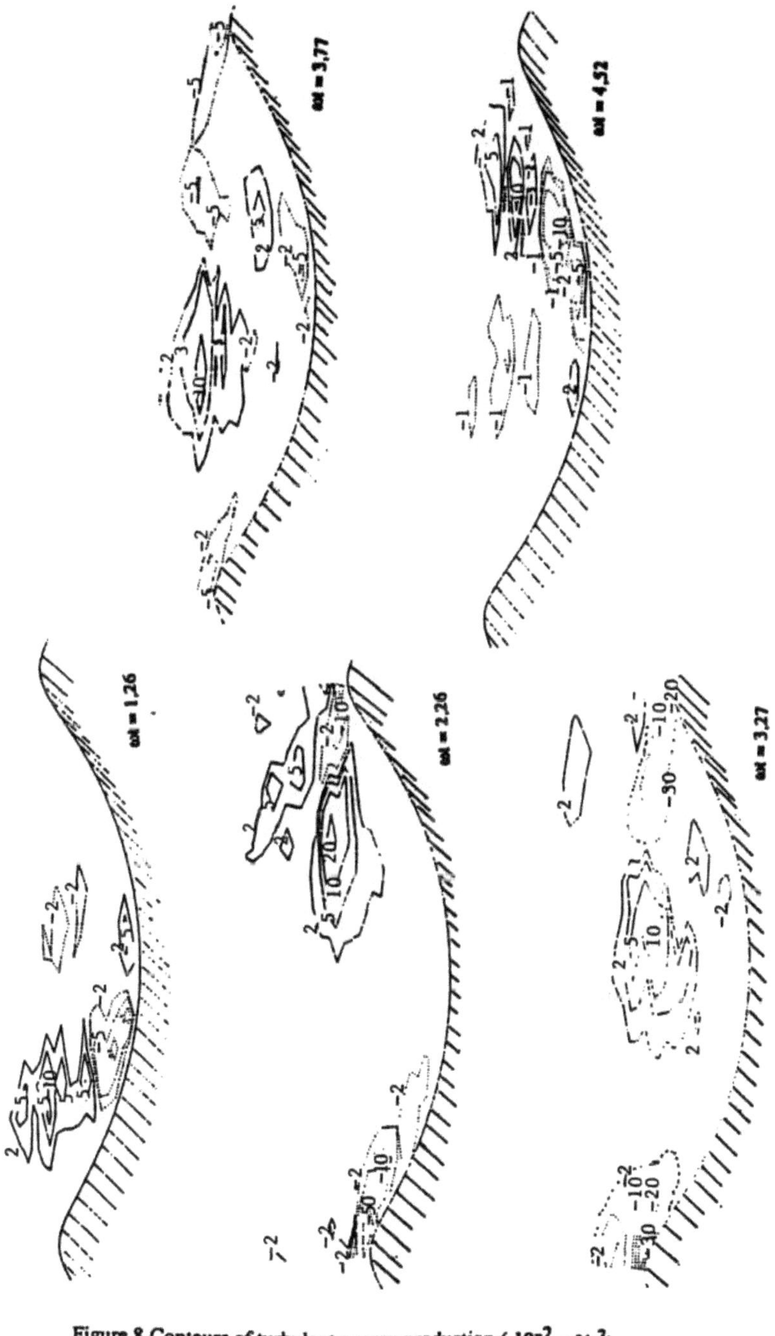

Figure 8 Contours of turbulent energy production ($.10^{-2}$ m^2/s^3) ;
(—) positive production, (---) negative production

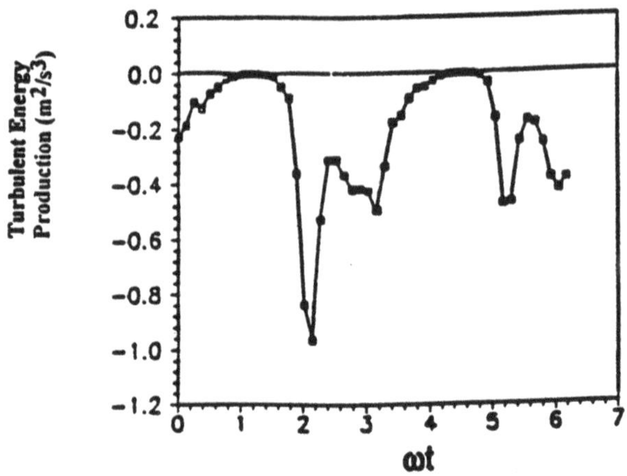

Figure 9 Temporal evolution of turbulent energy production at x=0, y=3.2 (m^2/s^3)

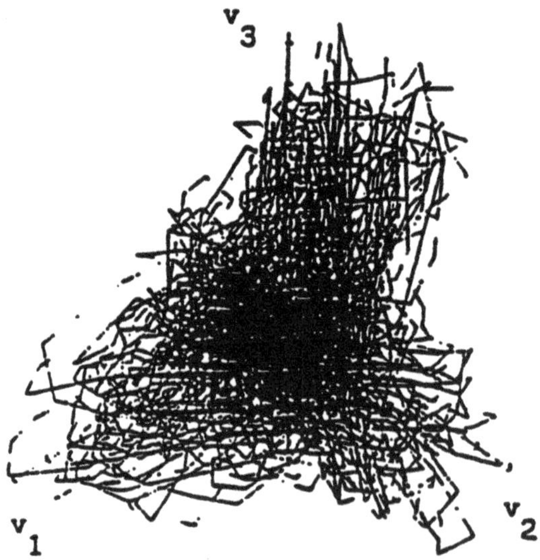

Figure 10 Projection of the attractor in a 3D embedding space (vertical component of velocity) ; R = 1400

10. References

1. R.A. BAGNOLD , *Motion of waves in shallow water. Interaction between waves and sand bottoms*, Proc. Roy. Soc., **A 187**, (1946), pp. 1-15.

2. P. BLONDEAUX, G. VITTORI, *Vorticity dynamics in an oscillatory flow over a rippled bed*, J. Fluid Mech. **226**, (1991), pp. 257-290.

3. C.G. DU TOIT, J.F.A. SLEATH, *Velocity measurements close to rippled beds in oscillatory flow*, J. Fluid Mech. **112**, (1981), pp. 71-96.

4. A.K.F.M. HUSSAIN, *Coherent structures, reality and myth*, Phys. Fluids **26**, 10, (1983), pp. 2816-2850.

5. N. LHUISSIER, F. MARIN, G. VITTORI, *Experimental investigation of the chaotic flow induced by surface gravity waves over a rippled bed : preliminary results*. Annual meeting of Italian Hydraulic Association (G.N.I.) on "turbulence", Rome, (1991).

6. S. SATO, N. MIMURA, A. WATANABE, *Oscillatory boundary layer flow over rippled beds*, Proc. 19 th Conf. on Coastal Engng, Houston, (1984), pp. 2293-2309.

7. S. SATO, K. SHIMOSAKO, A. WATANABE, *Measurements of oscillatory turbulent boundary layer flow above ripples with a laser Doppler velocimeter*, Coastal Engng in Japan **30**, (1987), pp. 89-98.

8. M. SAWAMOTO, T. YAMASHITA, T. KITAMURA, *Measurements of turbulence over vortex-ripple*, Proc. 18 th Conf. on Coastal Engng, (1982), pp. 282-296.

9. J.F.A. SLEATH, *Sea bed mechanics*, Wiley Ocean Engng Series, Mc Cornick Ed., (1984).

Mass Transport over Rippled Beds

F. MARIN [1] - J.F.A. SLEATH
Cambridge University, Engineering Department
Trumpington Street
Cambridge, England

ABSTRACT

Measurements of fluid velocity have been made with a laser Doppler anemometer in the vicinity of a rippled bed subjected to progressive waves in water of constant depth. The drift velocity was found to be reduced by the bed geometry. This reduction becomes less and less significant when the Reynolds number decreases and for small values of this parameter the drift velocity converges towards the theoretical value for a flat smooth bed. This effect of the bed geometry does not appear to be predicted by existing models.

1. Introduction

The drift currents induced by waves are of great importance for coastal protection. Although this steady drift is weak compared with the oscillatory component of velocity, it has a significant effect on the transport of sediment and pollutant in the sea.

There has been much work (both theoretical and experimental) on the wave-induced drift over flat beds (e.g. Longuet-Higgins, 1953, 1957, Lhermitte, 1958, Collins, 1963, etc.) but very little on drift over rippled beds. However, beds in nature are often rippled. Longuet-Higgins (1953) has developed a theory for the mass transport produced by waves over a smooth bed in laminar flow. His model takes account of the viscosity of the fluid and the solution is characterised by a forward mass transport near the bottom. Laboratory tests (e.g. Collins, 1963) have shown good agreement between theory and experiment in this case. Sleath (1984) measurements have shown that gravel increase the maximum drift velocity near the bed at low Reynolds numbers ; at high Reynolds numbers, there was no clear difference between the maximum mean drift velocities above rough and smooth beds even though the boundary layer was fully turbulent for the rough bed and laminar for the smooth bed. Analytical expressions have been derived by Sleath (1974a) for the mass transport velocity produced by waves in the boundary layer over a rippled bed. He has solved the two-dimensional equation of vorticity in terms of stream function.

[1]New address : Laboratoire de Mécanique des Fluides, Quai Frissard, B.P. 265, 76055 Le Havre Cedex, France.

This stream function has been expanded as a power series. The theoretical results are restricted to low Reynolds numbers (i.e. small wave heights). The effect of bed roughness was found to be to increase the mass-transport velocity, except for very long waves where the velocity could be decreased. A numerical model has been developed by Sleath (1974b) to study the effect of a wavy bed on the mass transport induced by waves near the bed. For beds whose wave-length is large compared with boundary layer thickness, the results have shown that the influence of the waviness of the bed depends critically on the Reynolds number and on the relative magnitude of the fundamental and the first harmonic in the flow outside the boundary layer. As far as the authors are aware, these models have not been experimentally tested.

The objective of the present work was to make new measurements of the drift velocity over a rippled bed.

2. Experimental layout

The experiments were carried out in the 17.5 m wave flume of the Cambridge University Engineering Department. Waves were generated at one end of the flume by a paddle and absorbed at the other end by a pebble beach of slope 1:20.

Table 1 shows the test conditions. In this table T is the period ($T=2\pi/\omega$), H is the wave height, d is the mean water depth, k is the wave number, a is the orbital amplitude of the fluid just outside the boundary layer at the bed and L is the wave length of the ripples. In no test did wave breaking occur in the test section.

Table 1 Test conditions

Test No.	T (s)	H (mm)	d (mm)	kd	a/L
1	1.35	29	200	0.72	0.38
2	1.38	48	200	0.70	0.59
3	1.33	64	200	0.73	0.74
4	0.72	31	200	1.68	0.12
5	0.73	52	200	1.64	0.21
6	0.70	65	200	1.73	0.23

Velocity measurements were made in the vicinity of the bed. The horizontal component of fluid velocity was measured with a laser Doppler anemometer operated in forward scatter mode. At each position the velocity was sampled 100 times per cycle for 50 cycles. Ensemble-average was then determined at each phase. Time-mean drifts were computed in two independent ways : by an average of sampling and by filtering the output of the tracker. Results of each of these methods are very similar. For each test, the velocities have been measured over the ripple crest at four locations across the wave tank and over the ripple trough.

The bed profile was the same for all tests. This was obtained by allowing the waves (test N° 3 conditions) to act on an initially plane bed of sand ; the grain diameter was 0.4 mm. Ripples formed on the bed and when they had reached their equilibrium condition they were stabilised with a thin powdering of cement. These ripples were of length 45 mm and height 8 mm.

3. Test results and discussion

It is convenient to work in terms of the curvilinear coordinates (ξ, η), which are defined in terms of Cartesian coordinates (x,y) as

$$\xi = x + h/2 \exp(-k\eta) \sin k\xi$$

$$\eta = y - h/2 \exp(-k\eta) \cos k\xi$$

where y is measured from the mean bed level as shown by Figure 1. This coordinate system maps the bottom profile into the line $\eta = 0$.

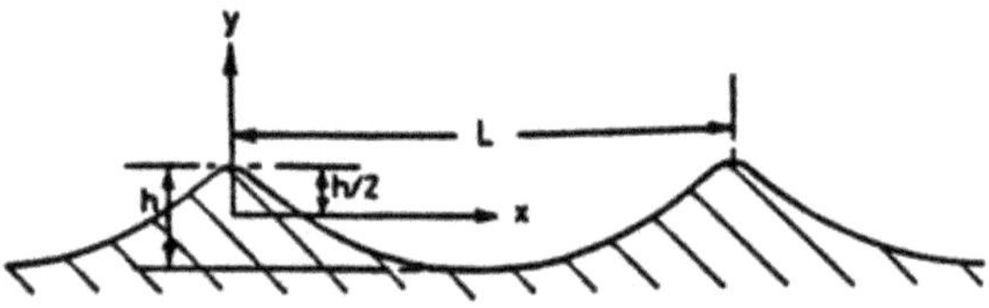

Figure 1 Definition sketch for ripple profile

Figure 2 shows a comparison between the amplitude of the measured and theoretical velocities U_0 just outside the boundary layer at the bed. In this figure, $(U_0)_{meas}$ is the amplitude of the fundamental component of velocity obtained by Fourier analysis of the velocity record and $(U_0)_{th}$ is that calculated from third-order Stokes wave theory. The agreement between theory and experiment is quite good.

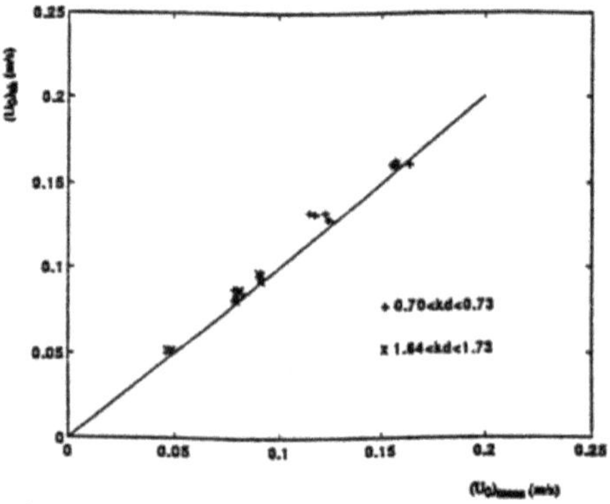

Figure 2 Comparison of measured and theoretical values of the velocity amplitude just outside the boundary layer

Visualisations by injection of dye in the water have shown the basic role of the parameter kd on the drift profile of the flow. This is illustrated on figure 3 which shows the drift profile in the body of the fluid for two values of kd. For kd = 0.7 the drift profile is oriented in the direction of the wave propagation at half height of the mean water level. This suggests that the mean drift velocity just above the boundary layer at the bed is in the opposite direction to that of wave propagation. For kd = 1.7 the drift velocity is oriented in the opposite direction. Measurements of fluid velocity have been made close to the rippled bed with the laser Doppler anemometer for each of these values of kd and for three values of the ratio a/L.

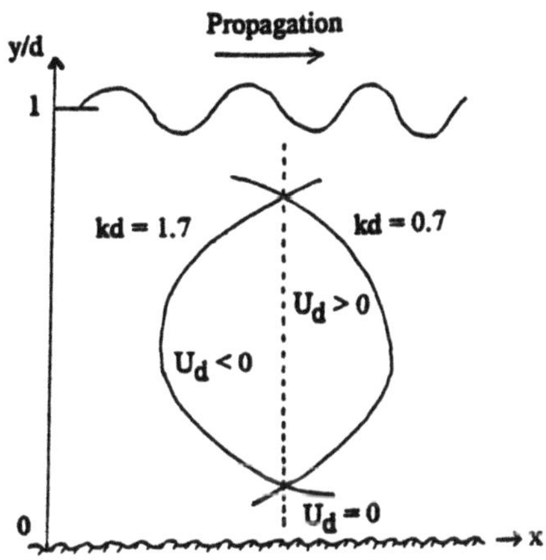

Figure 3 Sketch of drift profiles in body of fluid for wave height H = 50 mm

As no significant differences were found between velocities measured at four locations across the tank above the ripple crest, the drift velocity has been averaged across the wave flume for each test.

Figures 4,5,6,7 show profiles of the measured Eulerian drift velocity U_d in the vicinity of the bed. In these figures δ is the Stokes thickness ($=(2\nu/\omega)^{1/2}$ where ν is kinematic viscosity). The effect of a/L is very marked. The trends are clearest at kd=0.7 (Figures 4, 5). Over the trough (Figure 5) it is possible to detect a peak in the drift profile, at about $\eta/\delta = 0.6$. This peak is in the direction of wave propagation (positive values of U_d). We will refer to this as the "max positive" peak. For Test Nos. 1 and 2 it is possible to detect, even closer to the bed, a peak in the opposite direction. We will refer to this as the "max negative" peak. Bearing in mind the values of a/L listed in Table 1, we see that as a/L increases the max positive peak becomes progressively smaller (or more negative) and so does the max negative peak.

Similar trends are seen over the ripple crest (Figure 4). For Test Nos. 1 and 2 the max positive peak is found at about $\eta/\delta = 2$ but for the largest value of a/L (Test No. 3) the drift has become so strongly negative close to the bed that the positive peak has disappeared.

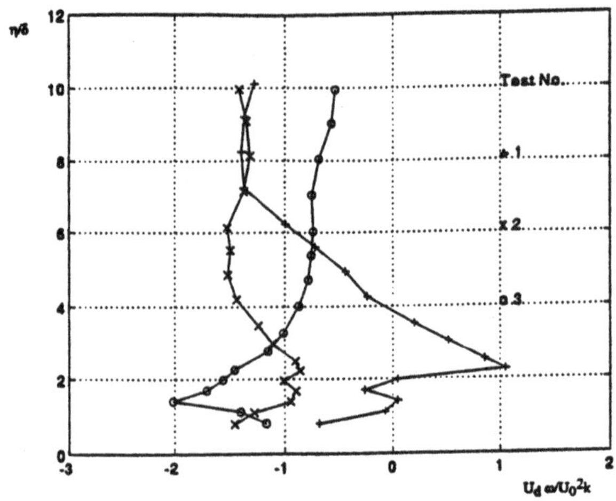

Figure 4 Variation of the drift velocity U_d with height above the ripple crest (kd=0.7)

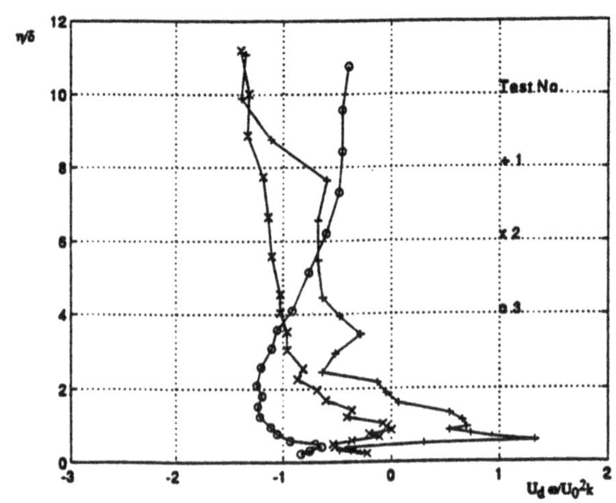

Figure 5 Variation of the drift velocity U_d with height above the ripple trough (kd=0.7)

For kd = 1.7 (Figures 6,7) the trends are less clear. Over the trough (Figure 7) there is still a well-defined forward peak but over the crest (Figure 6) the trends are very confused. To some extent this is probably due to the physical limitations of the experimental equipment. At period as small as about 0.7 second there is a tendency to develop cross-waves in this channel. Also the results at the smallest wave height (Test No. 4) may have been affected by the negative drifts observed by Sleath (1973) in the vicinity of the bed. For these reasons the tests at kd = 1.7 should be treated with caution.

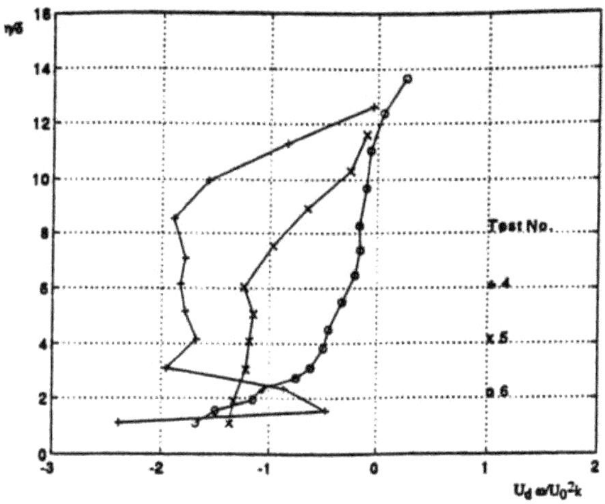

Figure 6 Variation of the drift velocity U_d with height above the ripple crest (kd=1.7)

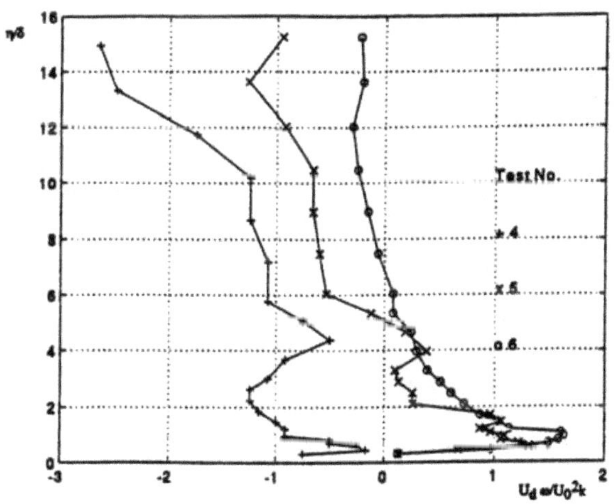

Figure 7 Variation of the drift velocity U_d with height above the ripple trough (kd=1.7)

The way in which the magnitudes of the max positive peak and of the max negative peak vary with a/L are summarised in Figure 8 for kd = 0.7 and for kd = 1.7 when well defined peaks exist in the drift profiles. At very small values of a/L the max positive peak is equal to or a bit larger than the theoretical value for a flat bed suggested by Longuet-Higgins (1953). This is in qualitative agreement with Sleath's (1974a) analytical model for rough beds. For the present experimental conditions this model predicts a maximum mass transport velocity near the bed which is 77 % greater than Longuet-Higgins' value in the limit as a/L $\rightarrow$ 0.

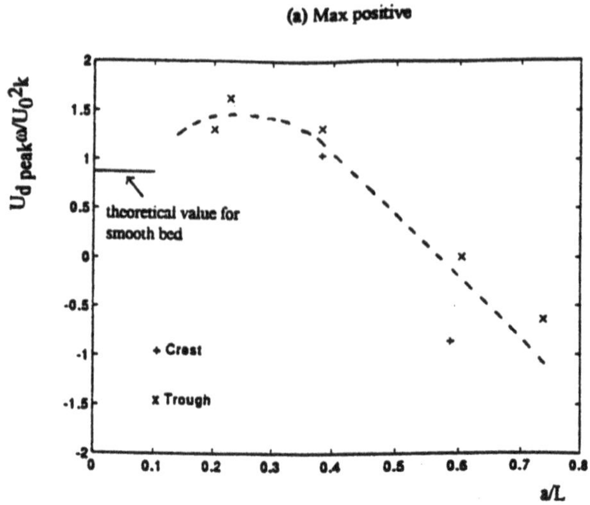

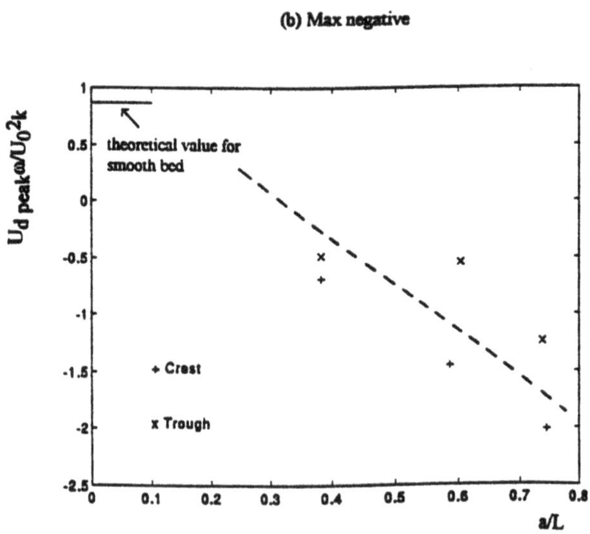

Figures 8 Variation of the peaks of the drift velocity $U_{d\ peak}$ with a/L

As a/L increases, both the max positive and max negative peaks become progressively more negative. This is probably due to the effects of the vortices formed behind the ripple crests during each half cycle. The vortices are strongest at about a/L = 0.7 or 0.8. At smaller values of a/L they become progressively weaker and probably disappear altogether below about a/L = 0.2. It was not possible with the present apparatus to obtain values of a/L much larger than 0.8. However, the numerical model of Sleath (1974b) suggests that the change in drift velocity due to vortex formation would become smaller again as a/L increases beyond 0.8.

Sleath's (1974b) numerical model shows opposing trends. If higher harmonics are neglected the effect of bed roughness is to increase the near-bed drift velocity in the direction of wave propagation. However, the effect of the first harmonic is to encourage a drift velocity in the opposite direction. If the first harmonic is relatively important the net effect of vortex formation linked to bed roughness would be to decrease the forward drift and, possibly, to reverse it completely. On the other hand, if the first harmonic is unimportant the net effect of bed roughness is to increase the forward velocity near the bed. This may explain why the presents results show a decrease in forward velocity whereas the measurements of Sleath (1984) with gravel showed an increase at equivalent values of a/L. In the 1984 measurements the waves at such low values of a/L were almost perfectly sinusoidal whereas the present measurements (with the much larger value of L) have a significant first harmonic for all of the measurements with kd = 0.7. However, although Sleath's numerical model provides a qualitative explanation for the observed trends it does not show quantitative agreement. This may be because the computation were carried out for somewhat different conditions. Also, the bed profile is different and the model does not allow for turbulence.

Finally, it is obvious that the reversal in near-bed drift velocity observed for values of a/L approaching 0.7 or 0.8 has significant implications for sediment transport in the near-shore region. The value of a/L for natural ripples is frequently close to 0.7 or 0.8.

4. Conclusion

The laser Doppler anemometer measurements in a wave flume indicate significant effect of ripples on the drift velocity near the bed. At large values of a/L, the drift velocity is reduced by the bed geometry and is in the opposite direction of wave propagation. As far as a/L decreases, the drift velocity increases and converge towards the theoretical limit for flat bed.

This work has been supported by the European Community, under programme 'Human Capital and Mobility'.

5. References

1. J.I. Collins, *Inception of turbulence at the bed under periodic gravity waves*, J. Geophysical Research, Vol. **68**, No. 21, 1963, pp. 6007-6014.

2. P. Lhermitte, *Contribution a l'etude de la couche limite des houles progressives*, C.O.E.C. No. 136, Imprimerie Nationale, Paris, 1958.

3. M.S. Longuet-Higgins, *Mass transport in water waves*, Phil. Trans. Roy. Soc. Series A, Vol. **245**, No. 903, 1953, pp. 535-581.

4. M.S. Longuet-Higgins, *The mechanics of the boundary layer near the bottom in a progressive wave*, Proc. 6th Conference on Coastal Engineering, Miami, 1957, pp. 184-193.

5. J.F.A. Sleath, *A second approximation to mass transport by water waves*, J. Marine Research, Vol. **30**, No. 3, 1972, pp. 295-304.

6. J.F.A. Sleath, *Mass transport in water waves of very small amplitude*, J. Hydraulic Research, Vol. **11**, No. 4, 1973, pp. 369-383.

7. J.F.A. Sleath, *Mass transport over a rough bed*, J. Marine Research, Vol. **32**, No. 1, 1974(a), pp. 13-24.

8. J.F.A. Sleath, *A numerical study of the influence of bottom roughness on mass transport*, Numerical Methods in Fluid Dynamics, Edit. C.A. Brebbia & J.J. Connor, Pentech Press, London, 1974(b), pp. 482-493.

9. J.F.A. Sleath, *Measurements of mass transport over a rough bed*, International Conference on Coastal Engineering, Houston, 1984, pp. 1149-1160.

LABORATORY EXPERIMENTS ON SEDIMENT EROSION INDUCED BY PARTIALLY STANDING WAVES

VINCENT REY
Laboratoire de Sondages Electromagnétiques de l'Environnement Terrestre
Université de Toulon et du Var
BP 132, 83957 LA GARDE Cedex, FRANCE

NATHALIE CESSELIN, MAX BELZONS
Département de Physique des Systèmes Désordonnés, SETT
Université de Provence, centre de Saint Jérôme
case 161, 13397 MARSEILLE Cedex 20, FRANCE

ABSTRACT

Dynamic interactions between partially standing waves and erodible beds have been examined in a small-scale wave tank. In presence of a solid rectangular obstacle, the bottom evolution, including both ripples and dunes formation and development of local scour in the vicinity of the obstacle was measured by using an ultrasonic system. The associated evolution of the surface waves was also monitored using an ultra-sonic sensor with a determination of the harmonic composition of the wavefield. The capabilities of the experimental set-up in describing in situ phenomena was validated by measurements of the threshold for sediment motion, and the ratio between ripple length and heigth, in good agreement with those of previous laboratory or in situ investigations. Ripples and bars formations and evolution of rate of standing wave upstream the solid obstacle do not depend on the shape of its corners, which means that both vorticy and current due to the solid obstacle are local phenomena and have a small influence far away from the bar. On the contrary, the amplitude of the local scour depends strongly on the corners' shape and on the wave phase, and the volume of sand involved in this process was chosen to be the physical parameter. At short times, the ratio between the displaced volume and time is proportional to the difference between the near bed velocity and the corresponding threshold value, an asymptotic behaviour for the volume of sand displaced is observed at long times.

In previous experimental studies carried out in Marseille, it has been shown that the main features of the interaction between surface waves and regular or irregular fixed bed profiles (reflecting power of varying topographies, Bragg resonances, vorticity and nonlinearity effects) can be accurately observed in a small-scale wave tank (Belzons et al, 1988; Belzons et al, 1991; Rey et al, 1992; Guazzelli et al, 1992).

The capabilities of the experimental set-up have been extended in order to examine the dynamic interactions between partially standing waves and erodible beds. In a first stage, initially flattened beds were concerned, with the study of the influence of ripples and bars (or dunes) formations on the surface wave behaviour (Rey, 1991; Rey et al, 1994). We then studied the evolution of such topographies in presence of solid rectangular obstacles and more particularly the local scour around these obstacles (Cesselin, 1992). Indeed, artificial obstacles are widely used for coastal protection, and estimation of their possible influence on sediment transport is of great interest.

In the present paper, we present the main findings of our study concerning erodible beds, after a brief recall of the experimental facilities and of the model used for calculations of the fluid movements.

The experiments were performed in a glass-walled wave tank, 4.7 m long and 39 cm wide. The mean water depth was H = 6 cm and the wave frequency 1.5 Hz. A very accurate (precision of order 100 µm), rapid, ultrasonic detection system was used to monitor the evolution of an initially flattened sand bed, including both the formation of ripples and dunes, and the development of local scour in the vicinity of a rectangular solid bar. The associated evolution of the surface waves was also monitored using an ultra-sonic sensor (precision 100 µm), with which the harmonic composition of the wavefield was determined (see figures 1 and 2).

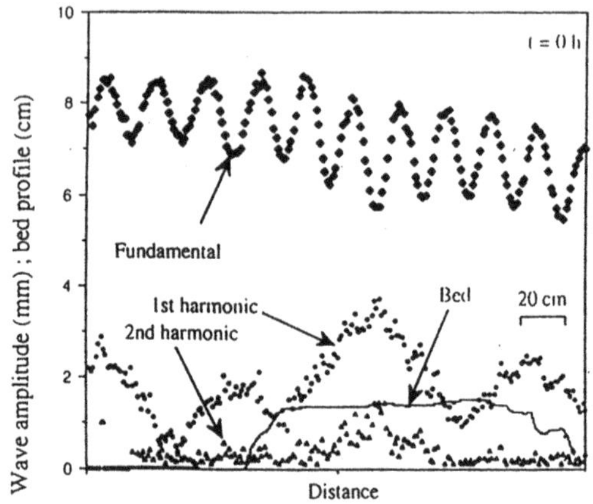

figure 1 : measured wave and bed profile at t = 0 h

The capability of a model based on the linear potential theory (Rey, 1992; Rey, 1993) in describing the experimental wave field has been carefully tested for both the wave envelope and the reflection coefficients. For a relative small incident wave amplitude, i.e. in the vicinity of the erosion threshold, the experimental data obtained are recovered from the model to within some 5%. For the largest incident wave amplitudes, the discrepancy does not exceed 20%. Consequently, we used the numerical model to calculate the magnitude of the whole flow characteristics and in particular the near bed velocity for each given experimental condition.

In the absence of rectangular solid bars, small-scale ripples with wavelength λ and heigth η formed rapidly on the initially flattened bed. The rate η/λ (see Rey, 1991; Rey et al, 1994) was in good agreement with those found in previous in situ experiments or in laboratories (Nielsen, 1979). The threshold for the ripple formation was found to be slightly above the threshold for sediment motion on a flat bed given by the criteria of Bagnold (1946) and Komar & Miller (1975) for the present experimental conditions. Both results confirm the capability of such a small-scale experimental set-up to study accurately dynamic interactions between surface waves and erodible beds. Ripple sizes varied as bars formed during the experiment, with larger ripples ultimately becoming established on the bar crests and smaller ripples in the troughs. In the ultimate bed profile, the average ripple slope was nearly the static angle of repose of sediments in water.

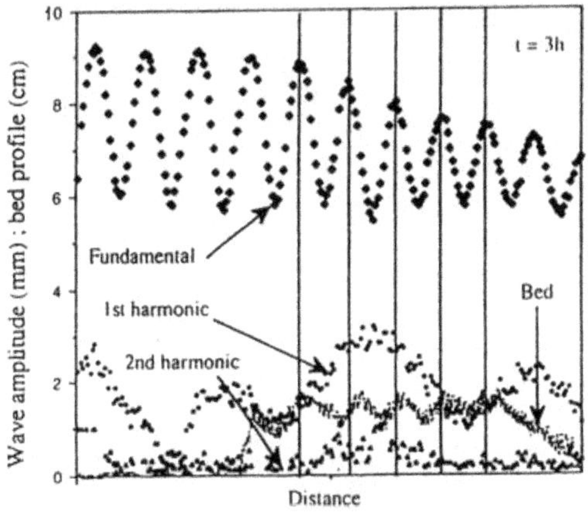

figure 2 : measured wave and bed profile at t = 3 h

From figure 2, it appears that the distance between two successive nodes (i.e. half the surface wavelength) is approximately equal to the wavelength of the surface of the bed. So, sand bars satisfy the Bragg condition. The bar crests formed slightly down-wave of the antinodes of elevation, as a result of a net, near-bed, sediment transport due to vortex shedding from the (asymmetrical) ripples. As the bars grew, the up-wave reflection coefficient increased, as expected for this positioning of bars relative to the wave envelope (O'Hare and Davies, 1990). It is thought that the phase matching revealed through the Bragg condition was the visible sign of a dynamic adaptation between the surface and the bed profile.

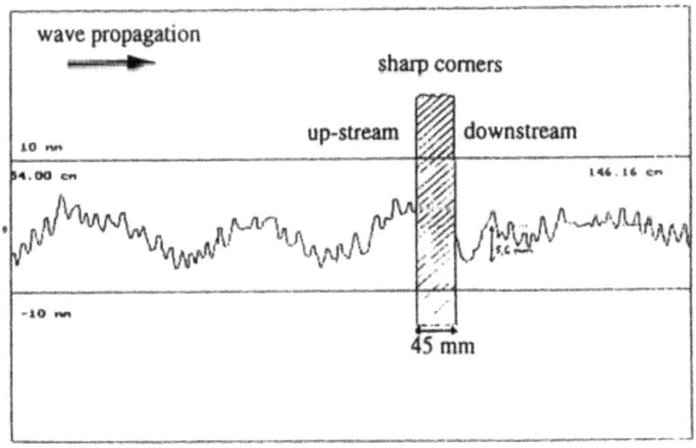

figure 3

In the presence of a submerged or half submerged bar (4.5 cm long, 2.5 cm high and 39 cm wide), ripples and dunes formed far away from the bar, while typical scour occured in the vicinity of the bar. We have investigated the threshold and evolution of the local scouring induced by the rectangular bar. This local scour was due to the near field i.e. strongly dependent on both vortices generated at the bar edges, and current generated under the bar for the half submerged bar.

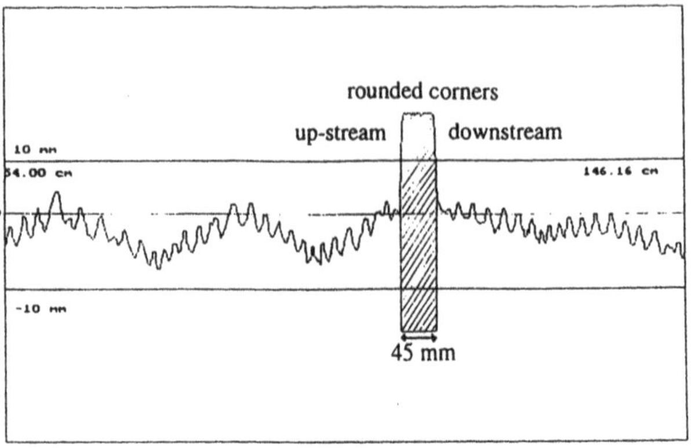

figure 4

Figures 3 and 4 show the influence of the corners' shape of the submerged rectangular obstacle on the bed topography evolution, initially partly submerged in the sand (depth 0.5 cm). We can observe that ripples and bars formations up-stream the obstacle do not depend on the shape of the corners, what means that both vorticies and current due to the rectangular obstacle are local phenomena and have a small influence far away from the bar. On the contrary, the amplitude of the local scour is strongly increased for the sharp corners case, what means a vorticity very sensitive to the shape of the bar corners. These results confirm the results of Rey et al (1992) on both far field and near field effects of a rectangular bar on a solid bottom.

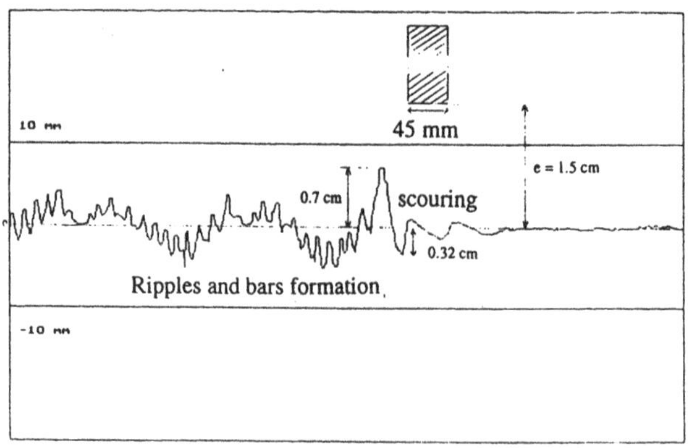

figure 5:vortex formation and local scour induced by a submerged bar (e = 1.5 cm)

We have investigated the threshold and evolution of the local scouring induced by the rectangular bar submerged at a distance e from the initially flat sand bed (figure 5). The bar was located either at a node or an antinode of the partially standing wavefield induced by the beach reflection. For each given bar location, the near bed velocity threshold U_{th} of the local scouring was estimated by introducing in the numerical model the corresponding experimental threshold of the wave amplitude. This threshold velocity increases significantly with the gap e; it is far less than the flat bed value for small e (0.5 cm) and approaches it for large e (1.5 cm). This is thought to reveal the action of the vortices shedded from the bar corners on the local scouring.

The dynamics of the local scouring was then investigated versus the time t and for a given difference between the near bed velocity and the threshold value, $U - U_{th}$. The volume of sand V involved in this process was chosen as the physical parameter. This volume was estimated from the bed profile over a distance from the bar beyond which the standard erosion is recovered. The results are plotted on the figure 6.

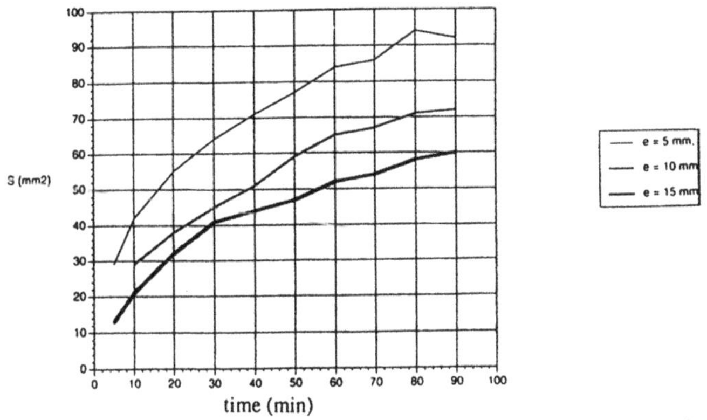

figure 6: volume of sand per unit width involved in local scouring: bar located at a node of the partially standing wave

We observed an asymptotic behaviour for all the distances from the threshold which is correctly fitted with the function $V = V_0\,[1 - \exp(-\alpha t + \beta t^2 + \gamma t^3)]$. At short times ($t \ll 1/\alpha$), the time evolution of the displaced volume is given by $V/t \approx C\,(U - U_{th})$, with $C \approx$ constant whatever e or the bar location. Experimental results reveal in a first stage the development of an instability, the physical parameter was chosen to be $(U - U_{th})$. Such a bahaviour for the flux of sediment transport over sandy beds under the action of waves was proposed by Sleath [1978], the physical parameter was there $(\Psi - \Psi_{th})$ where $\Psi = (\rho U^2)/[(\rho_s-\rho)gD]$ is the Shieds' parameter (ρ_s = sand density, ρ = water density, D = mean grain diameter). For large values of time, evolution of each case studied in our laboratory have shown an asymptotic limit of the volume of sand displaced near the obstacle. For larger waves' amplitudes i.e. far from the threshold conditions, further experiments in a bigger-scale wave tank are necessary to estimate quantitatively the scouring and its possible evolution toward an equilibrium state.

References:

Bagnold, R. A. 1946 Motion of waves in shallow water: Interaction between waves and sand bottoms, *Proc. Royal Society of London* , series A, **187**, 1-8.

Belzons, M., Guazzelli, E. and Parodi, O. 1988 Gravity waves on a rough bottom: experimental evidence of one-dimensional localization, *J. Fluid Mech.* **186**, 539-558.

Belzons, M., Rey, V. and Guazzelli, E. 1991 Subharmonic Bragg resonance for surface water waves, *Europhysics Letters* **16**, 189-194.

Cesselin, N. 1992 Modélisation de l'érosion d'un fond sableux par la houle littorale: application à la formation des barres et à l'affouillement, *Rapport de stage de DEA Sciences de l'environnement marin, Université Aix-Marseille II.*

Nielsen, P., 1979 Some basic concepts of wave sediment transport. Technical University of Denmark, Institute of Hydrodynamics and Hydraulic Engineering, Series Paper n° 20, 160 pp.

Guazzelli, E., Rey, V., and Belzons, M. 1992 Higher-order Bragg reflection of gravity surface waves by periodic beds. *J. Fluid Mech.* **245**, 301-317.

Komar, P. D., and Miller, M. C. 1974 Sediment threshold under oscillatory waves. *Proceedings of the 14th International Conference on Coastal Engineering* American Society of Civil Engineers, 756-775.

O'Hare, T. and Davies, A. G., 1990 A laboratory study of sand bar evolution, *Journal of Coastal Research*, **6** (3) , 531-544.

Rey, V. 1991 Propagation d'ondes de gravité au dessus de fonds solides ou constitués de sédiments : application à l'étude d'interactions dynamiques ondes-sédiments, thèse de l'Université de Provence, Marseille.

Rey, V. 1992 Propagation and local behaviour of normally incident gravity waves over varying topography, *European Journal of Mechanics B/Fluids*, **11**, 213-232.

Rey, V., Belzons, M. and Guazzelli, E., 1992 Propagation of surface gravity waves over a rectangular submerged bar. *J. Fluid Mech.* **235**, 453-479.

Rey, V., Davies, A. G., and Belzons, M. 1994 On the formation of bars by the action of waves on an erodible bed: a laboratory study. *Journal of Coastal Research* , in press

Rey, V. 1993 Propagation d'ondes de gravité d'incidence oblique en présence d'obstacles cylindriques par profondeur finie, *Colloque "Les méthodes informatiques de la conception industrielle"*, ESIM, Marseille, 17 june 1993.

Sleath, J. F. A. 1978 Measurements of bed load in oscillatory flow,. *J. Waterw., Port Coastal Ocean Div. Am. Soc. Civ. Eng., 19th,* 2310 - 2321.

Bed shear stress study in the neighbourhood of a near-wall cylinder in waves

A. Sakout, A. Jarno-Druaux & M. Bélorgey

Laboratoire de Mécanique des Fluides, Université du Havre

B.P. 265 - 76055, Le Havre cédex - France

ABSTRACT

This paper presents an analysis of the wall proximity influence on the bed shear stress distribution in the vicinity of a cylinder in waves. The bed shear stress is very sensitive to the change in flow regime noted when the cylinder is /or not partly immersed in the wall boundary layer. Some particular variation of the bed shear stress versus time are analyzed. Then, the present experiments conducted with a flat non erodible bed give informations on the initial conditions for the erosion process.

1. Introduction

The flow around a cylinder beneath waves is very much influenced by the vicinity of a wall. This induces both important modifications of the hydrodynamic forces and of the bed shear stress.

In waves, the experimental works dealing with the behaviour of the bed shear stress in the vicinity of near-wall cylinders are rather limited. Some works are based on observations which can give qualitative informations on the scour process around cylinders. For example, Leeuwestein [1], has observed the erosion process around a cylinder and noticed that it occurs essentially under the cylinder giving rise to a tunnel-type erosion. Other authors have studied the interaction between the cylinder and the wall, analyzing its influence on the bed shear stress. In Sumer *et al* study [2], the bed is non erodible and the Keulegan-Carpenter number, calculated on U_m, the maximum horizontal velocity at the cylinder elevation measured in absence of the cylinder is equal to 10, 20 and 100. The bed shear stress is calculated from the measured velocity field. Their results indicate that when the e/D, gap-to-diameter ratio is equal to zero, there is an increase in the bed shear stress of about twice its undisturbed value in the downstream region of the cylinder and that the increase reaches four times the undisturbed value under the cylinder when $e/D = 0.05$.

The objective of this study is to characterize the bed shear stress through the analysis of the velocity field measured in the vicinity of the cylinder and to improve our knowledge of the erosion process in the initial stage. Two series of tests were conducted. In one case, part of the cylinder is immersed in the bed boundary layer and in the other one, the cylinder is out of the direct influence of the wall boundary layer but always interacts with the bottom. The two gap-to-diameter ratios are respectively 0.09 and 0.5.

This work makes part of a more complete study concerning the effect of waves on near-wall cylinders (Sakout [3]).

2. Details of Experiments

The experiments were performed in the wave flume of the Fluid Mechanics Laboratory of Le Havre University. It is 9 meters long and 0.8 meter wide. The wave characteristics are indicated on figure 1. The cylinder is smooth with an external diameter D of 0.02 meter.

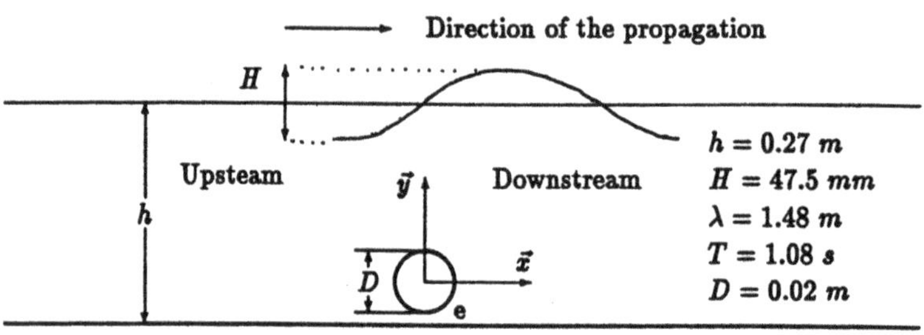

Figure 1: Experimental conditions

The wave propagation generates an Eulerian current for which intensity and direction depend on depth. Near the wall, this current has the same sense than the wave propagation and its amplitude is 8% of the maximum horizontal velocity measured in this zone. The flow generates a laminar oscillatory boundary layer on the bottom for which the maximum thickness measured at the wave crest is δ. The δ/D parameter defined for this study is equal to 0.15. For the two gap-to-diameter ratios tested (e/D = 0.09 and e/D = 0.5), the center of the cylinder is located in a region where U_m, the maximum velocity measured without the cylinder remains nearly constant ($U_m = 0.09 m/s$). The Reynolds number ($Re = U_m D/\nu$) and Keulegan-Carpenter number defined respectively by $Re = U_m D/\nu$ and $K.C = U_m T/D$ are then constant and their value are fixed to 1800 and 5 for the two positions tested. The τ_w, bed shear stress is based on its turbulent expression, where the non-slip condition has been added. It is then given by

$$\tau_w(t,x) = \mu(\frac{\partial \overline{U}}{\partial y} + \frac{\partial \overline{V}}{\partial x})_{wall} \tag{1}$$

where $\overline{U}$ and $\overline{V}$ represent the mean horizontal and vertical components of the velocity and (x,y) are respectively the horizontal and vertical directions.

The estimation of the bed shear stress was obtained by using near-bed velocity measurements. The laser velocimetry technique was used and the minimum distance from the wall was evaluated to 0.2 mm, which represents half the diameter of the measuring volume.

The results of bed shear stress presented are issue from a statistical determination of near-bed velocities, obtained through ensemble averages calculated over more than ten waves.

The bed shear-stress calculations were conducted for $e/D = 0.09$ and $e/D = 0.5$ for eight downstream distances from the cylinder, in the region extending from $x/D = 0.75$ to $x/D = 3$.

3. Results and Discussion

3.1 Evolution of the Measured Velocity Field for the two tests e/D = 0.09 and e/D = 0.5

A complete flow analysis is presented in Sakout study [3]. The vortex development process differs markedly from a case to the other. In the near-wall case ($e/D = 0.09$), two attached vortices issuing from the upper part of the cylinder develop in a preponderant way, one downstream during the first half-period and the other upstream during the second half-period. The streamlines obtained for the two t/T time instants when the vortices are preponderant are given on figure 2 (case (a)). Two small wall-side vortices are also generated during the two half-periods. But, under the effect of a jet flow which takes place under the cylinder, these vortices are rapidly shed and vanish. A vortex shedding from the wall-side of the cylinder is then maintained for a very small value of the gap-to-diameter ratio. This finding is consistent with the results of Sumer *et al* [4] in oscillating flow.

In the $e/D = 0.5$ case, a major development of the wall-side vortices is observed (figure 2, (b)). These vortices are convected towards the wall where they vanish. The development of the two wall side-vortices is quite identical. Nevertheless we can observe that the vortex which develops up the cylinder remains at a shorter distance form the cylinder than the one developing down the cylinder. It suggests that the Eulerian current influence the vortex convection and favour the downstream displacement of the vortices.

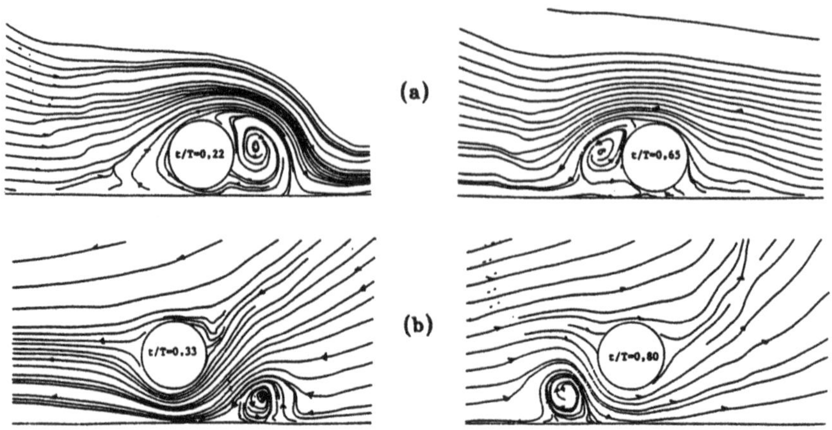

Figure 2: Evolution of the streamlines (a) $e/D = 0.09$, (b) $e/D = 0.5$

4. Bed Shear Stress Measurements for the two tests e/D = 0.09 and e/D = 0.5

The bed shear stress is presented as a function of t/T for different X/D, in figure 3. The two tested cases are represented on the same figure.
Some general remarks can be made for the two tests.

The $(\frac{\partial \overline{V}}{\partial x})$ term is negligeable and the bed shear stress can be evaluated with a laminar form of τ_w

$$\tau_w(t,x) = \mu(\frac{\partial \overline{U}}{\partial y})_{wall} \tag{2}$$

Furthermore, it can be noted that the x-extend of the downstream perturbation induced by the cylinder is inferior to 2.5D. Out of the cylinder influence, the bed shear stress has a sinusoidal variation versus t/T and can be written

$$\tau_w(t/T) = \tau_0 + \tau_1(t/T) \tag{3}$$

where τ_0 denotes the positive constant component of the bed shear stress due to the existence of the eulerian drift in the propagation sense and τ_1 is the monochromatic sinusoidal component of the bed shear stress.

The bed shear stress predicted by the small-amplitude wave theory in laminar flow (Batchelor [5]) is given by

$$\tau_w(t/T) = \mu U_\infty \beta \sqrt{2} \ cos(2\pi t/T - \frac{\pi}{4}) \tag{4}$$

where $\beta = (\frac{\omega}{2\nu})^{1/2}$ and $\omega = \frac{2\pi}{T}$.

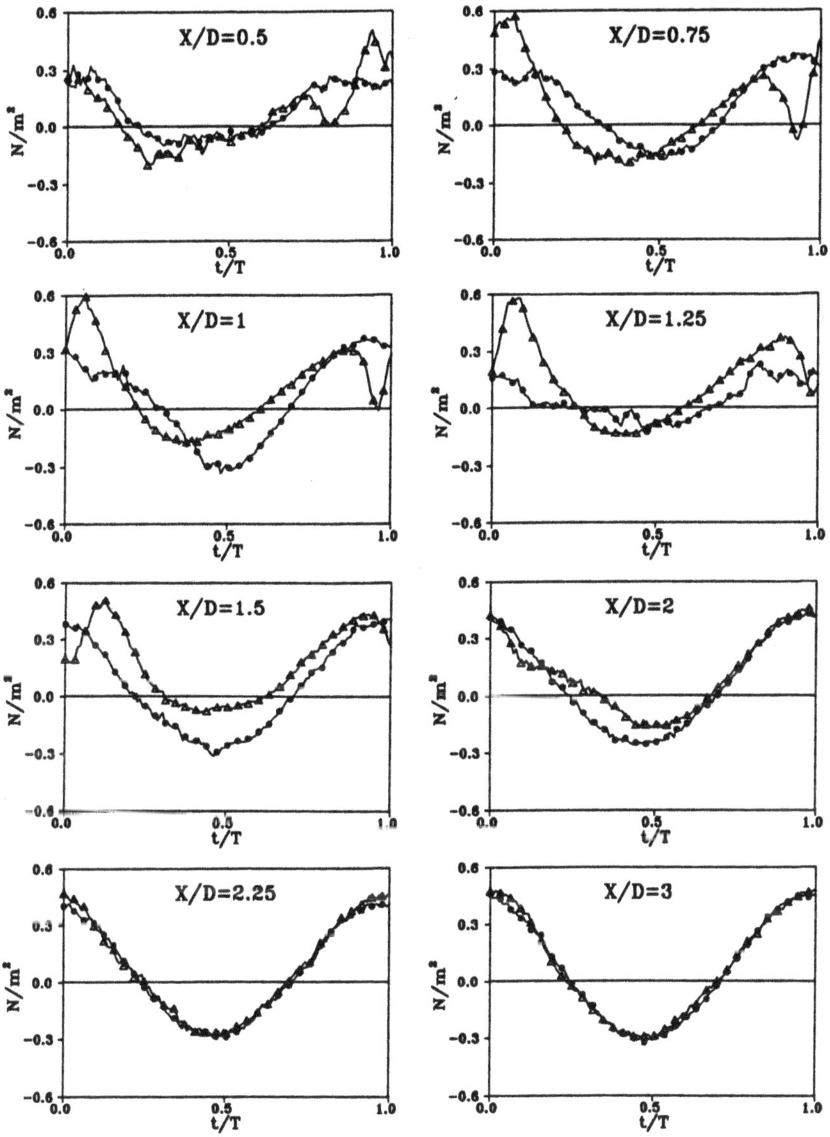

Figure 3: Variations of the bed shear stress versus t/T for different x/D horizontal positions: (Δ) $e/D = 0.09$, ($\bullet$) $e/D = 0.5$

The phase of τ_{max}, the maximum bed shear stress is 45° ahead that of the free stream velocity. The value of the maximum bed shear stress is equal to $0.22N/m^2$ and may be compared with the experimental $\tau_{1_{max}}$ value equal to $0.37N/m^2$. It can be noted that the experimental value of $\tau_{1_{max}}$ exceeds the predicted value. The flow generated by waves cannot certainly be considered as a purely oscillatory flow. Furthermore, the measuring conditions near the wall are not good, due to reflexions on the wall, difficulties to evaluate accurately the distance from the wall. The bed shear stress values obtained give an approximation of the bed shear stress and may be analyzed more by comparison than absolutely.

Another remark concerns the maximum bed shear stress. In the studied region, the maximum bed shear stress never exceeds its undisturbed value for $e/D = 0.5$.On the contrary, when $e/D = 0.09$, the bed shear stress slightly exceeds it. This relative increase in bed shear stress occurs at the beginning of the wave period and stays limited in its duration.

Another form of the bed shear stress is given on figure 4. The amplification factor α, defined by

$$\alpha = \frac{\tau_{max}}{\tau_{1_{max}}} \tag{5}$$

is given versus the x/D, horizontal position in which $\tau_{1_{max}}$ is the maximum value of the undisturbed bed shear stress and τ_{max} is the maximum bed shear stress at a fixed x/D. The variations obtained in the two tests slightly differ. When $e/D = 0.5$, no amplification of the bed shear stress is noted in the downstream region of the cylinder ($0.5 \leq x/D \leq 2.5$). On the contrary, when $e/D = 0.09$, τ_{max} becomes larger than $\tau_{1_{max}}$ in the $0.5 \leq x/D \leq 2$ region.

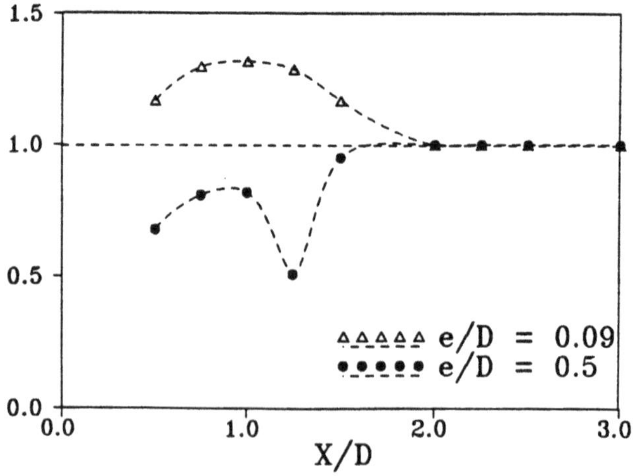

Figure 4: Variation of the $\alpha = \tau_{max}/\tau_{1_{max}}$ amplification coefficient with x/D

Some particular bed shear stress evolutions can be interpreted with the flow previously studied.In figure 5, it can be seen that, in the $e/D = 0.09$ case and $x/D = 0.75$, the bed shear stress can undergo rapid variations. It can also be noticed that for the time instants when the bed shear stress rapidly decreases ($t/T = 0.85$), a jet flow develops in the bottom clearance and that it is responsible for the creation of a stagnation region near the wall.

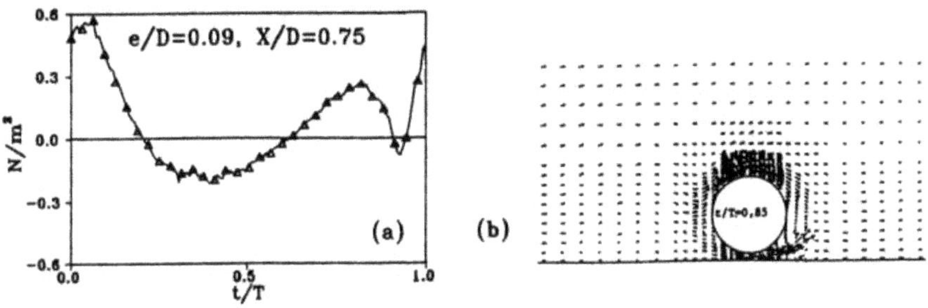

Figure 5: Evolution of the bed shear stress versus t/T for $e/D = 0.09$ and $x/D = 0.75$ (a) and the associated velocity field at $t/T = 0.85$ (b)

Some variations of the bed shear stress with wave position are greatly disturbed. For example, in the $e/D = 0.5$ test, for a horizontal position given by $x/D = 1.25$ represented in figure 6, the bed shear stress becomes zero when t/T varies between 0.15 and 0.35. The vortex with positive vorticity convected towards the wall is, for these instants, adjacent to the wall in the x region around 1.25D. The vortex and the vorticity generated in the layer closed to the wall are of opposite signs. It results a great reduction of both vorticity near the wall and bed shear stress.

This example suggests that the convection of a vortex towards a wall is not always accompanied with an increase of the bed shear stress.

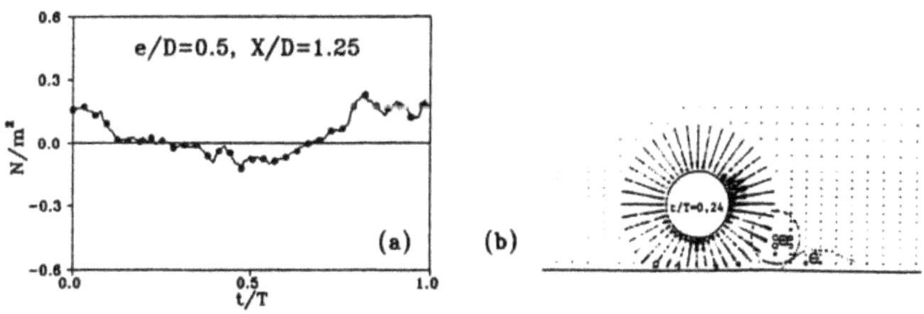

Figure 6: Evolution of the bed shear stress versus t/T for $e/D = 0.5$ and $x/D = 0.25$ (a) and the associated rotational field at $t/T = 0.24$ (b)

5. Conclusions

When a cylinder is placed in the neighbourhood of a wall, in fixed conditions of Keulegan-Carpenter number and Reynolds number (K.C=5 and Re=1800), it undergoes different flow regimes depending on the distance between the cylinder and the wall. Two typical flow regimes have been identified. In one case ($e/D = 0.09$), part of the cylinder is immersed in the wall boundary layer and, in the other case, it is outside of the wall boundary layer ($e/D = 0.5$). There is consequently a very different bed shear stress distribution along the wall in the two tested cases and in particular in the downstream region of the cylinder. The influence of both, the jet flows which develop in the gap between the wall and the cylinder when $e/D = 0.09$ and the vortices convected towards the wall when $e/D = 0.5$ are detected in the bed shear stress. Some rapid variations of the local bed shear stress in the case of a flat bed may favor the erosion process in the region where this phenomenon is noted. Furthermore, the initial conditions for the erosion process strongly depend on the relative distance of the cylinder to the wall and can lead to different evolutions of the bed profile.

7. References

1. W. Leeuwestein, *Scour around submarine pipelines*, **1602**, (1984), Def. Inf. Int., p. 16.
2. B.M. Sumer, C. Pedersen, D. Yu and J. Fredsoe, *Bed shear-stress measurements in the vicinity of a pipeline in waves*, (1990), Prog. Rep. 71, Inst. Hydrodyn. and Hydraulic and Eng. Tech. Univ. Denmark, p. 61.
3. A. Sakout, *Analyse des actions de la houle sur un cylindre à proximité d'une paroi plane*, (1992), Thèse de Doctorat de l'Université du Havre, France.
4. B.M. Sumer, B.L. Jensen and J. Fredsoe, *Effect of a plane boundary on oscillatory flow around a circular cylinder*, **225**, (1991), J. Fluid Mech., p. 271.
5. G.K. Batchelor, *An introduction to fluid dynamics* (Cambridge University Press, 1967).

DIAMETER PARTICLE MEASUREMENTS BY USING SPECTRAL INTERPOLATION METHOD APPLIED TO EXTREMA DETECTION

E. VASSELIN
Laboratoire de Mécanique des Fluides, Le Havre University
B.P. 265 Quai Frissard Le Havre, France1[1]

and

R.D. RAJAONA
Groupe d'Energetique et de Mécanique , Le Havre University
B.P. 265, Quai Frissard Le Havre, France

ABSTRACT

Particle diameter is studied by using Mie theory. Numerical simulations have been developed on micro computer to analyze the extrema number of the scattered light intensity. A biunivoque relation is evidenced between the extrema number and the particle diameter. A specific instrumentation has been developed to validate the theory prediction. The second part of the paper is devoted to a spectral interpolation method applied to the measurement of the extrema number. It is shown that the method is more efficient in the measurements of the particle diameter for a diameters wider range compared with a classical FFT technique.

1 Introduction

The knowledge of particle diameter involved in industrial and research and development laboratory environement owing to the sedimentation process is an important subject. The diameter measurement gives further informations to complete those related to the velocity, shape, size, concentration of the particles. A classical non intrusive technique is based upon the exploitation of the physical characteristics of the light intensity. A number of authors has theoretically described the application of light in particle diameter measurement (BOHREN & al. , HUFFMAN[1], KERKER [5], VAN DE HULST [7]). However those experimental techniques require the so-called Rayleigh, Raleigh-Gans or Raleigh-Debye approximations theories or more frequently the Fraunhofer diffraction theory. Those classical techniques have been recently improved (HAMELIN [4], KOUZELIS [6], nevertheless some limitations are still necessary conditions for a successful application of the methods e.g. particle size and the incident light wavelegnth ratio, measurement range, concentrations or predefinite distributions hypothesis .

---

[1]New address: L.A.C.O.S. Université du Havre 76600 Le Havre France

A solution of the electromagnetic wave scattering problem over a spherical particle has been proposed for a plane wave by integration of the Maxwell equations by Lorentz (1890) and Mie (1908). Recently the related techniques have been applied to the case of other wave types (GOUESBET and GREHAN [3]) and various measurement techniques have been developped by directly using the Mie theory (CORBIN [2]).

From these basis, we have developed a new method of measuring the particle diameter using the Mie theory and an efficient spectral interpolation method has been introduced in order to improve the measurement range of the experimental technique (RAJAONA [10] and RAJAONA and SULMONT [12]).

2 Extrema Detection

Let us consider an arbitrary shape particle lighted by a plane wave. The onset light propagation direction is the the z axis. The particle center is the origin of the axis (Oxyz). The scattering direction and the wave propagation direction are defined as the diffusion plane. The diffusion directions are defined by the diffusion angle θ and the azimuth .

$$\begin{pmatrix} E_{IIscat} \\ E_{\perp scat} \end{pmatrix} = \frac{e^{-ik(r-z)}}{ikr} \begin{pmatrix} S_2 & S_3 \\ S_4 & S_1 \end{pmatrix} \begin{pmatrix} E_{IIscat} \\ E_{\perp scat} \end{pmatrix} \tag{1}$$

The [S] matrix is the scattering matrix and its elements Sj are complex functions of the θ and Φ. We have:

$$I_{dif}(r,\theta,\phi) = \frac{f(\theta,\phi)}{k^2 r^2} I_0 \tag{2}$$

where f(θ,Φ) is the diffusion function and I_{scat} is the intensity of the scattered light.

The solution of the problem of the diffusion of a electromagnetic plane wave past a spherical particle has been derived following the Lorentz and Mie theories. The final result is expressed in terms of the matrix [S] coefficients and derived by KERKER. The far field solution is a transverse one and we have $S_3 = S_4 = 0$

The only non zero amplitude functions are S_1 and S_2 and we have:

$$S_1 = \sum_{n=1}^{\infty} \frac{2n+1}{n(n+1)} \left[a_n \pi_n(\cos\theta) + b_n \tau_n(\cos\theta) \right] \tag{3.a}$$

$$S_2 = \sum_{n=1}^{\infty} \frac{2n+1}{n(n+1)} \left[a_n \tau_n(\cos\theta) + b_n \pi_n(\cos\theta) \right] \tag{3.b}$$

The functions π_n and τ_n are Legendre functions, the terms a_n and b_n are the diffusion coefficients:

$$a_n = \frac{\Psi_n(\alpha)\Psi'_n(\beta) - m\Psi'_n(\alpha)\Psi_n(\beta)}{\xi_n(\alpha)\Psi'_n(\beta) - m\xi'_n(\alpha)\Psi_n(\beta)} \tag{4.a}$$

$$b_n = \frac{m\Psi_n(\alpha)\Psi'_n(\beta) - \Psi'_n(\alpha)\Psi_n(\beta)}{m\xi_n(\alpha)\Psi'_n(\beta) - m\xi'_n(\alpha)\Psi_n(\beta)} \qquad (4.b)$$

$\alpha=\pi d/\lambda$ is the first size parameter, $\beta= m\alpha$ is the second size parameter of the particle, $m=n(1-ik)$ is the complex refraction indice, Ψn are th Riccatti-Bessel functions of the first order, ξn are the hankel functions, and $\Psi'n$ and $\xi'n$ are the first derivative of these functions related to their arguments.

The POYNTING theorem is used to compute the energy flux corresponding to the scattered lightening:

$$S = \frac{1}{2}\mathrm{Re}(E_\theta H_\phi^* - E_\phi H_\theta^*) \qquad (5)$$

the asterisk symbols indicate the complex conjugate for the functions H and E.

For the case of a plane wave , the incident lightening is $\frac{1}{2}\left(\frac{\varepsilon}{\mu}\right)^{1/2} E_0^2$. Therefore for an incident lightening equal to 1, one can write:

$$I_{scat} = I_\theta + I_\phi \quad \text{where}$$

$$I_\theta = \frac{\lambda^2}{4\pi^2 r^2} i_2 \cos^2\phi \quad \text{and} \quad I\phi = \frac{\lambda^2}{4\pi^2 r^2} i_1 \sin^2\phi \qquad (6)$$

The lightening functions i_j are given by $i_j = |S_j|^2$.

The results obtained for the case of a spherical air bubble of a 13µm diameter in water lighted by a plane wave light for which the wavelength is $\lambda = 488$ nm are reported on the figure 1.

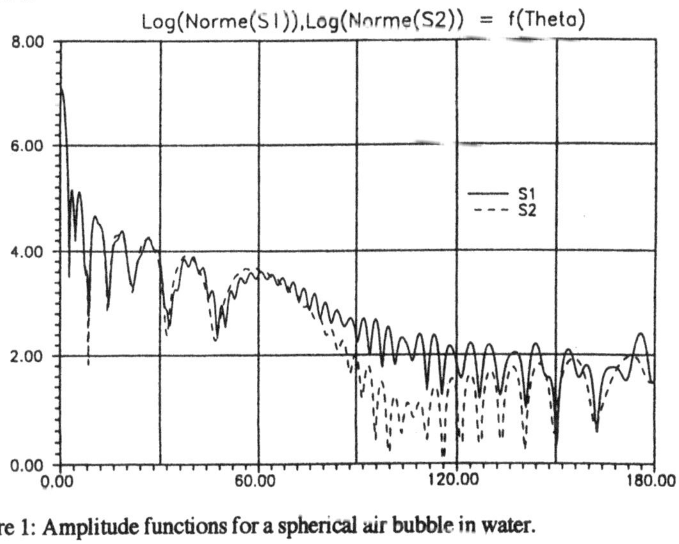

Figure 1: Amplitude functions for a spherical air bubble in water.

These results have shown that the number of the lobes are related to the diameter of the bubble. The number of extrema has been computed by using the Mie theory versus the diameter of the particle. The results of this numerical simulation are illustrated in Fig. 2. We have shown that three ranges can be used to predict the particle diameter once the number of lobes is known. These results are related to a spherical glass of refraction index of 1.545 and lighted by plane wave for which the wavelength is 632.8 nm. The angular sector analyzed [θ_1, θ_2] is [5°,15°] and the azimuth angle is zero.

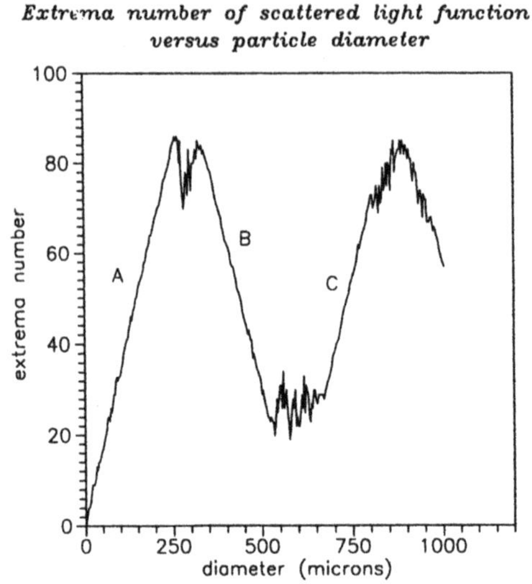

Figure 2: Number of extrema for glass spheres in water
λ=632.8 nm, n = 1.545

The interesting observation is that for these three ranges noted (A, B, C), a linear evolution of the graph shows that there is a biunivoque relation between the particle diameter and the number of extrema. It is then possible to make use of this curve in order to predict the diameter of a particle.

3 Experimental Settings

A specific experimental arrangement has been developed in order to validate these numerical results. The experimental setting consists of the measurements of the diffused light intensity in a adequate angular sector. The measurements are performed by using a linear CCD that allow the recording of the intensity of the light diffused by the small spheres of glass lighted in air by a laser beam. The use of a laser beam allows a high light intensity level. However it should be noticed that in that case the wave form geometry corresponds to a Gaussian distribuiton and not a plane. It has been observed that despite this situation the results are in good agreement with the plane theory provided that a specific diameter value of the laser beam is used .It is assumed that in this case the local deformation of the light wave is negligeable. The whole experimental setting is illustrated in figure 3

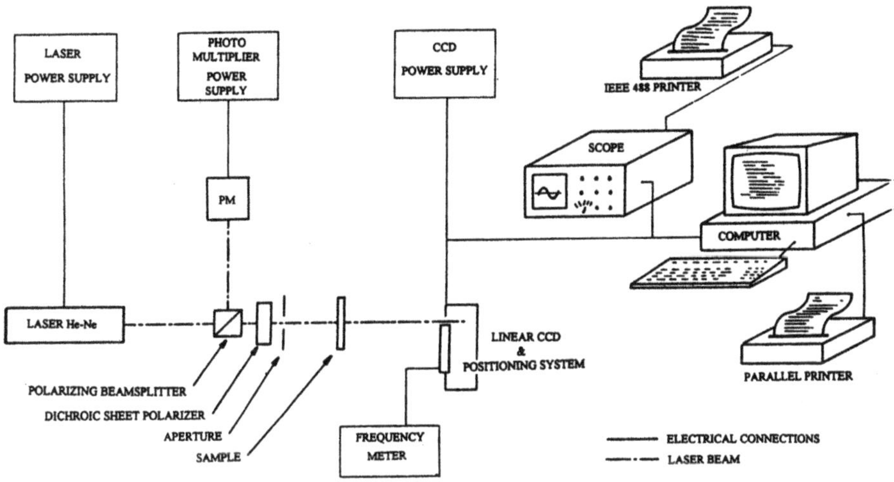

Figure 3: Experimental setting.

4 Preliminary Experimental Results

The CCD sensor converts the fluctuations of the scattered light in a waveform variation of an electric signal. This electric signal is then digitized and recorded in a microcomputer memory. The number of extrema is then computed by using a classical FFT technique. The experimental values have been obtained by two methods. The first method is based upon a direct measurement of the diameter by using a microscop for a range of 170 microns adn 250 microns (these results correspond to the continuous line) . The second method is our method based upon Mie theory . These results correspond to the asterisk symbols. The values obtained by our method are scattered but the mean values are in good agreement with the theoretical values specifically for high diameters for which the number of extrema can be estimated with a higher accuracy .

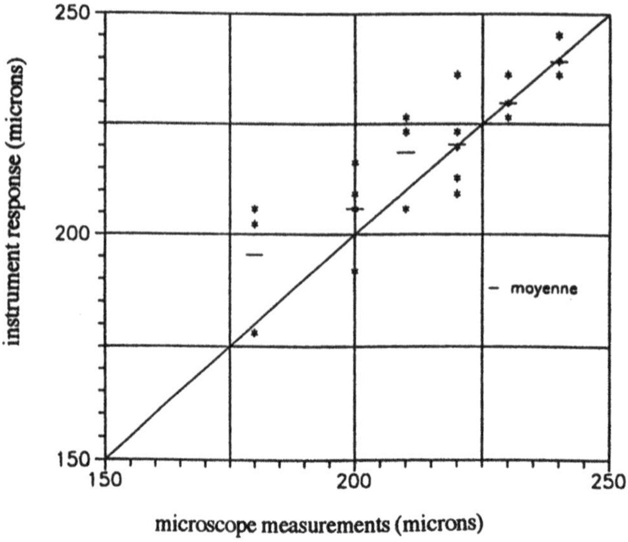

Figure 4 Comparison between particle diameter measurements by using a microscope and the extrema detection method.

5 Use of the Spectral Interpolation Method

In order to improve the efficiency of the present method an interpolation method developed by RAJAONA [10] and RAJAONA and SULMONT [11] is proposed. As we have previously seen when the diameter of the particle is low the direct FFT technique fails down or leads to biaised values. The interpolation technique has been used to improve the accuracy of the DFT calculation.

5.1 Background

Let us consider the light intensity g(t) sampled at regular space interval $t_r = rT/N$, where T is the width of the observed signal in meters, N is the number of samples. And we have $0 \le r \le N-1$. In other words T is the width of the window function which will be used for the analysis of the complete signal g(t) defined in the time interval $-\infty < t < +\infty$. Furthermore we define :

$$g_r = g(t_r) \tag{7}$$

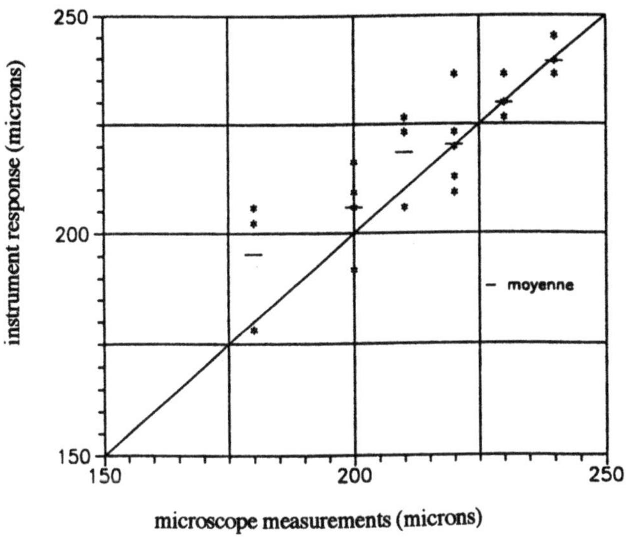

Figure 4 Comparison between particle diameter measurements by using a microscope and the extrema detection method.

5 Use of the Spectral Interpolation Method

In order to improve the efficiency of the present method an interpolation method developed by RAJAONA [10] and RAJAONA and SULMONT [11] is proposed. As we have previously seen when the diameter of the particle is low the direct FFT technique fails down or leads to biaised values. The interpolation technique has been used to improve the accuracy of the DFT calculation.

5.1 Background

Let us consider the light intensity g(t) sampled at regular space interval $t_r = rT/N$, where T is the width of the observed signal in meters, N is the number of samples. And we have $0 \leq r \leq N-1$. In other words T is the width of the window function which will be used for the analysis of the complete signal g(t) defined in the time interval $-\infty < t < +\infty$. Furthermore we define :

$$g_r = g(t_r) \tag{7}$$

These sampled data are supposed to satisfy the Shannon condition (sampling frequencies are greater than twice the largest frequency of the signal) in such a way that no aliasing phenomenon is present [10].

The discrete Fourier transform of the function g(t) corrresponding to the frequency interval 1/T is given by the equation (8) :

$$G_j = \frac{1}{N}\sum_{r=0}^{N-1} g_r \exp(-i2\pi rj / N) \qquad (8)$$

In general, the values G_j do not provide the exact spectrum of the function g(t) ; particularly if periodic components are present in the signal, the peaks of the spectrum are not equal to the amplitude of the periodic components except for very special cases for which the frequency peak is an integer multiple of 1/T ; this has been shown to be a result of the leakage effect [11]. Thus it is not possible to determine the amplitude and the spatial frequency of the various components in the signal from the knowledge of the G_j values .The spectral interpolation described in [10] is useful to compute the values of amplitudes and frequencies.

5. 2 Definition of the pseudoperiodic signals

Furthermore, the signal is assumed to consist of a superposition of a finite number of well distinct and weakly damped periodic components. We thus have:

$$g_r = \frac{1}{2}\sum_{r=0}^{N-1} A_k \exp(i\Omega_k t_r) + \overline{A}_k \exp(-i\overline{\Omega}_k t_r) \qquad (9)$$

for a real signal and it is useful to associate the following complex signal :

$$g_r = \sum_{r=0}^{N-1} A_k \exp(i\Omega_k t_r) + w(t_r) \qquad (10)$$

where we have :

$$A_k = a_k + i\, b_k \qquad (11)$$

$$\Omega_k = i\,\lambda_k + \omega_k \qquad (12)$$

with a_k, b_k, λ_k, ω_k real and $|\lambda_k| << 1$.

$w(t_r)$ represents a low level noise.

As was shown in [10], the integer k takes values between 0 and N-1; however only some values of A_k are nonzero since the signal is assumed to consist of distinct components. For sake of simplicity, we only present the calculation related to the complex associated signal. For the actual light wave signal the analysis needs to be slightly modified but the procedure for numerical calculation is essentially identical; specifically, the amplitude of the actual signal as computed by using an algorithm based on the input of a complex signal only needs to be multiplied by two.

The use of weighting functions different from the rectangular window function is expected to provide more accurate values of the pseudoperiodic components ; specifically by reducing the so-called leakage effect as described in HARRIS [13] . This paper concerns only the effect of a Hanning window although the method can be applied to other weighting functions .We thus define the weighted signal as :

$$g_r^h = h_r g_r \tag{13}$$

where

$$h_r = 1 \tag{14}$$

for a rectangular window

$$h_r = 1\text{-}0.5[\exp(i2\pi r/N) + \exp(-i2\pi r/N) \tag{15}$$

for a Hanning window .

5.3 Error computation.

5.3.1 Rectangular window.

For a rectangular window, Equations (8) (10) and (14) yield:

$$G_j = \frac{1}{N}\sum_{k=0}^{N-1} A_k \frac{1-\exp[-\lambda_k T + i(\omega_k T - 2\pi j)]}{1-\exp[-\lambda_k T + i(\omega_k T - 2\pi j)]/N} \tag{16}$$

Equation (16) shows that there is no obvious direct relation between that values G_j and A_k except for those components for which the frequencies ω_k are integer multiples of the frequency interval $\Delta\omega = 2\pi/T$. *For these particular components* with j=K and $\omega_K = K\, 2\pi/T$, we have:

$$G_K = A_K \frac{1-\exp(\lambda_K T)}{\lambda_K T} + \frac{1}{N}\sum_{k=0}^{N-1} A_k \frac{1-\exp[-\lambda_k T + i(\omega_k T - 2\pi j)]}{1-\exp[-\lambda_k T + i(\omega_k T - 2\pi j)]/N} \tag{17}$$

provided that λ_K is small.

It can be seen that if all the values λ_k vanish for k running from 0 to N-1 we have:

$$G_K = A_K + O(\frac{1}{N}) + W(K) \tag{18}$$

where O(1/N) is the contribution of the remaining components and W(K) is the noise contribution.

In the latter case, G_K thus directly yields the value of A_K disregarding the contribution of the noise.The term O(1/N) shows that the contribution of the remaining components is of order 1/N. More generally, the frequency of a component can be related to two successive integer multiples of $\Delta\omega$ as follows:

$$K\Delta\omega \le \omega_K \le (K+1)\Delta\omega \qquad (19.a)$$
$$(K-1)\Delta\omega \le \omega_K \le K\Delta\omega \qquad (19.b)$$

In practice K corresponds to the values of j for which G_j is a peak in the spectrum despite the leakage effect. Equations (19) can be expressed as follows :

$$\omega_K = (K+\varepsilon_K)\frac{2\pi}{T} \qquad \text{with } 1 \ge |\varepsilon_K| \qquad (19.c)$$

The interpolation method consists in determining ε_k from which ω_k is then obtained by using Eq. (19.c). We can isolate the contribution of the peak of order K in equation (8) and Eq. (16) yields:

$$G_j = \frac{1}{N}A_K\frac{1-\exp[i2\pi(F_KT-j)]}{1-\exp[i2\pi(F_KT-j)/N]} + \frac{1}{N}\sum_{k=0,k\neq K}^{N-1} A_k\frac{1-\exp[i2\pi(F_kT-j)]}{1-\exp[i2\pi(F_kT-j)/N]} \qquad (20)$$

where we have :

$$F_k = \Omega_k/2\pi = i\frac{\lambda_k}{2\pi} + f_k = i\frac{\lambda_k}{2\pi} + (k+\varepsilon_k)\frac{1}{T} \qquad (21)$$

for k running from 0 to N-1.
W(j) = the discrete Fourier transform of the noise $w(t_r)$
For the values K of the integer j , the contribution of the peak of order K is dominant compared with that of the remaining components. We thus have :

$$G_K = A_K\frac{1-\exp z_K}{-z_K} + O_K(\frac{1}{N}) + W(K) \qquad (22)$$

where we have :

$$z_K = i2\pi(F_KT-K) = -\lambda_KT + i2\pi\varepsilon_K \qquad (23)$$

$$O_K(\frac{1}{N}) = \frac{1}{N}\sum_{k=0,k\neq K}^{N-1} A_K\frac{1-\exp[i2\pi(F_KT-j)]}{1-\exp[i2\pi(F_KT-j)/N]} + W(K) \qquad (24)$$

Similarly for j = K+1 and j =K-1 , we have :

$$G_{K+1} = A_K\frac{1-\exp z_K}{-z_K+i2\pi} + O_{K+1}(\frac{1}{N}) + W(K+1) \qquad (25)$$

$$G_{K-1} = A_K\frac{1-\exp z_K}{-z_K-i2\pi} + O_{K-1}(\frac{1}{N}) + W(K-1) \qquad (26)$$

The lineshape fitting method described in [14] needs the calculation of the differences

$$S^{+} = G_{K} - G_{K+1} \tag{27}$$

$$S^{-} = G_{K} - G_{K-1} \tag{28}$$

These quantities involve the differences of terms as $O_K(1/N) - O_{K-1}(1/N)$. It can be easily shown that for sufficiently high value of N we have :

$$O_K\left(\frac{1}{N}\right) - O_{K+1}\left(\frac{1}{N}\right) = i\frac{2\pi}{N^2}\sum_{k=0,k\neq K}^{N-1} A_k \frac{[1-\exp z_k]\left[\exp i\frac{2\pi}{N}(k-K+\mu_k)\right]}{\left[1-\exp i\frac{2\pi}{N}(k-K+\mu_k)\right]\left[1-\exp i\frac{2\pi}{N}(k-K+\mu_k-1)\right]} \tag{19}$$

where $\mu_k = z_k / i2\pi$

Eq. (29) shows that the contribution of the components of order k different of K is of order $1/N^2$ in the computation of S^+. A similar derivation leads to a similar result for the value of the difference S^- . The essence of the procedure consists in determining the parameter ε_k, from which ω_k is then obtained by using Eq (19.c). The value of ε_k is deduced from that of the complex number z_k defined in Eq. (23) and (30) :

$$\frac{S^+}{S^-} = \frac{G_K - G_{K+1}}{G_K - G_{K-1}} = \frac{i2\pi + z_K}{i2\pi - z_K} \tag{30}$$

The Equations (20), (27),(28),(29) and (30) show that the contribution of the components of order k different from K,-that are the so called " remaining components " - is of order $1/N^2$ and we have :

$$z_K = i2\pi\frac{G_{K+1} - G_{K-1}}{2G_K - G_{K+1} - G_{K-1}} \tag{31}$$

It should be noticed that the contribution of the noise has been partly eliminated in the calculation of z_K as was shown in [10] provided that its Fourier transform is gradually varying.

5.3.2 Use of Hanning window

The effect of the Hanning window function on the error computation involved in the method [10] is analyzed in detail in [12] . It is shown that further calculations lead to Eq. (32)

$$G_K = A_K\frac{1}{i2\pi}P(\mu_K)\left[1 - \exp(i2\pi\mu_K)\right] + \sigma(K) \tag{32}$$

where we have :

$$P(\mu_K) = \frac{1}{\mu_K(\mu_K + 1)(\mu_K - 1)} \tag{33}$$

and

$$\sigma(K) = O_K(\frac{1}{N}) - \frac{1}{2}O_{K-1}(\frac{1}{N}) - \frac{1}{2}O_{K+1}(\frac{1}{N}) \tag{34}$$

where $\mu_k = z_k / i2\pi$ (35)

and z_k is the complex frequency shift as defined in the case of rectangular window.

Specifically , σ_k is of order $1/N^3$ while its value is of order 1/N in the case of rectangular window. Once again the calculation of G_{K+1}, G_{K-1} leads to the computation of the frequency by using the ratio $(G_K - G_{K+1}) / (G_K - G_{K-1})$ in order to partly eliminate the noise contribution ; we thus have :

$$\frac{G_K - G_{K+1}}{G_K - G_{K-1}} = -\frac{i4\pi + z_K}{-i4\pi + z_K} = -\frac{\mu_K - 2}{\mu_K + 2} \tag{36}$$

We can now determine the complex value z_k . It represents a complex value of the complex frequency shift similar to that was introduced by Feit and Fleck [14] .
We have :

$$z_K = i4\pi \frac{G_{K+1} - G_{K-1}}{2G_K - G_{K+1} - G_{K-1}} \tag{37}$$

Eqs. (32) (33) (36) (37) show that the improvement expected by using a Hanning window function is achieved. The previous development also illustrates the mechanism of reduction of the error by using the contiguous values of the spectrum. This is an answer to a question arised by Feit and Fleck described in [14].

6. Computation of the components characteristics

From Eqs. (21), (31) and (37) the complex frequency can be determined by the following formulae:

$$z_K = i2\pi\nu \frac{G_{K+1} - G_{K-1}}{2G_K - G_{K+1} - G_{K-1}}$$

$$\lambda_K = \frac{-1}{T}\mathrm{Re}\{z_K\} \tag{38}$$

$$\varepsilon_K = \frac{1}{2\pi}\mathrm{Im}\{z_K\} \tag{39}$$

$$F_K = (\varepsilon_K + K)\frac{1}{T} + i\frac{\lambda_K}{2\pi} \tag{40}$$

where we have $\nu = 1$ for a rectangular window
$\nu = 2$ for a Hanning window

The complex amplitudes are derived by using the values of z_K and $2G_K - G_{K+1} - G_{K-1}$. This technique is a local interpolation as described in [3] , since the contiguous values of G_K G_{K+1} and G_{K-1} have been used . It can be shown that this leads to the following results:

$$A_K = (2G_K - G_{K+1} - G_{K-1})\frac{z_K(z_K^2 + 4\pi^2)}{8\pi^2(1 - \exp z_K)} \tag{41}$$

for a rectangular window and :

$$A_K = (2G_K - G_{K+1} - G_{K-1})\frac{i\pi(4 - \mu_K{}^2)}{6P(\mu_K)(1 - \exp z_K)} \tag{42}$$

for a Hanning window function .

The accuracy of these values depends upon the accuracy of z_K and μ_K values and the efficiency in eliminating the noise contribution. An alternative method for computing the complex amplitudes consists of resolving the linear system related to amplitudes values once the frequencies and damping are computed. This technique is a global method [8] and it is used in the time domain methods .

7 Application of the interpolation method to diameter measurement

The spectral interpolation method has been applied to the extrema detection in order to illustrate its efficiency. It should be noted that that the spectral resolution is inversely proportional to the width of the window . So, the sampled light intensity signal analyzed is similar to those used for Fig. 4. The results are reported in Fig. 5 . It is shown that the range of detected diameters of glass spheres is wider than that related to the direct FFT. It was necessary subtract the mean value of the signal in order work with a centered signal [13]. It is clearly demonstrated that the particle of diameter value of 120 microns is widely biaised when the direct FFT is used. The result obtained by using the interpolation is greatly improved

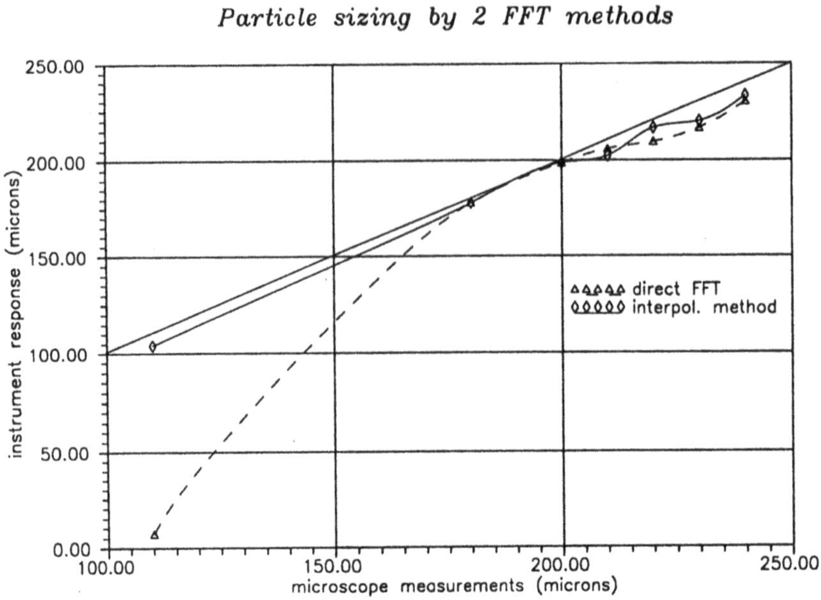

Figure 5: Comparison between the actual diameter values measured by using a microscope and the extrema detection coupled with interpolation method.

8. Conclusions

The main conclusions of this works are:

- A new optical method is used to measure particle diameter by the detection of the extrema number of the light scattered intensity. The theoretical basis are derived from the Mie theory applied to fixed spherical particles and plane waves.

- An experimental setting has been built up to validate the theoretical theory. The corresponding results needed the use of a laser beam for which the light intensity distribution is gaussian. However the reults are encourageing and agree with the actual values measured by a microscope.

- The FFT technique applied to the extrema detection is coupled with a spectral interpolation method based upon the determination of the complex frequency shift. The light intensity waveform is assumed to be a pseudoperiodic signal with lossy components. The accuracy of the method is demonstrated to be in order of $O(1/N^3)$ when a Hanning function is involved to weight the original light intensity signal. The practical improvement of the method is to increase the measurement range of the extrema detection method.

9 References

1. C.F. BOHREN and D.R. HUFFMAN Absorption and scattering of light by small particles (John Wiley and Sons, New York, 1983)

2. F. CORBIN Développement d'un vélocimètre-granulomètre à faisceau laser corrigé. Application aux écoulements dispersés. Thèse Université de Rouen 1989

3. G. GOUESBET and G. GREHAN Sur la généralisation de la théorie de Lorentz-Mie J. Optics vol 13, N°2 pp 97-103 (1982)

4. P. HAMELIN Application de la diffusion lumineuse à la métrologie des particules en écoulement diphasique dispersé. Thèse Institut Polytechnique de Toulouse (1986)

5. M. KERKER The scattering of light. Academic Press, Orlando (1969)

6. D. KOUZELIS Mesure de taille de particules par une méthode optique utilisant la diffraction de lumière laser. Thèse Ecole Centrale de Paris (1988)

7. H.C. VAN DE HULST Light scattering by small particles. Dover Publications, New York (1981)

8. E. VASSELIN Analyse des phénomènes de diffusion lumineuse- Développement d'un granulomètre laser. Thèse. Université du Havre 1993)

9. E. VASSELIN and R.D. RAJAONA Extrema detection and spectral interpolation-Application to particle diameter measurement. EUROMECH 310 Le Havre (1993)

10. R.D. RAJAONA Pseudoperiodic Phenomena and Fluid Flows past a circular cylinder . Thèse d'Etat Université de Nantes(1987)

11. R.D. RAJAONA and P. SULMONT, J. Computational Physics, Vol 61, pp186 (1985)

12. R.D. RAJAONA and P. SULMONT, J. Computational Physics, Vol 97, N°1, pp 30-52(1991)

13. F.J. HARRIS On the use of windows for harmonic analysis with the Discrete Fourier Transform, Proceedings of IEEE, Vol 66(1), pp. 51-83 (1978)

14. M.D. FEIT, J.A. FLECK Mode properties of optical fibers with lossy components by the propagating beam method Applied optics Vol 20(5), pp. 848-856 (1981)

THE BEHAVIOUR, INTERACTION AND EFFECT OF DEPOSITING GRAINS.

M.V.C. Wheeler,
Engineering Department,Aberdeen University,
Fraser-Noble Building, Old Aberdeen,
AB9 2UE, Scotland

and

B.B.Willetts,
Engineering Department,Aberdeen University,
Fraser-Noble Building, Old Aberdeen,
AB9 2UE, Scotland

Abstract

This flume based study, investigated the effect that feeding material has upon the movement of mixed grain size sediment. Three 30 hour, constant flow rate experiments were completed. The first run without sediment feed acted as a control. Two separate runs were then each fed with single size material classed respectively as fine and modal in size with regard to the original mixture. Transported material was continually collected within a trap, allowing sediment transport rates to be calculated and bedload compositional changes to be recorded. Using laser doppler anemometry velocity profiles were taken over the developing bed surface, while extensive longitudinal bed surface profiling highlighted bed degradation/aggradation sequences. Results showed that feeding and the size of fed material directly affected the flow, the rate at which material was transported and the bedload composition. In particular, feeding fine material enhanced the transport rate and time-averaged near-bed velocity, while feeding modal material reduced near-bed velocities substantially.

Introduction

In a heterogeneous bed mixture it has been observed that with time, a bed surface becomes water worked and selectively eroded, leaving a coarsened or armoured surface layer (Harrison, 1950; Proffitt, 1980). General bed load equations have been adapted to incorporate hiding functions to account for surface coarsening, but fieldworkers (Hollingshead, 1971; Laronne & Carson, 1976; Klingeman & Emmett, 1982; Brayshaw, 1983) have observed that these empirically based flume derived formulae (Einstein, 1950; Ackers & White, 1973) are unreliable because the assumption that there is a unique relationship between transport rate and contemporary flow strength is unfounded. Their studies each reported that bedload variations were also partially attributable to bed microtopography which depends on the size distribution and arrangement of grains. Hence it was seen to be important to consider grain availability and bed texture when calculating expected transport rates.

Dietrich et al, 1988; Iseya & Ikeda, 1987; Parker, Dhamotharan & Stefan 1982; Schiller & Rowney, 1984 all completed laboratory sediment fed experiments that showed armouring, bed texturing and subsequent transport rates to be dependent on the size and rate at which material was fed into their systems but little is known about their effect on the flow regime.

This paper reports on three sediment fed laboratory experiments. They support and extend the understanding of physical transport mechanisms through the use of laser doppler anemometry measurements and continuous bedload sampling.
Results show that bedload transport is significantly affected by material feed. A vigorous and unsteady bedload transport was recorded when fines were fed into the system whilst, introduction of coarse sediment reduced near bed velocities and stifled grain movement.

Experimental set-up and procedure

Experiments were undertaken in a recirculating,12m long, 300mm wide glass-walled tilting flume with an initial bed slope set at 0.004 (Fig.1). The original bed material had a size distribution shown in Fig. 2; it was placed into the flume and scraped flat to a similar depth for each run. This provided a well mixed smooth reproducible bed surface. A uniform flow rate capable of moving all grain sizes was maintained in each experiment. The water delivery rate was set at 9 l/s which produces an approximate shear velocity of 0.04m/s.
Three 30 hour runs were completed. The first experiment was <u>unfed</u>. It acted as a control for later fed runs. Two separate runs were each fed with dyed material from a single size fraction. These two fractions were fine and modal in size with respect to the original bed material. Each material was fed, 3m from the upstream end of the flume using an automated rotating feeding drum which could deliver sediments at a pre-set rate. The purpose of the experiment was not to overload the system but to observe grain interaction and to relate differences in behaviour directly to the size of fed material. Feed rates were set for both input sizes at values below their equilibrium transport rates. Feeding commenced once uniform flow conditions had been established and continued for 4 hours. Experimental parameters are shown in Table 1.
Sediment transport rate calculation and analysis of bedload composition were made possible by the collection of transported material within a non-intrusive bedload trap, 7.8m downstream of the feeder. The design of the trap (Tait, Willetts & Maizels, 1992) was such that it permitted continuous sampling without disruption of either the bed surface or flow. The bedload samples were dried, weighed and then sieved. All the sediment samples were analysed using a series of sieves at 0.5ϕ intervals. Water and bed level variations were monitored with vernier depth gauge measurements. Taken at two hourly intervals, levels were recorded every 90mm across the flume width and at 1m intervals along the flume length, from 1.5m to 8.5m
Using laser doppler anemometry, velocity profiles were taken at three hourly intervals 6.8m downstream of the feeder over the developing bed surface. Profiles were taken through the flow depth in 0.5mm increments on the flume centre line. With a sampling frequency of 100Hz, several turbulent events were recorded as well as mean velocities. Shear velocities were then calculated using the logarithmic law of the wall equation.

At the site of the velocity profiles, bed rugosity was measured using a laser displacement sensor. The sensor, traversing (in x and y) in a plane parallel to the mean bed surface, recorded bed surface profiles over a 100 x 100mm grid. Although not reported in this paper, work is being carried out characterising these bed surfaces and relating them to variations in near bed velocities and bedload fluctuations.

Table 1. Experimental parameters

Experiment	Feed rate	Input size fraction (mm)	Feed initiation	Feed termination	Flow depth	Mean shear velocity (m/s)
Control	N/A	N/A	N/A	N/A	60mm	0.04
Fine	47(g/min)	1.00–0.85mm	20 mins	255mins	58mm	0.04
Modal	46(g/min)	3.36–2.36mm	60 mins	300mins	57mm	0.04

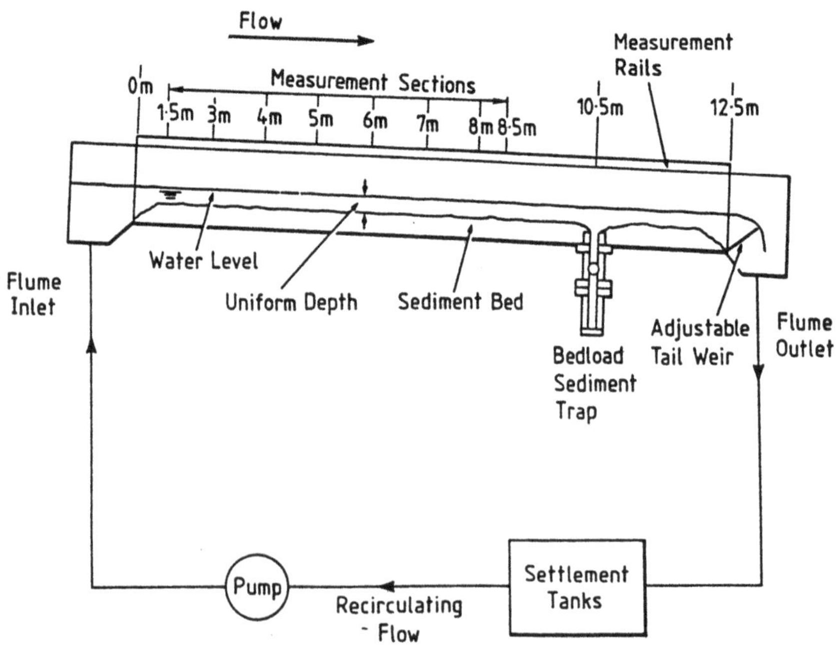

Material fed in at 3m
Velocity profiles taken at 9·8m

Figure 1. Sketch of experimental layout

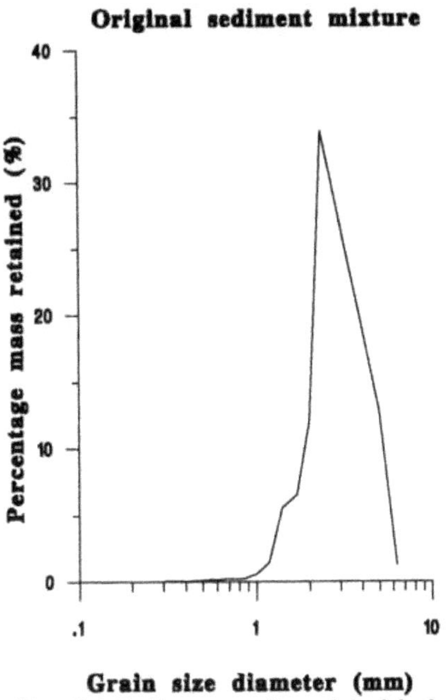

Figure 2. Size distribution curve for original bed material

Results

Sediment transport rates.

Data gathered from the unfed experiment exhibited a pattern which was similar to that found in previous depletion investigations, (Tait, Willetts & Maizels, 1992) with the sediment transport rate first rising and then falling in dramatic fashion (Fig. 3a). When the systems were fed the transport rate responses were similar in an overall sense, but each had distinctive characteristics (Fig. 3b).

Feeding modal sized material reduced the transport rate (Fig. 3c) and bedload movement was sporadic. Although reduced, transport was sustained for a longer duration (8 hours) than in the unfed case (4 hours).

With the feeding of fines bedload transport was significantly increased for the duration of feeding but then dropped dramatically when feeding ceased (Fig. 3d). Whilst the system was fed, transport was intermittent. Pulses migrated down the flume with a period which increased with time from 11 mins to 30mins, creating distinct layers within the trap. These layers were composed of normally mixed bedload separated by bands of dyed input material. No clear evidence of pulsing was seen in either of the other two runs.

Observation of the bed during the fine run showed that a section of bed 1.2m downstream of the feeder was being significantly eroded. Longitudinal bed levels (Fig. 4) confirmed this and also indicated material accumulation on the bed surface during the modal run where material was deposited over a region extending almost 3m. downstream of the feeder. Bed slopes also varied (Table 2.) .

Table 2. Bed slope comparison

	Bed slope (%)		
Elapsed time (mins)	Fine fed run	Modal fed run	Unfed run
240	0.46	0.38	0.40
1300	0.41	0.35	0.39

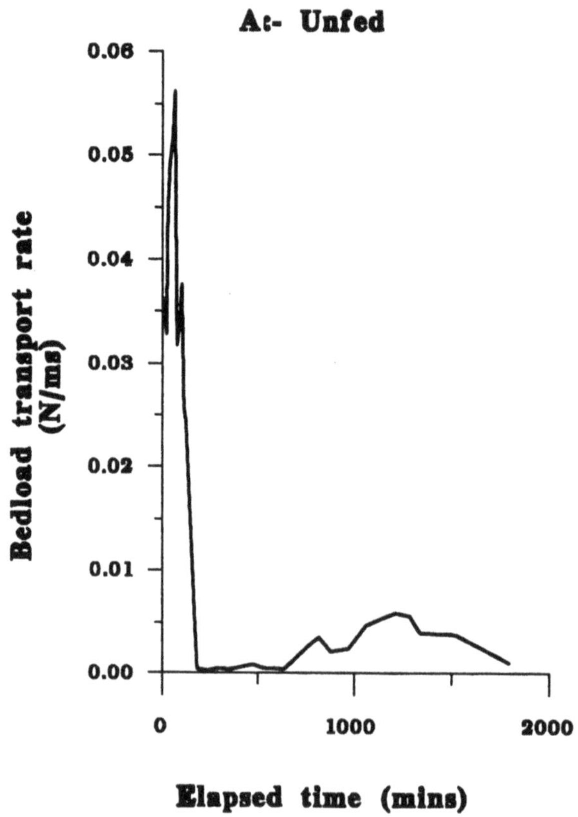

Figure 3a.: Bedload transport rate for the unfed experimental run

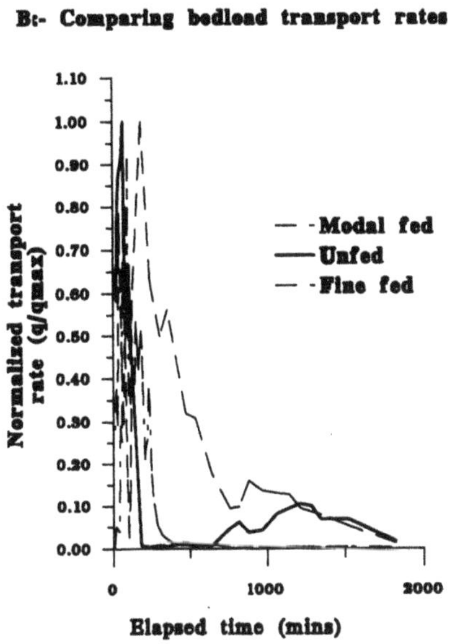

Figure 3.b:- Comparing transport rates for each run

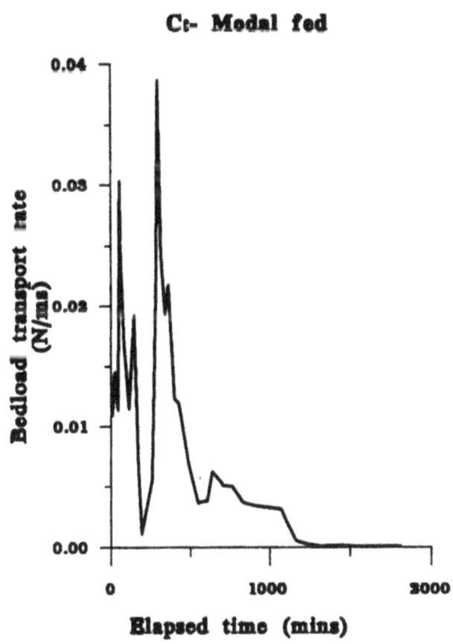

Figure 3c:- Bedload transport rate for modal fed run

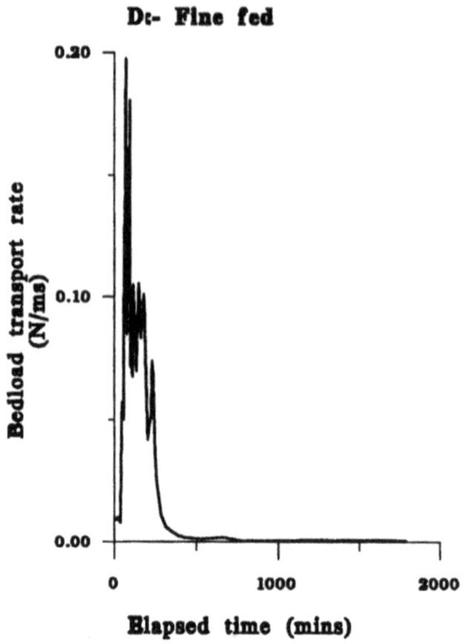

Figure 3d:- Bedload transport rates for fine fed run.

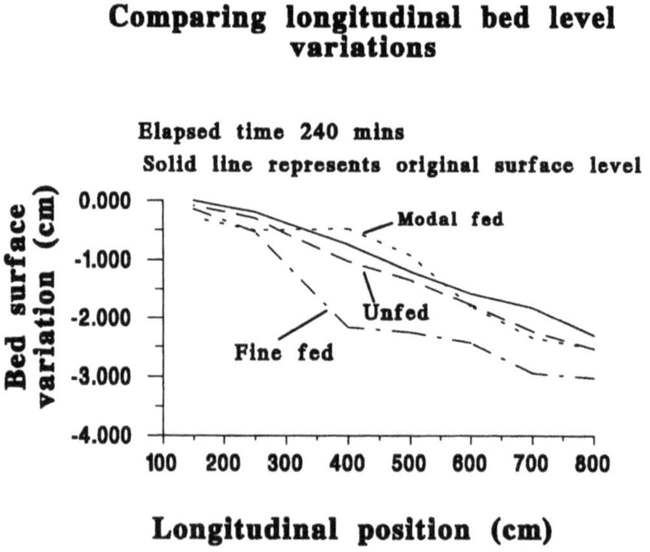

Figure 4. Longitudinal bed profiles. See also Fig. 1 & Table 2

Bedload composition.

Overall, bedload compositions varied temporally, and the degree to which individual size fractions moved was quite different in each run.

Bedload size distributions for the fine fed run showed the composition to become distinctly bi-modal, one mode corresponding to the original bed mixture and the other reflecting the input size (Fig. 5a). Bi-modality persisted throughout, suggesting that input grain infiltration had occurred because input grains were still being released long after feeding ceased. Physical recovery of bed surface samples, to measure grain infiltration was impossible during experimental run time. However, a strong suggestion of infiltration was seen in the time difference recorded for dyed grains to arrive in the trap. Significant proportions of modal input material were collected in the trap only 12 mins after initial introduction whilst it took 40 mins before notable proportions of fine input material were collected. After the completion of the run, surface layers at several locations were removed revealing dyed fine tracer material in the underlying sub-layers. Although quantitative measurements of infiltration were not made, all this evidence suggested that infiltration was an important bed mechanism.

In the unfed run, the concept of equal mobility (Parker et al., 1982) was tenable when grain size distributions at the start and end of the run were compared (Fig 5b). However, at an intermediate stage, after 240 minutes had elapsed, the proportion of some of the finer size fractions increased and the proportion of the modal size fraction decreased compared with proportions expected on the basis of equal mobility. Although the sediment transport rate had fallen, the prevalence of these finer fractions was attributed to longitudinal sorting of the bed material, this having occurred during the preceding period of high sediment transport. After slow pulse-like motion along the flume these sorted areas were eventually caught in the trap. These results were of importance as they highlighted the necessity for continual bedload sampling. Based solely on the evidence of bedload size distributions at the start and end of the run, selective sorting and resulting compositional changes on the bed surface would have gone undetected. Also even when the sediment transport was low, bedload sheets moved through the system. Without thorough bedload analysis these material pulses would have gone unnoticed as there was no evidence of pulsing in the sediment transport rate (Fig. 3a).

In the modal fed case, dyed grains were clearly visible on the bed surface and no infiltration occurred. The resulting bedload size distribution (Fig. 5c) distinctly narrowed about the input size with a reduction of the finer fractions. When comparing the aggregated bedload masses (Table 3.) feeding modal sized sediment suppressed overall grain movement with respect to the unfed experiment, whereas feeding fine material enhanced yields for all sizes, but especially for the finer fractions.

For the modal run 11.5kg of 2.36m material was input 60 minutes after the start of the run over a 240 minute period. The transported material of this size collected, within the trap after 1500mins was 9.42kg; only 4.4kg more than was collected in the equivalent period in the unfed run.

This implied, that 7.5kg of input material was retained by the bed. During the fine fed run, 11.5kg of 0.85mm material was input 60 minutes after the start of the run over a 240 minute period. In this case, after 1500 mins 12.1kg more, of this size material was collected when compared with that collected in the unfed experiment, which implied that an extra 0.6kg of material was released. As separating dyed grains from normal bedload grains proved too difficult, it was impossible to determine the percentage of input material actually moving through the system but it was clear that the transport processes were different in the three cases.

Table 3: The aggregated masses of collected bedload for each size fraction after elapsed time1500mins

	Cumulative collected bedload masses (g)		
Grain size (mm)	Fine fed run	Unfed run	Modal fed run
6.30	0.00	0.00	0.00
5.00	148.00	209.00	212.00
3.36	2515.00	2950.00	2529.00
2.36	5889.00	5422.00	9421.00
2.00	3834.00	2744.00	2156.00
1.71	2350.00	1412.00	908.00
1.40	2509.00	981.00	618.00
1.18	829.00	165.00	106.00
0.85	12168.00	63.00	32.00
0.60	694.00	3.00	1.68

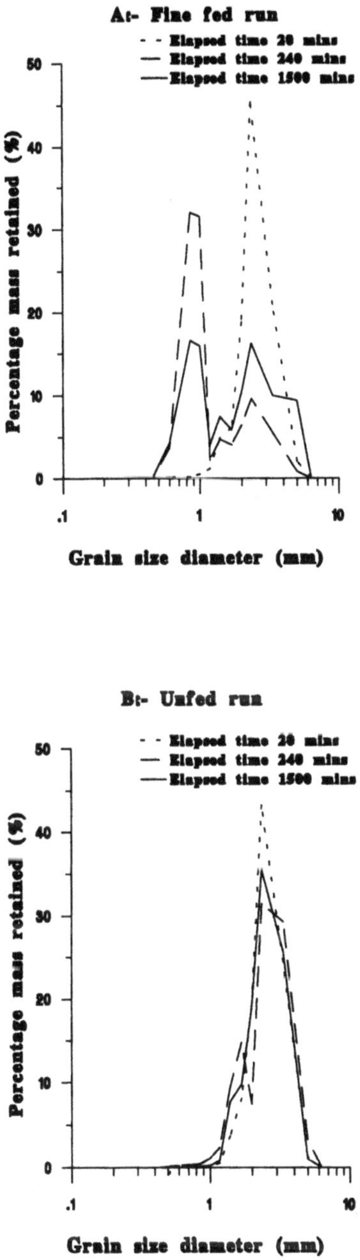

Figure 5.a,b:- Temporal Variation of bedload size distributions

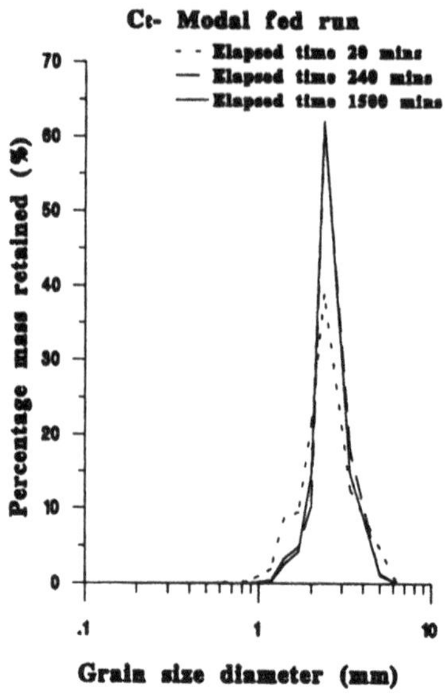

Figure 5.c:- Temporal Variation of bedload size distributions.

Velocity Profiles

Analysis of velocity profiles confirmed that feeding and the size of fed material directly affected the flow. At elapsed time 100 minutes (Fig. 6a.), the profile shape for each run was distinct. The near-bed velocities associated with the fine fed run were larger than those measured during the unfed and modal fed runs, by almost a factor of 2 compared with the modal fed case. Profiles taken at elapsed time 1400 minutes (Fig. 6b.) showed much smaller differences, which suggested that irrespective of the fed size, the resulting bed surface had a similar effect on the flow. At the earlier times the profiles exhibited two distinct regions separated by a kink. It was postulated that these kinks separated an upper layer of constant shear stress from a lower roughness layer. Within the roughness layer the mean flow was three dimensional and shear stresses were thought to be carried in part by the moving grains and partly by bed surface irregularities. Due to the nature of the flow in the roughness layer, shear stresses were only calculated using the law of the wall equation in the fully turbulent, upper constant shear layer (Table 4.). These calculations showed the most hydraulically rough bed encountered by the flow occurred when bed activity was high and modal material was added. The bed with least drag resistance resulted from feeding fine material. After 1400 minutes the transport rate had dropped and the shear stresses were attributed solely to bed roughness and were found to be almost the same for all these cases.

Table 4: Calculated shear velocities and shear stresses at times of differing bed activity

	Upper profile portion					
	Fine run		Modal run		Unfed run	
Elapsd time (min)	V^* ($m.s^{-1}$)	Shear stress ($N.m^{-2}$)	V^* ($m.s^{-1}$)	Shear stress ($N.m^{-2}$)	V^* ($m.s^{-1}$)	Shear stress ($N.m^{-2}$)
100	0.034	1.121	0.050	2.460	0.040	1.563
1400	0.034	1.121	0.030	0.924	0.030	0.927

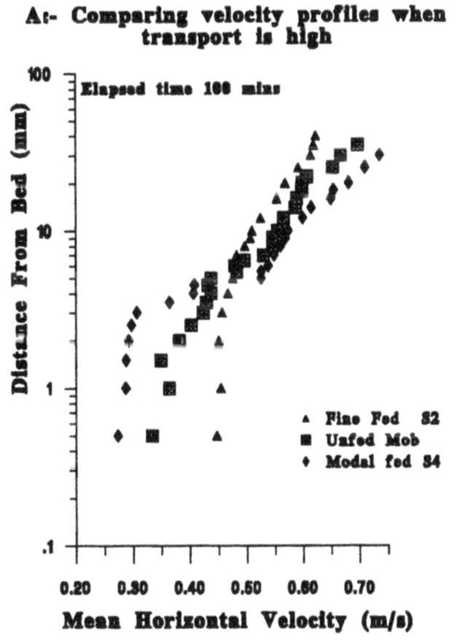

Figure 6.(a,b) Velocity profiles over differing bed surfaces

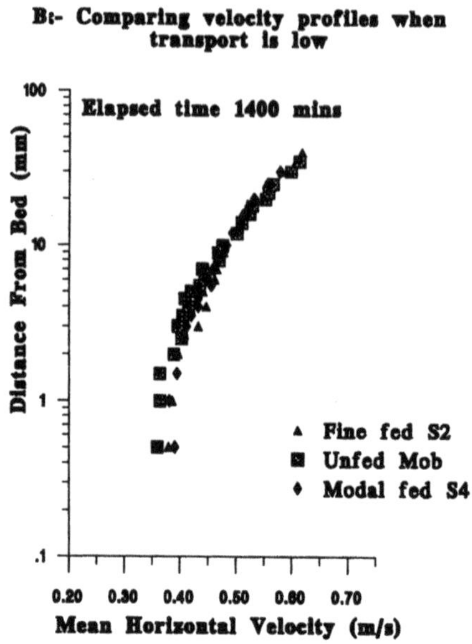

Figure 6.(a,b) Velocity profiles over differing bed surfaces

Discussion

As suggested by other authors (Di Silvio, 1992; Andrews and Parker, 1987), the bed is best described as a series of layers. The uppermost layer here termed bedlayer, contains moving particles as bedload, with individual grains constantly interchanging with an underlying surface or exchange layer. Below these two interactive layers lies a generally stable sub-surface, of similar composition to the original mixture. It is by using this layering idea that the effects of input grains can best be described.
Added fines were observed to interact with the bed, infiltrating down into the exchange layer, filling interstices between grains and eventually saturating the layer with fines. Within these saturated zones grain dislodgement is thought to be suppressed as a result of grain interlocking. This causes a continuity imbalance and directly downstream of these zones sediment supply is greatly reduced. This imbalance is further accentuated by lower bed shear stresses measured over these smoothed well packed areas which also contributes to reduced grain entrainment. With a reduction in bed drag, near-bed velocities increase and this, together with a supply defecit arising from reduced entrainment, causes areas downstream of these zones to suffer severe bed erosion. This exposes underlying well mixed subsurface layers making additional material available for transport. Bed surface levels (Fig. 4) show bed degradation to be more pronounced and sudden towards the feeder where fine infiltration is more active. Ikeda, 1984 suggested that bedload transport resulting from fine infiltration was related to particle collision rather than fluid entrainment.

This conforms to some extent with these findings, but substantial evidence would be difficult to gather, involving the use of sophisticated video equipment. Once the bed near the feeder becomes saturated with fine material, further added fines are thought to be moved downstream to another infiltration site leading to longitudinal aggrading-degrading sequences. The pulses of bedload are thus attributable to longitudinal sediment sorting arising from these sequences. In essence the fines act as a catalyst for sediment transport. Without input the bed is not degraded further and the exhausted layers cannot provide material, so the transport rate rapidly drops shortly after feeding is terminated.

The mechanisms for the modal fed system are very different. The grains were seen to roll over the surface for sustained distances but then become trapped and remain inactive for considerable periods. This sporadic movement and deposition leads to a random array of roughness elements which sit in the exchange layer, reducing the near bed velocities and lessening the flow's potential to move grains.

The addition of coarser material therefore surpresses grain movement as a result of increased hiding and near bed flow retardation.

The similarity between the velocity profiles taken at a time of low transport is quite puzzling. The values for bed shear stress are all alike (Table 4.), which implies that feeding different sized material has no affect on the final hydraulic roughness. This does not necessarily mean that each bed surface would respond to a flood event in the same manner because differing bed mechanisms formed each final bed surface. The sediment depleted bed associated with the fine fed run could be more stable at higher flows than the artificially roughened modal fed bed.

Conclusions

Input fines integrate with the bed material and act as a catalyst for sediment movement. Coarser input surpresses movement through increased bed hiding and near-bed flow retardation. Fine fed pulses are attributed to longitudinal size sorting arising from aggrading-degrading bed sequences. Velocity plots show that the size of fed material alters near-bed velocities and bed shear stresses at times of high bed activity. When bedload transport is low the velocity profiles are very similar, suggesting that final bed hydraulic roughness is unaffected by feeding.

References

Hollingshead, A.B, 1971: *Sediment transport measurements in a gravel river.* J Hydrau. Div.ASCE 97:1817-1834

Laronne, J. B. and Carson, M.A., 1976: *Interrelationships between bed morphology and bed-material transport for a small, gravel-bed channel.* Sedimentology 23:67-85

Klingeman, P. C. and Emmett, W.W., 1982: *Gravel bedload transport processes. Gravel bed rivers,* John Wiley & Sons. Hey, R.D., Bathurst, J.C. and Thorne, C.R. (eds)

Brayshaw, A.C., 1983: *Bed microtopography and entrainment thresholds in gravel-bed rivers.* Geological Society of American Bulletin, **v96**, p.218-223

Einstein, H.A., 1950: *The bedload function for sediment transportation in open channel flows.* U.S. Department of Agriculture Technical Bulletin 1026

Ackers, P. and White, W.R., 1973: *Sediment transport: New approach and analysis.* American Society of Civil Engineers. v99, p2041-2060

Dietrich, W., Kirchner, J.W., Ikeda, H., Iseya, F., 1988: *Sediment supply and the development of coarse surface layer in gravel-bedded rivers.* Nature 340 215-216

Iseya, F. and Ikeda, H., 1987: *Pulsations in bedload transport rates induced by a longitudinal sediment sorting: A flume study using sand and gravel mixtures.* Geografiska Annaler, 69 p15-27

Parker, G., Dhamotharan, S., Stefan, H., 1982: *Model experiments on mobile paved gravel bed streams.* Water Resourses Res 18:1395-1408

Schiller, E. and Rowney, C., 1984: *Stream-bed armouring under known conditions of upstream sediment input.* Can. J. Earth Sci 21. 1061-1066

Tait, S.J., Willetts, B.B. and Maizels, J.K., 1992: *Laboratory observations of bed armouring and changes in bedload composition. Dynamics of gravel bed rivers,* John Wiley & Sons. Billi, Hey, Thorne and Tacconi (eds)

Parker, G., Klingeman, P.C. and McLean, D.G., 1982: *Bedload and size distribution in paved gravel-bed streams.* J Hydrau. Div. ASCE 108:544-571

Di Silvio, 1992: *Modelling sediment transport .Dynamics of gravel bed rivers,* John Wiley & Sons. Billi, Hey, Thorne and Tacconi (eds)

Andrews, E.D. and Parker, G., 1987: *Formation of coarse surface layer . Sediment transport in gravel bed rivers.* John Wiley & Sons. Hey, R.D., Bathurst, J.C. and Thorne, C.R. (eds)

Ikeda, H., 1984: *Flume experiments on the causes of superior mobility of sediment mixtures.* Ann. Rep., Inst. Geosci. Uni. Tsukuba, no. 10, p53-56

BREAKING WAVES-PLUNGING POINT AND ITS EFFECTS

Yalçın YÜKSEL
Department of Hydraulics
Civil Engineering Faculty, Yıldız Techncial University
Istanbul, Turkey

and

M. Sedat KAPDAŞLI
Department of Hydraulics
Civil Engineering Faculty, Istanbul Technical University,
Istanbul, Turkey

ABSTRACT

This paper deals with the following aspects of periodic water waves breaking on rigid and sandy beaches: a qualitative description of wave breaking, wave height variation, effects of plunging points and erosion of sandy beach in the vicinity of wave plunging point.

Experiments were carried out in two different wave flumes for eight different incident wave conditions on 1:10 beach slope.

1. Introduction

Knowledge of the instantaneous variations of surface elevations and internal velocity fields in breaking waves in shallow water regions is of importance for investigation of the problem of sediment transport modelling in the surfzone and that of wave forces on structures in the surfzone.

New computer-based methods have made it possible to analyze the pre-breaking behavior and the initial wave overturning up to the point where the falling jet strikes the onrushing trough. After the instant the jet hits the surface, the wave undergoes further transformation, primarily due to the role that turbulence now plays in the flow processes. However a general understanding of mechanisms and processes governing the mean flow field in water waves just after breaking remains one of the outstanding unsolved problems in fluid mechanics. When the plunging jet of breaker hits forward face, the flow field becomes too complex.

Many investigators tried to understand the mechanism of the waves in surf zone and notable progress has been made in recent years, regarding our understanding of the anatomy of breaking waves (e.g. Y.Yüksel, (1993); D.V. Basco, (1985); W.J. Easson et al., (1988); W.J. Easson and C.A. Greated, (1984); A. Seyama and A. Kimura, (1988); F. Raichlen and P.Papanicolaou, (1988); H.Wang et al., (1982); M.J.F.Stive (1980); E.D. Cokelet, (1979); etc.).

Basco (1985) tried to give a qualitative description of wave breaking. According to Basco, both classic spilling and plunging-type breakers are found to have similar initial breaking motions, but at vastly different scales. He identified two primary vortex which are surface roller and plunger vortex generate turbulence but to varying degrees, rates and scales.

Yüksel (1993) observed both spilling and plunging type breaker and he concluded that plunging breakers show strong mixing properties. During breaking a jet is generated which strikes upon the forward slope of the wave generating a series of eddies that contribute to the rapid decay of the wave. Each eddy moves forward and downward upto the bottom yielding a strong mixing process. However, spilling breakers have less effective mixing properties because of the eddies of relatively small scale, which are confined to the near-water surface region and extend much below the trough level but not reach the bottom. He also measured the breaking wave forces on submerged horizontal cylinders, experimental data show that the peak forces on horizontal cylinders resting on sea bed occur at the plunging point in the surf zone.

They are almost four times at plunging point than breaking point and plunging type breaking wave forces are much bigger than spilling type.

Thus, the effect of breaker shape and breaker vortices is indicated in three applications; (1) bedforms in the surf zone, (2) impact pressures due to breaking waves, (3) the interaction of post -breaking bore and foreshore. This study waves undertaken with the aim to contribute to an understanding of the breaking waves- the plunging jet and its effects on bed movement and submerged inshore structures.

2. Wave Breaking

Before discussing the effects of breaking waves, some features of breaking waves will be considered and then compared to some breaking wave characteristics which were obtained from present experimental data.

There are a number of criteria that are used to define the onset of breaking. In this paper, it has been assumed that wave breaking coincides with the occurence of the maximum wave height showing the initiation of a bubble with attendant foam formation. At the later stage of plunging, the shape of the breaking wave is completely lost with the crest falling onto the forward moving water. It has been followed the definitions of the two points due to Battjes (1974) and Bijker et al. (1974). Thus surf zone is clearly divided into two regions as the transition zone which is between breaking point and plunging point, and the bore establishment zone which is between plunging point and coastline.

A number of empirical formulas are available to determine the breaking wave height. The most widely used one of these empirical formulas is the one derived by Weggel (1972) using data from a large number of experiments.

$$(H/d)_b = \frac{b}{1+\dfrac{ad}{gT^2}} \tag{1}$$

where H and d are wave height and still water depth at the breaking point, T is period, g is accelaration of gravity and subscript b denotes breaking point and

$$a = 43.75\ (1-\exp\ (-19\ S))$$

$$b = \frac{1.56}{1+\exp(-19s)}$$

$$s = beach\ slope$$

After a wave has been broken as a spilling or plunging breaker, a transition occurs. For the spilling the surface roller grows, and the wave height decreases rapidly. For the plunging breaker the jet of water that plunges down pushes up a very turbulent mass of water which continues the wave - breaking process. In both cases the wave is transformed into a bore - like broken wave, and this inner part of the surf zone can be described as a series of periodic bores (Svendsen et al., (1978)).

While the shoaling process is characterized by a very small energy loss, the wave-breaking is associated with a very large loss of wave energy. The surf zone along the beach is where the wave energy flux from offshore is dissipated to turbulence and heat. Due to the strong energy dissipation, the wave height decreases towards the shore in the surfzone.

Nadaoka and Kondoh (1982) worked with LDV to measure the velocity field structure in the surfzone and they found that turbulent intensity remains small around the breaking point but it attains its maximum at $d/d_b=0.7$, i.e. around plunging point, where d is still water depth and d_b is depth of breaking point.

Nadaoka and et al., (1988) also described "obliquely descending eddies" which bring highly intermittent turbulence to the bottom with the large onshoreward momentum. The eddies hit the bottom and then lift up the sediment into suspension, yielding the spot-like sediment cloud in accordance with the 3-dimensional eddy structure.

3. Experiments

First set of the experiments was carried out in a 18.50 m long and 1.20 m wide wave tank in which periodic waves were produced by a paddle at one end. A smooth and impermeable beach was modelled by using plywood at the other end, which was laid on a slope of 1:10.

Six different test conditions were considered in the first set of experiments for rigid bed condition (Table I). For every test condition, typical recordings of wave profiles from deep water to shore line were made. Reflection coefficient was around 3% during the experiments. These recordings of surface elevation in the near-breaking and the breaking region shows that the non-breaking wave motion is a purely periodic quantity, and that the breaking wave motion is a fluctuating quantity with a quasi-periodic character.

Table I. Laboratory Test Conditions On Rigid Beach

Case	Incident Wave Height H_o(cm)	Incident Wave Length L_o(m)	Wave Period T(s)	Incident Wave Steepness H_o/L_o	Breaker Type
1	17.14	3.95	1.92	0.043	TR
2	17.14	4.81	2.32	0.035	PL
3	13.71	3.20	1.66	0.043	TR
4	13.71	3.98	1.92	0.035	PL
5	9.71	3.81	1.90	0.025	PL
6	9.71	5.00	2.32	0.020	PL

PL: Plunging breaker

TR: Transition between plunging and spilling breaker

As it is also known from the previous studies, the depth at breaking point d_b, the wave height at breaking point H_b, and the form of the breaking wave depend upon the deep water wave steepness H_0/L_0 and the beach slope. The results are compared to the previous results, e.g. Weggel (1972) and Méhauté and Koh (1967), and it is found that they fit well with the presented results (Figure 1).

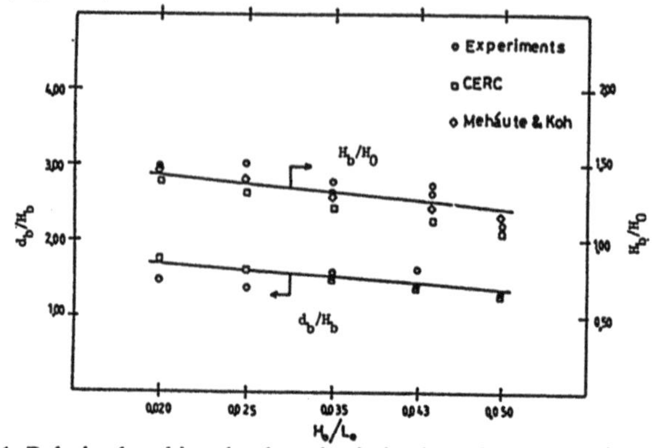

FIG.1. Relative breaking depth and relative breaking wave height

At the breaking point the wave profile (Figure 2) has only one wave crest but after the plunging point the broken wave has double hump since a very strong front surface roller occurs when the plunging jet hits the water and after a distance only one hump has been observed described as a bore wave.

Experimental measurements typically shows a concave curve in graphs of H versus d, indicating a relatively large dissipation of energy immediately after breaking (Figure 3). Althought wave height variation is not shown from this figure due to difficulties of measurement after plunging point to the shoreline it is observed that a progressively smaller rate of dissipation towards shorewards has been occured. Smith and Kraus (1987) also described this concave profile as a power law for the wave height decay. In Figure 3 wave height measured from the toe of beach to shore line is shown. This figure also shows that breaking wave height depends not only on incident wave steepness but also on incident wave height.

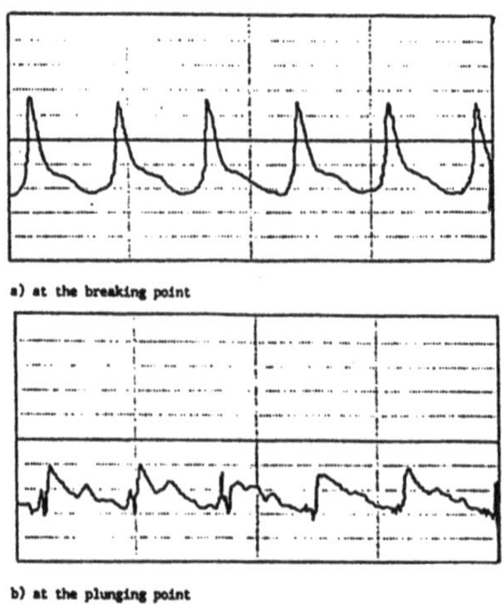

FIG. 2. Breaking wave profile (raw data)

After a short distance, roughly several times the depth at breaking, the wave adopts a steady, well organised profile which is more-or-less independent of the initial breaking behavior. Figure 4 shows that the ratio between wave height at plunging point (H_p) and depth of plunging incident point (d_p) is almost constant and seems to be independent of the incident wave steepness while breaking wave height depends upon the wave steepness. Where wave height at the plunging point was measured between the top point of the curling jet and the trough of the wave in front of it. However Figure 5 shows that the wave heights at plunging point increase with increasing of the incident wave height (H_o).

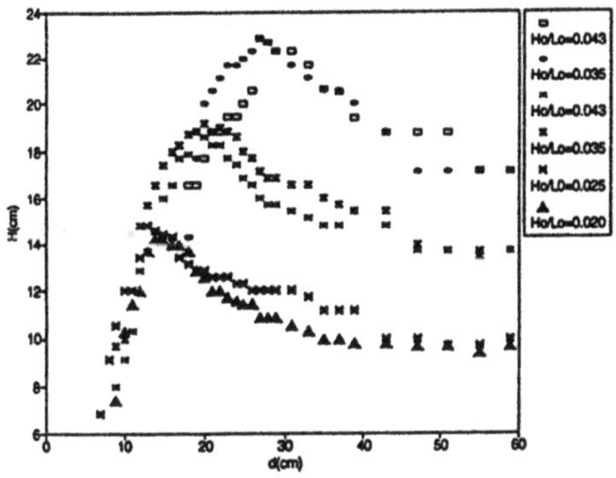

FIG. 3. Wave height variation along a beach (s=1/10) for different values of incident wave steepness (H_o/L_o)

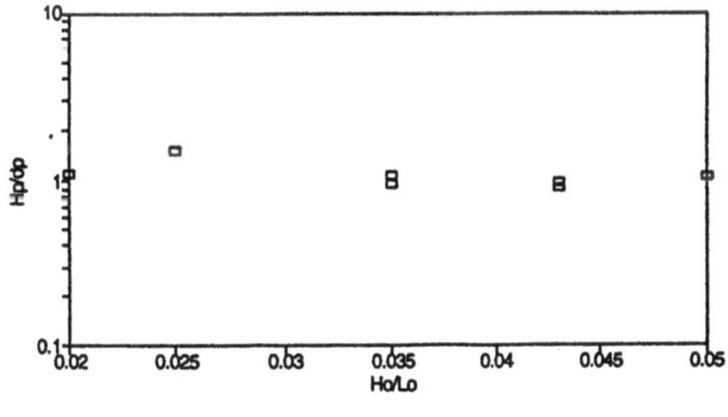

FIG. 4. Dimensionless plunging wave height ratio (H_p/d_p) versus incident wave steepness (H_o/L_o)

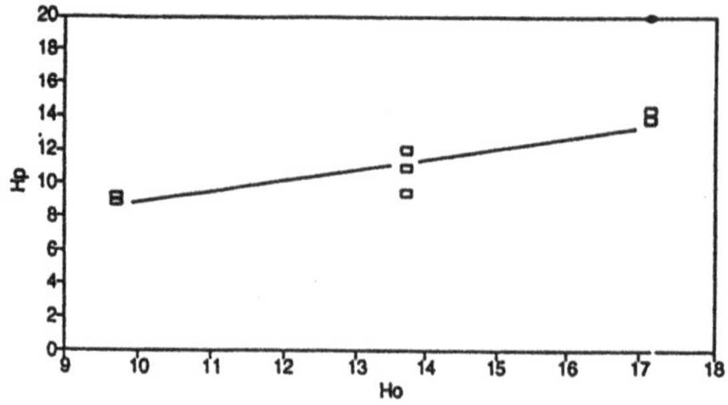

FIG. 5. Wave height at plunging point (H_p) versus incident wave height

During the experiments wave height at the plunging point was observed as being 40% smaller than the breaking wave height, (Figure 6). Basco (1985) also showed that the breaking wave has 30-40% smaller amplitude at the plunging point. This can be interpreted to mean that almost 60% of the wave energy is dissipated at the transition region in surf-zone (between breaking point and plunging point). Most of this wave energy is dissipated to turbulence in the transition region.

Second set of experiments was carried out in a flume with glass sidewalls which is 23 m long, 1 m wide and 0.5 m high with regular waves by using two different values of wave steepness for living bed conditions (Table II).

For a plunging breaker the crest of the wave moves forward and falls down at the trough in front of it as a single structured mass of water or a jet. The impact of the jet generates a splash-up of water which continues the breaking process and creates large coherent vertices which can reach the bottom and stir up considerable amounts of sediments. The flow caused by entrained air further spreads the sediment over the vertical, and clouds of suspended sediment are often observed near the plunging point (Fredsöe and Dergaard, (1992).

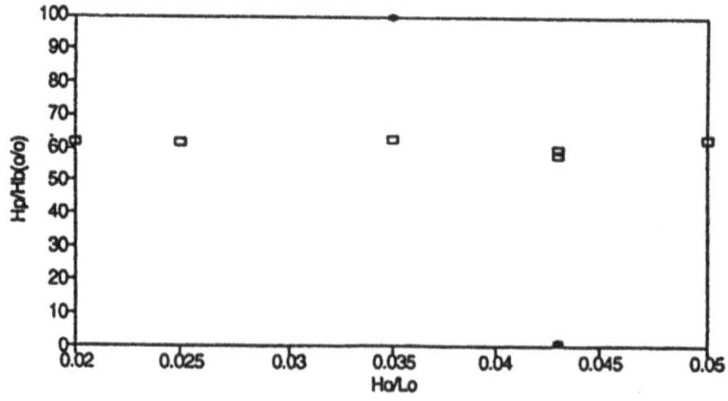

FIG. 6. Dimensionless wave amplitude ratio of plunging wave height (H_p) to breaking wave height H_b versus incident wave steepness (H_o/L_o).

Table II. Laboratory Test Conditions on Sandy Beach

Case	Incident Wave Height H_0(cm)	Incident Wave Length L_0(m)	Wave Period T(s)	Incident Wave Steepness H_0/L_0	Breaker Type
1	5.3	2.44	1.125	0.021	PL
2	4.7	1.35	0.93	0.035	PL

As a wave starts to break initially at a plane sloping bottom, its height decreases rapidly and after that a bore-type wave is generated which continues to break during propagation to the shore. The ratio of the wave height and water depth (after breaking) remains approximately constant, because there is a continuously decreasing water depth, when a sand bar is present and the larger waves break at the bar location. The breaking process tends to cease at some distance landward from the bar, because the water depth after the bar will first increase. As a result the waves tend to reform. The wave height to water depth ratio of the reformed waves is reduced to about $H/d=0.3$ to 0.5 (Hotta and Mizaguchi, (1980).

In the present study, second test condition has been carried out on both rigid rough and movable sand beaches which were laid on 1/10 slope. It was found that breaking wave height was not affected by bottom friction when compared with the results for smooth beds. The variation of wave heights on movable beach has been measured and it was found that after the beginning of the beach erosion, bottom slope changes very rapidly to become steeper. For this reason incident waves begion to break before reaching first breaking point i.e. offshore side of it. Moreover wave heights ware again increased just offshore the bar because of the reflection effects (see Figure 7).

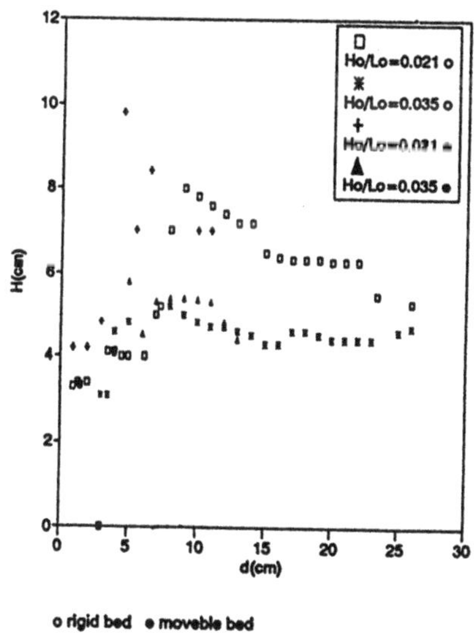

FIG. 7. Wave height variation on sandy beach

After a small number of waves attack, beach profile and bar formation reach a dynamically stable condition in which there is still very strong sediment movement but no change in beach profile. During the experiment dynamically stable condition has been reached in four hours (See Fig. 8). On this stable beach profile, incident waves break over the bar crest, that is the second breaking point, and there is deep scour inshore the bar because of the plunging of the waves. This phenomenon shows that, the estimation of the first and second breaking and plunging points is very important in order to design the artificial beach for protection of the shoreline.

From the observations, the first beginning point of erosion was found to occur at the plunging point. Shibayama yand Horikawa (1980) also reached the same conclusion. When sediments were brought into suspension the obliquely descending eddies which were defined by Nadaoka (1988) sweeped the suspended sediments in the offshore direction and bar formation occured near the plunging point.

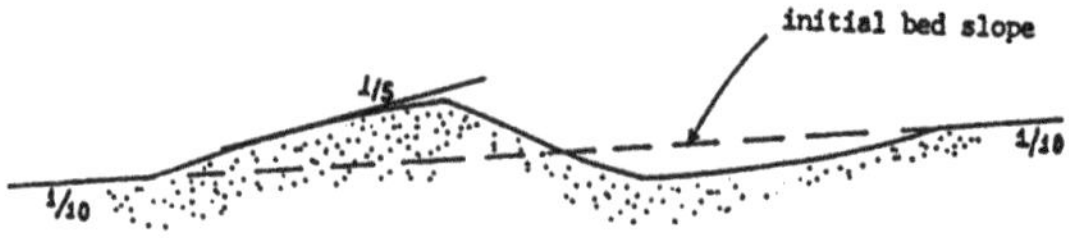

FIG. 8. Dynamic stable condition of bar formation

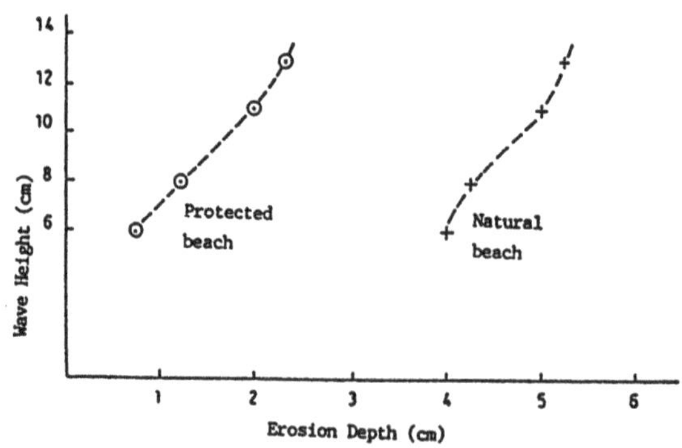

FIG. 9. Reduction of the erosion depth

This phenomenon is extremely important for dynamic stability of coastal zone. Kapdaşlı (1990) has shown experimentally that when the bar around the plunging point is protected by an armour layer of stones, erosion rate can be decreased up to 90%. Figure 9 shows that the erosion depths for natural beach and for protected beach are highly different. These results have clearly shown that most fragile part of the beach is the vicinity of the plunging point in which the sea bed is under the influence of quite strong dynamic fluid forces.

4. Conclusions

Laboratory experiments were carried out of plunging waves.

It has been concluded that;

1- The wave profiles show that at the breaking point wave form has only one wave crest but after the plunge point the broken wave has double hump due to very strong front surface roller formation.

2- Wave height (H) variation with local depth (d) along the beach shows a typical concave curve.

3- Breaking wave height depends not only on incident wave steepness but also on incident wave height.

4- Relative plunging wave height (H_p/dp) does not depend on incident wave steepness whereas breaking wave height depends upon the incident wave steepness.

5- Wave height at the plunging point is 40% smaller than that at the breaking point.

6- Sea bottom is more affected by plunging of waves due to plunging jet hitting onto the forward moving water and creating a coherent vortex formation rather than by the conditions at the breaking point. The first beginning point of erosion was found to occur at the plunging point.

7- Breaking wave height is not affected by sea bottom friction.

8- When beach profile and bar formation reach a dynamically stable condition after a time period there is stillvery strong sediment movement without changing the bar formation. However the first breaking point location at the plane bed will move to the second location which is the bar crest.

9- When the beach is protected by an armour layer at the plunging point, the beach erosion can be reduced.

5. References

1. Basco, D.V., A qualitative description of wave breaking, J. Water Ways, Harbours and Coastal Eng. Div., ASCE, Vol. III, pp. 171-188, (1985).
2. Battjes, J.A., Surf similarity, Proc. Int. Coastal Eng. Conf. 14th, Copenhagen, pp.466-480, (1974).
3. Bijker, E.W. Kalkwijk, J.P., and Pieters, T., Mass transport in Gravity waves on a sloping bottom, Proc. Int. Coastal Eng. Conf. 14 th, Copenhagen, pp. 447-465, (1974).
4. Cokelet, E.D., Breaking waves, Ed; Shav, T.L., Mechanics of wave- Induced forces on cylinders, (Pitman Advanced Pub. Program (1989)), pp. 287-301,(1979).
5. Easson, W.J., Griffiths, M.W.P. and Greated, C.A., Kinematics of breaking waves in coastal regions, Proc. Coast. Eng. pp. 871-883, (1988).
6. Easson, W.J. and Greated, C.A., Breaking wave forces and velocity fields, Coastal Eng. 8, pp. 233-241, (1984).
7. Fredsöe, J. and Deigaard, R., Mechanics of coastal sediment transport, (Advanced series on Ocean Eng. Vol. 3, World Scientific, Singapore, (1989)), p.369.
8. Hotta, S., and Mizugchi, M., A field study of waves in the surf zone, Coastal Eng. in Japan, Vol. 23, pp. 59-79, (1980).
9. Kapdaşlı, S., Low Cost Shore Protective Structures, (I.T.Ü. Water and Marine Science Research Center, Istanbul, (1990)), p. 130.
10. Le Mehaute, B. and Koh, R.C.Y., On the breaking waves arriving at an angle to the shore, J. of Hyd. Res. 5(1) pp. 67-88, (1967).
11. Nadaoka, K. and Kondoh, T., Laboratory measurements of velocity field structure in the surf zone by LDV, Coastal eng. in Japan, Vol. 25, pp. 125-145, (1982).

12.Nadaoka, K. Ueno, S. Igavaghe, T., Sediment suspension due to large scale eddies in the surf zone, Int. Coastal Eng. Conf. Malaga, Spain, pp. 1646-1660, (1988).
13.Raichlen, F. and Papanicolau, P., Some characteristics of breaking waves, Proc. Coast. Eng. pp. 387-392, (1988).
14.Seyama, A. and Kimura, A., The measured properties of irregular wave breaking and wave height change after breaking on the slope, Proc. Coast. Eng. pp. 419-432, (1988).
15.Shibayama, T. and Horikawa, K., Sediment suspension due to breaking waves, Coastal Eng. in Japan Vol. 25, pp. 163-176, (1980).
16.Smith, J. M. and Kraus N.C., Longshore current based on power low wave decay, ASCE Speciatlity Conf. on Coastal Hydrodynamics, Univ. of Delaware USA, (1987).
17.Stive, M.J.F., Velocity and pressure field of spilling breakers, Proc. 17 th CoastalEng. Conf. Sydney, pp 547-566, (1980).
18.Svendsen, I.A., Madsen, P. and Hansen, J.B., Wave characteristics in the surf zone, Proc., 16 th Coastal Eng. Conf., Hamburg, pp. 520-539, (1978).
19.Wang, H., Sunamuro, T. and Hwang, P.A., Drift velocity at the wave breaking point, Coastal Eng. 6, pp. 121-150, (1982).
20.Weggel, J.R., Maximum breaker height, J. Waterways, Harbours and Coastal Div. ASCE, Vol. 98, No WW4, pp. 529-548, (1972).
21.Yüksel, Y., Breaking wave forces on bottom seated horizontal cylinders in surf zone, (post doctoral reasearch report, UMIST, Manchester, UK, (1993)), p.70.

Ripple Dynamics

GRAIN SORTING OVER RIPPLES: PRELIMINARY RESULTS OF AN EXPERIMENTAL INVESTIGATION

ENRICO FOTI
Hydraulic Institute - University of Genoa
Via Montallegro 1, 16145 Genova, Italy

ABSTRACT

Sediment transport processes in the presence of mixtures show a tendency to select different grain sizes. This sorting process is usually enhanced by the presence of bedforms. In this paper, the sorting effect related to ripple formation under sea waves is experimentally investigated along with the critical conditions for ripple appearance in the presence of mixtures and ripple geometric characteristics. A comparison between the theoretical results by Foti & Blondeaux (1992) and the present experimental findings shows good qualitative agreement.

1. Introduction

Coastal sediments are usually characterized by a wide range of grain sizes. Yet, in the past, investigations into the development of bedforms and more generally morphodynamic studies were based on the assumption of uniform material. Only recently have both the language of statistics and physical considerations to describe the interaction of different sizes been introduced (see for instance Parker, 1991).

One of the main implications of considering a sandy bottom composed of a mixture is that selective transport may take place, which is enhanced by the presence of bedforms. For instance, selective transport acting on coastal ripples tends to accumulate the coarser fraction along the crests and the finer fraction in the troughs (Mei, 1989).

In an attempt to understand the mechanism underlying this grain-sorting process, Foti & Blondeaux (1992) carried out a linear stability analysis of a flat sandy bottom subject to an oscillatory flow taking into account the mixed character of the sediment. They found that the mechanism leading to ripple formation is similar to that characteristic of a uniform material. In other words, ripple growth is due to the steady streamings produced by the interaction of the oscillatory flow induced by sea waves with the bottom perturbation. These steady streamings usually tend to move the sediment from the troughs towards the crests of the perturbation, thus causing its growth. However a comparison of Foti & Blondeaux (1992) results with those described by Blondeaux (1990), who considered only a uniform material, shows that (i) the presence of a mixture has a stabilizing effect on the formation process, i.e. the amplitude of fluid oscillations which is necessary to trigger ripple formation is larger for a mixture than for a uniform material characterized by the same mean diameter; (ii) ripples in the presence of a mixture are longer than those found in the uniform case; (iii) in a mixture, a sorting process takes place which piles up large grain sizes at the crests leaving the fine fraction in the troughs.

The present paper deals with some preliminary results of an experimental investigation on ripple formation, when the bottom is composed of mixed size sediments, intended to check the theoretical predictions of Foti & Blondeaux (1992).

2. Experimental Method and Equipment

The experimental investigation was performed using a duct 290cm long with a rectangular cross section the width and height of which are equal to 40cm and 60cm respectively. The oscillatory flow inside the duct is provided by two rigidly connected pistons driven by a pneumatic motor. An electronic device allows both the amplitude and the period of the oscillations to be continuously varied. The duct ends are connected to two reservoirs, having a capacity of 0.55mc, which balance the volumes of fluid displaced by the pistons. At the bottom of the duct lies a fixed tray which can be filled with sediment.

The present investigation was carried out using four different mixtures obtained by mixing glass beads with a uniform diameter of 1.5mm with glass particles with a mean diameter of 0.65mm and a standard deviation σ_g equal to 1.4. The two materials were mixed as follows:

mixture (A): 80% of the coarse material and 20% of the fine material;
mixture (B): 70% of the coarse material and 30% of the fine material;
mixture (C): 60% of the coarse material and 40% of the fine material;
mixture (D): 50% of the coarse material and 50% of the fine material.

Moreover, some experiments were carried out using two almost uniform sands characterized by mean diameters equal to those of the mixtures (C) and (D), i.e. 0.99mm and 1.09mm respectively.

The experimental procedure was as follows. Starting from an initially flat bed, the pneumatic motor was switched on setting the desired values of the period T^* and amplitude a^* of the fluid oscillations. The time development of the bed configuration was then observed. By using different values of a^* and T^* it was possible to establish the critical conditions, i.e. the conditions discriminating between a stable flat bed configuration and a ripple configuration. The wavelength of the bedforms was then measured by means of a pointer, after about 20 cycles.

In order to measure the grain sorting effect produced by the presence of bedforms, the experiment was filmed. To improve the quality of the acquired images, the coarse and fine materials of the mixtures were painted by yellow and red respectively. Then the images were digitized, transferred to a workstation and analyzed using specially developed codes. The codes can reproduce each frame with only two colours corresponding to the two fractions present in the mixtures (i.e. the spurious colours are eliminated). The surface grain size distribution can then be obtained from the colour distribution. In principle the applied procedure allows the determination of local values of the size distribution to be obtained even though limits are posed by the grain size, the screen resolution and the size of the bottom area covered by the frame. The use of a sequence of images then allows a time series analysis to be performed.

3. Test Results

Most researchers, who have experimentally observed ripple formation in the presence of uniform sands (Sleath, 1984) consider the phenomenon as controlled by four dimensionless parameters:

- the Reynolds number of the flow in the boundary layer

$$R_\delta = \frac{\sqrt{2}U_o^*}{\sqrt{\nu\omega^*}} \tag{1}$$

- the sediment Reynolds number

$$R_d = \frac{U_o^* d^*}{\nu} \tag{2}$$

- the sediment Froude number

$$F_d = \frac{U_o^*}{\sqrt{(\frac{\varrho_s}{\varrho} - 1)gd^*}} \tag{3}$$

- the relative density of bottom material

$$s = \frac{\varrho_s}{\varrho} \tag{4}$$

In the above definitions, U_o^* denotes the amplitude of the velocity oscillations, ω^* the angular frequency equal to $2\pi/T^*$, ν the kinematic viscosity of the water, ρ and ρ_s the density of the water and sediment respectively, d^* the diameter of the sediment and g gravity.

Using mixtures, the language of statistics must be introduced to describe the sediment and the dimensionless parameters must be re-written in terms of the mean diameter d_m^* of the sediment. Moreover, because of the presence of mixtures, further dimensionless parameters appear. Natural sands are usually well described by a log-normal grain size distribution, hence the standard deviation σ_g is the only parameter needed to characterize it. In the present analysis, two size mixtures has been used; hence it is sufficient to specify the diameters of the two components along with their relative percentage.

A comparison between the critical conditions for uniform sands and for the mixtures characterized by the same mean diameter is given in table 1. The experiments show that a mixture is always more stable than a uniform sand. Indeed, table 1 shows that fixing the period of fluid oscillations, the critical amplitude of the mixtures is always larger than that for uniform sands.

bottom material	d_m^* (mm)	T^* (sec)	temp. (° C)	$(U_o^*/\omega^*)_{critical}$ (cm)
mix. (A)	0.8	3	18	15
mix. (A)	0.8	2	15	10
mix. (A)	0.8	4	17	20
mix. (B)	0.9	3	15.5	16
mix. (B)	0.9	4	15.5	23
mix. (B)	0.9	2	16	11
mix. (C)	0.99	2	14.0	12
mix. (C)	0.99	3	14.0	18
mix. (C)	0.99	4	14.0	24
mix. (D)	1.09	2	15.5	12.5
mix. (D)	1.09	3	15.0	18.5
mix. (D)	1.09	4	15.5	24.5
unif. sand	0.99	2	14.5	11
unif. sand	0.99	3	14.5	17
unif. sand	0.99	4	14.0	23
unif. sand	1.09	2	15.0	10
unif. sand	1.09	3	14.0	17
unif. sand	1.09	4	13.0	24

Table 1: Critical amplitudes of fluid oscillations for mixtures and uniform sands.

The results are summarized in figures 1a and 1b, which show the critical value of the sediment Froude number versus the ratio between the sediment Reynolds number and the Reynolds number of the flow. It is worth recalling that the sediment Froude number can be interpreted as the ratio between the destabilizing convective terms and the stabilizing gravity terms. Thus F_{dm} represents the key parameter which controls ripple formation (Blondeaux, 1990). It appears that the critical value of the sediment Froude number for mixtures is always larger than the critical value of the corresponding uniform sands.

The experiments just described also showed morphological differences of the bedforms. More precisely, it has been observed that by fixing the mean diameter and the characteristics of the fluid oscillations, ripple wavelength ℓ^* was larger for mixtures than for uniform sands. The results are reported in table 2.

The results on ripple geometry can be better inferred from figures 2a and 2b, where the ratio between the amplitude of the fluid oscillations and ripple wavelength, is plotted versus the dimensionless parameter

$$\frac{d_m^*}{(s-1)gT^{*2}} = \frac{F_{dm}^2 R_{dm}^2}{\pi^2 R_\delta^4} \tag{5}$$

which was proposed by Sleath (1984) as the key parameter controlling ripple geometry. It can be seen that, the conditions being equal (mean diameter, period, amplitude of the fluid oscillations), the values of a^*/ℓ^* for mixtures are always smaller than those

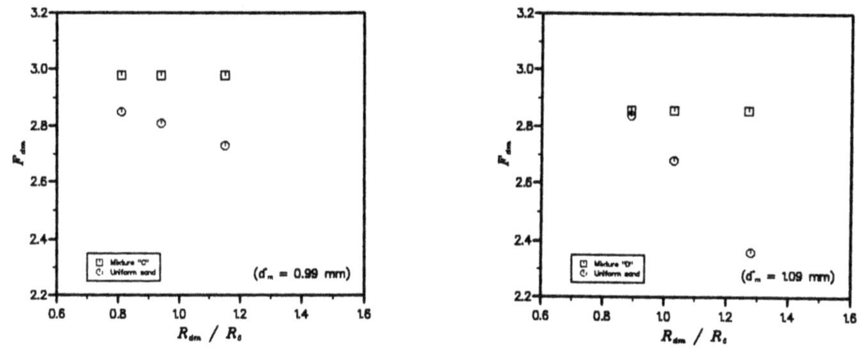

Figure 1: Critical sediment Froude number versus the ratio between R_d and R_δ. (a) Bottom material characterized by $d_m^* = 0.99$ mm (b) Bottom material characterized by $d_m^* = 1.09$ mm.

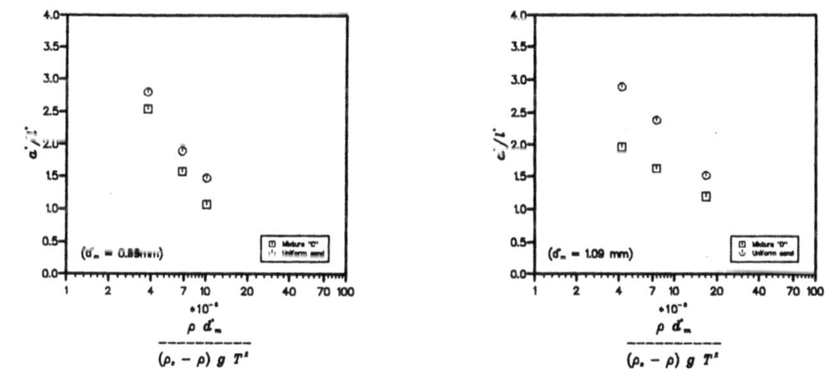

Figure 2: Ratio between the amplitude of fluid oscillations and ripple wavelength versus the dimensionless parameter $d_m^*/(s-1)gT^{*2}$. (a) Bottom material characterized by $d_m^* = 0.99$ mm (b) Bottom material characterized by $d_m^* = 1.09$ mm.

bottom material	d_m^* (mm)	T^* (sec)	Temp. (° C)	ℓ^* (cm)	U_o^*/ω^* (cm)
mix. (D)	1.09	2	11	11.7	14
unif. sand	1.09	2	14	9.2	14
mix. (D)	1.09	3	13	12.3	20
unif. sand	1.09	3	14	8.4	20
mix. (D)	1.09	4	13	13.2	26
unif. sand	1.09	4	14	9.0	26
mix. (C)	0.99	2	14.5	14(*)	15
unif. sand	0.99	2	13	10.2	15
mix. (C)	0.99	3	14.5	14(*)	22
unif. sand	0.99	3	13	11.7	22
mix. (C)	0.99	4	14	11(*)	28
unif. sand	0.99	4	14	10.0	28

(*) These wavelengths were obtained from the digitized images not from a direct measure.

Table 2: Comparison between the ripple wavelengths for mixtures and for uniform sands.

for uniform sands, thus implying larger wavelengths.

Before starting to analyze the sorting process some experiments were performed to obtain information on the critical conditions for each mixture (cfr. table 1). Once the critical conditions had been established, the experiments to determine the grain size distribution of the surface layer were performed in such a way that the actual sediment Froude number is about 1.15 times the critical value. The experimental conditions are given in table 3.

Photo 1 shows the top view of the bottom just before the start of experiment no. 4. The image analysis after the acquisition at the workstation and the colour filtering procedure gives an average density of the coarse and fine fraction equal to 31% and 69% respectively. The relative error of the measurements of the average density of the coarse and fine fractions in all the experiments turns out to be always smaller than 5% thus showing how well the technique measures grain size distribution.

Photo 2 is an image similar to that of photo 1 which shows the bottom configuration after about 20 cycles from the beginning of the experiment. The fluid oscillates from the top to the bottom of the picture and viceversa. The formation of ripples is clearly recognizable from the appearance of stripes, where the coarse fraction is more dense. These stripes oscillate in time around the crests of a waves pattern which is not evident from this top view, but is evident from a side view.

From a qualitative point of view, the present results confirm the grain sorting process detected during field observations with the coarse fraction observed at the crests of the ripples and the fine fraction in the troughs (Mei, 1989). However, it should be pointed out that large fluctuations of the grain size distribution are present during a cycle. Indeed, looking at figure 3, where the coarse sand concentration is plotted versus the progressive distances along a ripple at different times during a cycle, it can be seen that the concentration of the coarse fraction is characterized by

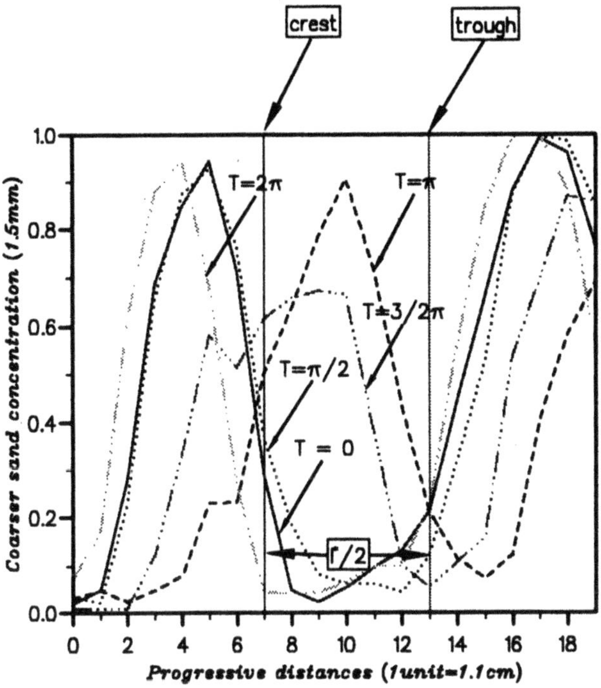

Figure 3: Coarser sand concentration along a bedform profile during one cycle. Experiment no. 4.

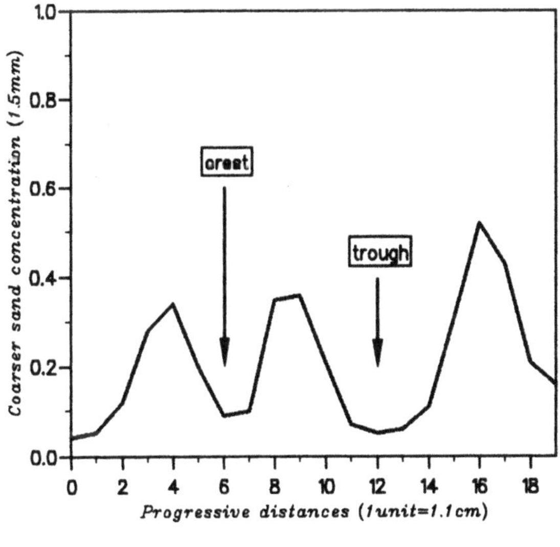

Figure 4: Averaged coarser sand concentration along a bedform profile during one cycle. Experiment no. 4.

exp. number	Bottom material	T^* (sec)	U_o^*/ω^* (cm)	Temp. °C
1	mix. (A)	3	18	18
2	mix. (A)	2	13	18
3	mix. (A)	4	25	18
4	mix. (B)	3	20	16
5	mix. (B)	4	25	16
6	mix. (B)	2	15	15.5
7	mix. (C)	2	15	14.5
8	mix. (C)	3	22	14.5
9	mix. (C)	4	28	14
10	mix. (D)	4	28	12
11	mix. (D)	2	15	12
12	mix. (D)	3	22	13

Table 3: Data concerning the experiences for which the sorting process has been quantified.

one peak for each ripple. This peak oscillates close to the crest, crossing it when the fluid velocity, is maximum and being almost at rest on the slope of the waviness. It is worth pointing out that the time t is chosen in such a way that the velocity far fron the bottom reverses its direction when t vanishes. This time behaviour is quite evident from figure 4 where the time average concentration is plotted. Indeed, two maxima at the side of the ripples can clearly be detected and the relative minimum at the crest is larger than that at the ripple trough.

This two-maxima behaviour becomes more evident by decreasing the percentage of the coarse fraction in the mixture (cfr. figures 5a, 5b, 5c).

4. Conclusions

In this paper an experimental investigation into ripple formation under an oscillatory flow in the presence of mixtures is presented.

The experiments show that the mixed character of sediments tends to stabilize the bottom thus delaying ripple appearance. Moreover, for fixed values of the amplitude and period of the fluid oscillations, the bedforms which appear in the presence of a mixture are characterized by longer wavelengths. By computer aided image analysis of the bottom time development, the grain sorting process which accumulates coarser sediments along the crests and the fine fraction in the troughs have been quantified.

Qualitative agreement of these experiments with the predictions by Foti & Blondeaux (1992) has been found. A quantitative comparison cannot be performed because the analysis carried out by Foti & Blondeaux (1992) assumes a laminar regime in the bottom boundary layer while in all the experiments turbulence was present. The extension of the Foti & Blondeaux (1992) analysis to cover the turbulent case is currently under investigation.

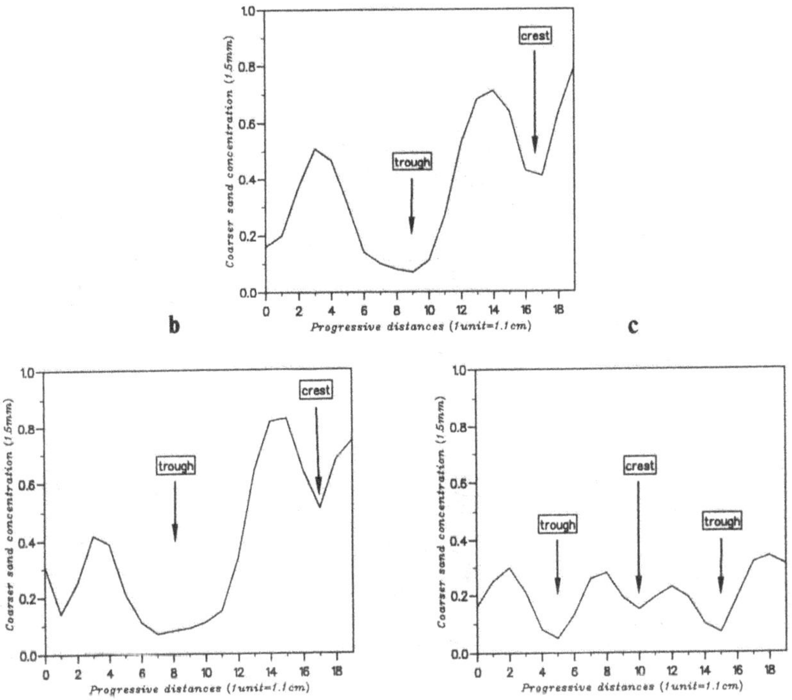

Figure 5: Averaged coarser sand concentration along a bedform profile during one cycle for different mixtures. (a) Experiment no. 12; (b) Experiment no. 8; (c) Experiment no. 1.

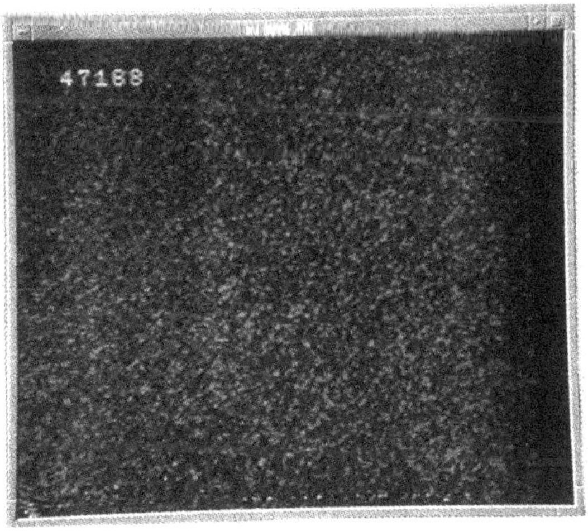

Photo 1. Top view of the flat sandy bottom before starting the experiment. Experiment no. 4.

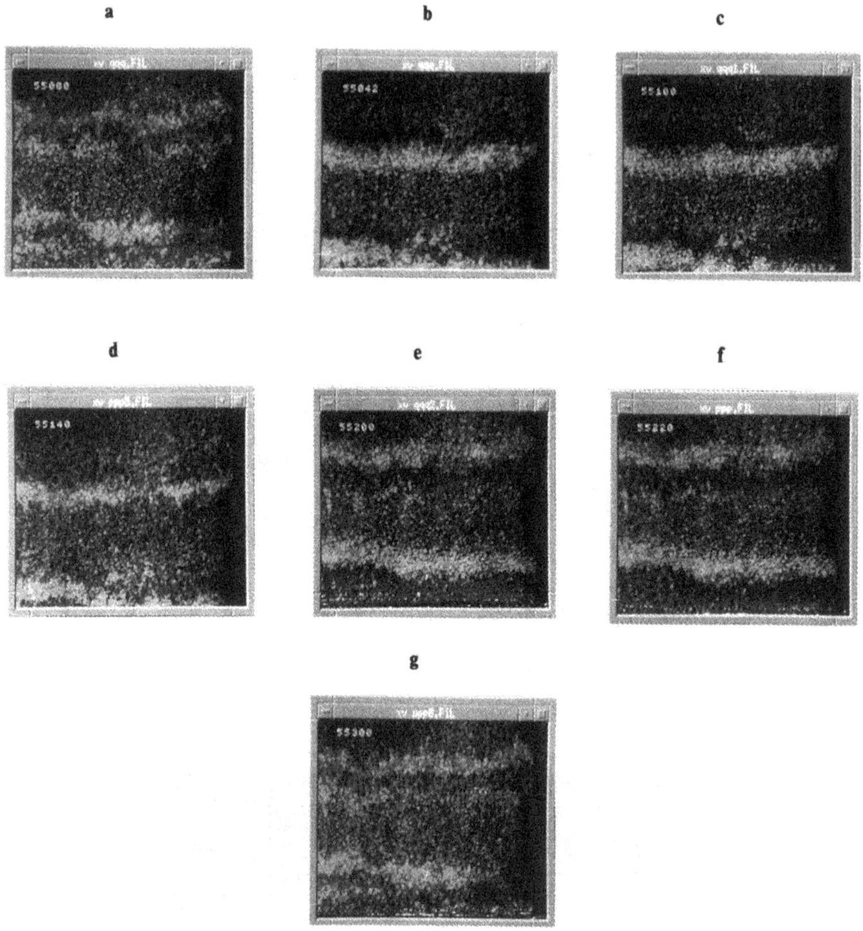

Photo 2. Top view of the sandy bottom (after 20 cycles) at different times during a cycle. Experiment no. 4. (a) $t/T = 0.0$; (b) $t/T = 0.14$; (c) $t/T = 0.33$; (d) $t/T = 0.47$; (e) $t/T = 0.68$; (f) $t/T = 0.73$; (g) $t/T = 1.0$.

5. Acknowledgements

Many thanks are due to P. Blondeaux for his suggestions on some points which arose during the development of the present investigation. The help of the technicians of the Hydraulic Laboratory of the University of Genoa and particularly of Mr. M. Galiani is also acknowledged. Financial support came from the "Ministero dell'Università e della Ricerca Scientifica e Tecnologica" and the Commission of the European Communities, Directorate General for Science, Research and Development under contract no. MAST2-CT 92-0027.

This work is part of the author's PhD Thesis to be submitted in partial fulfillment of his doctoral program.

References

[1] P. Blondeaux, *J. Fluid Mech.* **218** (1990) 1-17.

[2] E. Foti and P. Blondeaux, *Trans. XXIII International Conference on Coastal Engineering* **2** 2071-2084 (ASCE, Venice, 1992).

[3] C.C. Mei, *The applied dynamics of ocean surface waves* (World Scientific, 1989).

[4] G. Parker, *Grain Sorting Seminar* (Ascona, Switzerland, October 21-26, 1991).

[5] J.F.A. Sleath, *Sea-bed mechanics* (Wiley-Interscience Publications, 1984).

FORMATION OF SAND RIDGES: AN INSTABILITY MECHANISM

MARCO COLOMBINI
Hydraulic Institute, University of Genoa
Via Montallegro 1, 16145 Genova, Italy

ABSTRACT

The mechanism of formation of sand ridges is investigated. Such longitudinal stripes have been detected both in the laboratory and in rivers, they exhibit a spanwise periodic structure while being remarkably uniform in the streamwise direction.

A study of the stability of a uniform open-channel flow on an erodible bottom with respect to transverse bed disturbances is carried out. The analysis shows that the generation of sand ridges can be explained as an instability process.

Linear stability theory has been used successfully in the past to interpret the occurrence of a wide range of bedforms. Its ability to provide a simple and effective mechanism for the formation of sand ridges is confirmed here.

Although sand ridges can be obtained for the case of uniform sediment, variation in bed roughness also appears effective as a means of generation. Secondary flows can in fact be generated by alternate stripes of different roughness on the bed, with no need for variability of the bed elevation itself.

The development of a flow model capable of predicting correctly the linear response of the flow to small transverse perturbations either in elevation or in roughness represents a necessary step toward a complete linear stability analysis for the case of heterogeneous sediment. A second-order turbulence closure scheme is adopted here and is shown to yield a reasonable solution for the cellular secondary flows originated by both these mechanisms. The analysis thus fills the gap left open by previous works on the subject.

1. Introduction

Sand ridges are longitudinal bedforms periodic in the spanwise direction with typical wavelengths of the order of the flow depth. They are remarkably uniform in the streamwise direction, and this configuration resembles that of long ribbons placed on the bottom of the channel, so that they are often called "sand ribbons", too. Several field and laboratory observations suggest that the appearance of sand ridges is associated with the formation of stable pairs of counterrotating longitudinal vortices (see for example Culberston 1967).

Previous works on this subject suggest that the formation of sand ridges is related either to the effect of sidewalls or to the effect of non-uniformity in bed roughness associated with sediment sorting. Neither of these interpretations is of itself fully satisfactory.

Over a fixed bed the intensity of vortices induced by sidewalls decreases rapidly away from the sidewall (Nezu et al. 1985). Ridges cannot thus be assumed to be a passive response driven by the secondary flow induced by the sidewall. The mechanism proposed by Nezu et al. (1985) involves both the corner-induced vortices and an erodible bed. They postulated that sand ridges are generated with a cascade process that starts close to the sidewall, where the corner-induced vortex is responsible for the formation of a first ridge. When this ridge is sufficiently developed, it will induce a second vortex that, in turn, will generate a new ridge. The process is repeated until the whole section is covered with ridges. However, experimental results are not always in agreement with this hypothesis (Nezu & Nakagawa 1989).

Müller & Studerus (1979) and McLean (1981) showed that a non-uniformity of bed roughness of itself induces secondary currents. Since grain sorting is known to result in the accumulation of fine sand over the ridges, and since the near-bed secondary flow induced by changes in bed roughness is directed from the rough part towards the smooth one, it is reasonable to assume that a stronger destabilizing effect can occur if the hypothesis of uniform sediment is removed. Nevertheless, sand ridges have been obtained in the laboratory using uniform sediment (Ikeda 1981). Sediment heterogeneity does not seem to be necessary for the mechanism of formation of sand ridges, although it can be sufficient.

In the present contribution it is shown that sand ridges can be generated through an instability mechanism, without necessarily involving the presence of sidewalls or heterogeneity in the grain size of the bed sediment. The turbulence model of Speziale (1987) is used to perform a linear stability analysis of turbulent flow in an infinitely wide channel with an erodible bottom. Since part of the theory has already been published in a more complete form (Colombini 1993), most of the details are omitted here, in an attempt to concentrate more on the results of the theory and on its current (and future) developments. Results are also presented for the case of a fixed flat bed with a spanwise distribution of roughness.

2. Formulation of the problem

Let us consider a flow configuration like the one sketched in figure 1. We are going to assume that the undulations of the bed are of small amplitude, so that the problem can be studied as a perturbation of a uniform unidirectional flow in an infinitely wide channel. Due to the different timescales of flow and sediment evolution, it is possible to assume also that the flow instantaneously adapts itself to bed variations, so that time-derivatives can be neglected in the flow solution. Making all the variables dimensionless using the fluid density, the friction velocity and the flow depth for the basic uniform flow, the momentum and continuity equations can

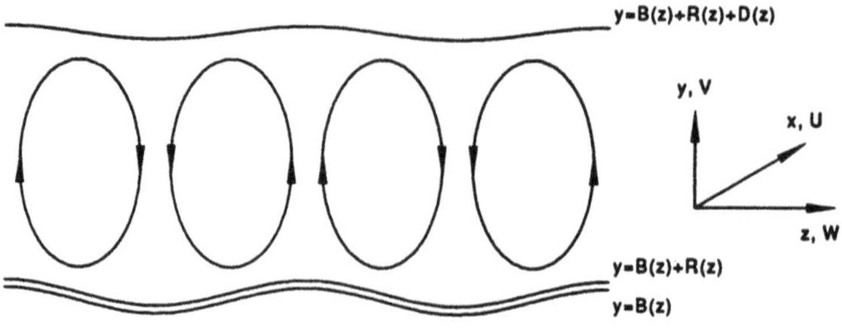

Figure 1: Sketch of flow configuration

be written as

$$V\frac{\partial U}{\partial y}+W\frac{\partial U}{\partial z}=1+\frac{\partial \tau_{xy}}{\partial y}+\frac{\partial \tau_{xz}}{\partial z} \tag{1}$$

$$V\frac{\partial V}{\partial y}+W\frac{\partial V}{\partial z}=-\frac{\partial P}{\partial y}+\frac{\partial \tau_{yy}}{\partial y}+\frac{\partial \tau_{yz}}{\partial z} \tag{2}$$

$$V\frac{\partial W}{\partial y}+W\frac{\partial W}{\partial z}=-\frac{\partial P}{\partial z}+\frac{\partial \tau_{yz}}{\partial y}+\frac{\partial \tau_{zz}}{\partial z} \tag{3}$$

$$\frac{\partial V}{\partial y}+\frac{\partial W}{\partial z}=0 \tag{4}$$

where τ_{ij} are the Reynolds stresses and P is the difference between the total and the hydrostatic pressure.

While at the free surface the dynamic and kinematic conditions hold, at the lower boundary the vanishing of the velocity vector is imposed at some small distance above the bed (namely R). The flow depth has been defined for convenience as the distance between the free surface and the reference level.

The continuity equation for sediment, assuming bedload only, reads

$$\frac{\partial B}{\partial t}=-\frac{\partial Q}{\partial z} \tag{5}$$

where Q is the dimensionless sediment flow rate in the spanwise direction and time has been scaled appropriately to follow the slow evolution of the bed topography.

The sediment discharge vector is assumed to be directed as the local direction of the bed shear stress, corrected for the effect of gravity as suggested by Ikeda (1982). The amplitude of this vector is specified through a load function, which relates the amplitude to the Shields parameter for the basic flow. The Meyer Peter–Müller load function has been used in the present calculations.

3. The turbulence closure

Any turbulence closure scheme that exhibits a linear dependence of the Reynolds stress tensor from the mean strain rate tensor, is unable to predict the appearance of the kind of cellular secondary vortices postulated here, which are often referred to as 'turbulence-driven' secondary flows. This concept is easily demonstrated. If a first-order turbulence closure is adopted, in the equation for the mean streamwise vorticity ω

$$V\frac{\partial\omega}{\partial y} + W\frac{\partial\omega}{\partial z} = \frac{\partial^2}{\partial y\partial z}(\tau_{zz} - \tau_{yy}) + \left(\frac{\partial^2}{\partial y^2} - \frac{\partial^2}{\partial z^2}\right)\tau_{yz} \tag{6}$$

all the terms are function of the secondary flow velocity V and W and no coupling exists with the streamwise momentum equation. The trivial solution of no secondary flow is therefore the solution of the problem of flow over a sinusoidal bottom, in strict analogy with the laminar case.

Several more refined turbulence closures can be invoked to overcome this problem. Solutions for the secondary flow driven by the sidewalls in a square duct have in fact been obtained with algebraic stress models (Naot & Rodi 1982; Demuren & Rodi 1984) and can in principle be obtained by a more general solution involving transport equations for each individual stress. Here, however, we seek a turbulence closure able to grasp the main features of the flow field without rendering the flow field solution too complex.

In this view the model of Speziale (1987) seems to be the appropriate choice, being explicit and relying on very simple and general constraints. Furthermore only two new constants need to be calibrated with experimental data, and the effect of boundaries can be introduced by means of assumed algebraic structure functions for the eddy viscosity and for the mixing length, both of which are well known for the case of uniform flow. These structure functions can then be slightly perturbed to study the present problem. The absence of separation due to the peculiar structure of sand ridges also encourages the assumption of local equilibrium implied by this choice.

Speziale derived a constitutive relationship as an higher-order approximation of the linear Boussinesq's closure. Assuming a generic functional dependence of the Reynolds stress tensor from the gradient of the velocity vector and imposing some general constraints (i.e. coordinate and dimensional invariance, positiveness of the turbulent kinetic energy and independence from rigid-body translation and rotation) a second-order turbulence closure is then readily derived as

$$\begin{aligned}\tau_{ij} = -\tfrac{2}{3}k\delta_{ij} + 2\nu_t D_{ij} \; &+ \; C_{\mathrm{D}}l^2(D_{im}D_{mj} - \tfrac{1}{3}D_{mn}D_{mn}\delta_{ij}) \\ &+ \; C_{\mathrm{E}}l^2(\widehat{D}_{ij} - \tfrac{1}{3}\widehat{D}_{mm}\delta_{ij})\end{aligned} \tag{7}$$

where $\widehat{D}_{ij}$ is the Oldroyd derivative of the mean rate of strain tensor D_{ij}, k is the turbulent kinetic energy, and ν_t and l are the eddy viscosity and the turbulence lengthscale respectively.

In a recent paper (Gatski & Speziale 1993) it has been demonstrated that a close relationship exists between second-order closures and algebraic stress models, so that the former can be thought as an explicit, generalised form of the latter. It is then not surprising that, once the linearization process has been completed, the structure of the Reynolds stresses has a close similarity to that obtained using an algebraic stress model (Naot & Rodi 1982).

As discussed in Colombini (1993) a value of 3.4 needs to be chosen for both the constants C_D and C_E in order to obtain a secondary flow of the correct intensity. This value is larger than the one originally proposed by Speziale but still of the same order of magnitude. A comparison with the constants of the algebraic stress model of Naot & Rodi, even if only possible in an approximated way, indicates that a somewhat higher value for these constants than the one originally proposed by Speziale should indeed be expected. The results presented herein have all been obtained with the above-mentioned choice for the values of the constants.

4. Flow field solution

In this section, following a procedure similar to that of Colombini (1993), a solution for the system (1-4) is sought using perturbation techniques. The zero-velocity reference level $y = B + R$ is assumed to be a periodic function of the lateral coordinate z, with amplitude small enough to allow linearization. Defining

$$E(t) = \exp(\sigma t) \qquad C(z) = \cos(\alpha z) \qquad S(z) = \sin(\alpha z)$$

where σ and α are the growth rate and the wavenumber of the disturbance respectively, a splitting of the kind

$$\begin{aligned}
(U, V, P) &= (U_0(y), 0, P_0(y)) + \epsilon(u(y), v(y), p(y))C(z)E(t) \\
W &= -\epsilon\alpha w(y)S(z)E(t) \\
(B, D) &= (0, 1) + \epsilon(b, d)C(z)E(t) \\
R &= R_0(1 + \epsilon r C(z)E(t)) \\
(\nu_t, l) &= (\nu_{t0}(y), l_0(y)) + \epsilon(\nu_{t1}(y), l_1(y))E(t)C(z) \\
(k_s, d_s) &= (k_{s0}, d_{so}) + \epsilon(k_{s1}, d_{s1})E(t)C(z)
\end{aligned}$$

can be substituted in the differential system (1-4). After linearization has been performed, collecting term at various order in the small parameter ϵ, leads to differential systems for the basic uniform flow and for the perturbations.

The non-orthogonal coordinate transformation

$$\eta = \frac{y - B(z) - R(z)}{D(z)}, \qquad \zeta = z,$$

which maps the cross-section on a rectangular domain, has been adopted. The distance above the bed where the boundary condition of zero velocity is applied is chosen to be proportional to the local roughness height, so that

$$R = \frac{k_s}{30} \tag{8}$$

where k_s is the relative roughness, again proportional to the non-dimensional grain size $d_s = d_s^*/D_0^*$. In the case of the logarithmic velocity profile for rough flow, this point corresponds to vanishing streamwise velocity.

As in De Vriend (1977) the mixing-length hypothesis is adopted so that

$$\nu_t = l^2 \frac{\partial U}{\partial y} \tag{9}$$

$$l = \kappa(y - B)\left(\frac{B + D + R - y}{D}\right)^{\frac{1}{2}}. \tag{10}$$

With these choices, the differential system for the basic flow is readily solved, recovering the usual logarithmic velocity distribution for a uniform flow.

At first order in ϵ a system of ordinary differential equations is eventually obtained that can be written in the general form

$$\mathcal{D}\mathbf{Z} = d\mathbf{D} + b\mathbf{B} + r\mathbf{R} \tag{11}$$

where d is considered as a parameter to be determined and

$$\mathbf{Z} = \begin{pmatrix} u \\ v \\ w \\ p \end{pmatrix}$$

The two terms involving b and r act as forcing terms. Two cases can then be identified; when $b = 1$ and $r = 0$, the response of the fluid to a bed perturbation is obtained, under the constraint of uniform bed roughness. On the other hand if $b = 0$ and $r = 1$ the case of flow over a spanwise distribution of roughness is recovered, under the constraint of a flat bed. While in the first case ϵ represents the amplitude of the bed disturbance, in the second one it becomes the amplitude of the roughness perturbation.

Let us first analyse some results for the case of uniform roughness (case A). The tangential component of the bed shear stress is expanded as

$$\tau_t = -\epsilon\alpha\tau_{t1}E(t)S(z)$$

In figure 2a the quantity τ_{t1} is plotted versus the wavenumber α, with the non-dimensional grain size d_{s0} as a parameter. This quantity, as it will be clear from the next section, plays a crucial role in controlling bed instability. One need only note here that a positive τ_{t1} is associated with a near-bed secondary flow directed from the trough towards the crest of a ridge. A higher τ_{t1} corresponds to a more intense vortex. The secondary flow generated by a bed perturbation is then directed in such a way as to amplify the bed perturbation itself for a wide range of wavenumbers. The maximum response is attained for a wavenumber around 4.8. The intensity of the vortex increases as the non-dimensional roughness of the bed increases; its behavior is almost unaffected by changes in the Froude number. The effect of the

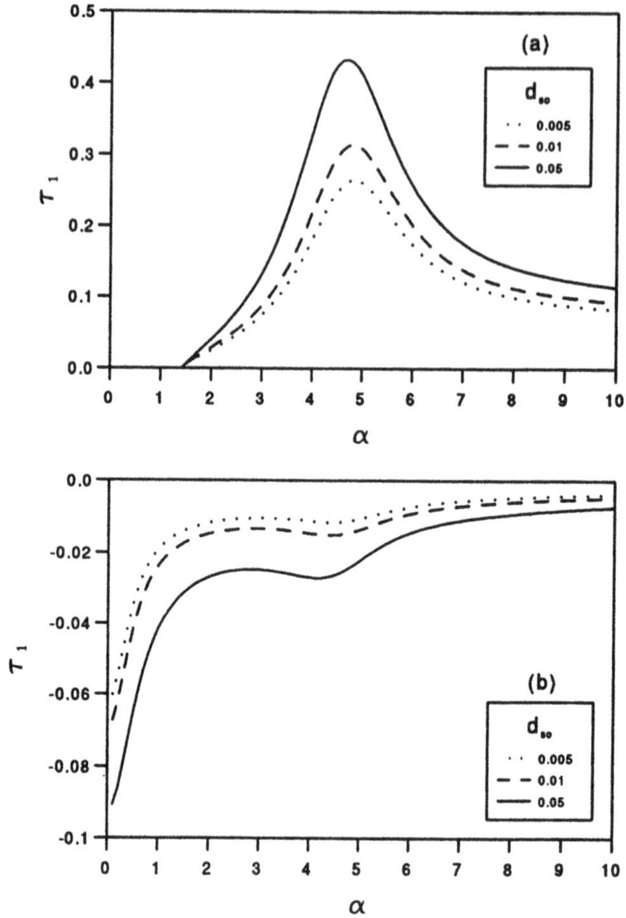

Figure 2: Tangential bed shear stress component τ_{t1} as a function of the wavenumber α for different values of the non-dimensional grain size d_{so}; (a) flow response to a perturbation of bed elevation; (b) flow response to a perturbation of bed roughness.

Froude number on the flow field is related only to the amplitude of the free-surface undulations associated to the bed perturbation. It is a very minor effect indeed, such that all the quantities are affected only on the third or fourth significant digit by changing the Froude number from zero to values of two or three.

As for the case of periodic roughness variation over a flat bed, the plot in figure 2b shows that the flow rotates in the direction opposite to the previous one. For all wavenumbers considered the near-bed secondary flow moves away from the coarser strip and toward the finer strip. This result is supported by experimental observations. It shows that no analogy can be cast in a facile way between the flow resulting from a variation in the elevation of the bed and the one obtained by similarly varying the roughness height.

5. Formation of sand ridges

We now proceed to expand the lateral sediment discharge as

$$Q = -\epsilon\alpha\Phi(\theta_0)\left(\tau_{t1} - \mu\theta_0^{-\frac{1}{2}}\right) E(t)S(z) \tag{12}$$

where θ_0 is the Shields parameter for the basic uniform flow, $\Phi(\theta_0)$ is the load function and μ is a constant related to the effect of gravity and taken here as 0.3. Linearization of (5) yields the following dispersion relationship for the growth rate of bed perturbations

$$\frac{\sigma}{\Phi(\theta_0)\alpha^2} = \tau_{t1} - \mu\theta_0^{-\frac{1}{2}} \tag{13}$$

The above equation clearly shows that the sign of the growth rate (and thus the amplification or decay of bed perturbations) is the result of a balance between a 'destabilizing' term, related to the bed shear stress component in the direction tangential to the bed, and the stabilizing effect of gravity.

The effect of gravity is to shift vertically the entire plot of figure 2a by the amount $-\mu\theta_0^{-\frac{1}{2}}$, so that a critical (in the sense of the instability) value for the Shield stress θ_{0c} can be identified when the action of gravity exactly balances that of the shear stress. Thus

$$\theta_{0c} = \left(\frac{\mu}{\tau_{t1}}\right)^2 \tag{14}$$

Results of the present theory show that sand ridges should appear only when the Shields stress for the basic flow θ_0 exceeds this value. This threshold is however unrealistically high, with characteristic values of order 1. Ridges can be easily identified on the bed when grains are just above the threshold for motion, and are often destroyed by other bedforms when the bed shear stress is increased. A possible explanation can be found in an underestimation of the tangential shear stress at the bed, or in the fact that, due to the very small amplitude of these bedform, the effect of gravity is overestimated by (13). If it were completely ignored, θ_{0c} would indeed devolve to the critical Shield stress for sediment motion.

We want to point out that in Colombini (1993) the same existence criterion was cast as a Froude criterion. Since the Froude number and the Shields stress for the basic flow are interrelated, this choice was formally correct. However it could be misleading since the role played by the Froude number in the analysis, as already stated, is a very minor one. A more feasible Shields criterion has then been presented here.

In figure 3 the behavior of the growth rate σ for a value of the relative roughness d_{s0} of 0.01 is presented as a function of the wavenumber α, with the Shields stress for the basic flow θ_0 as a parameter.

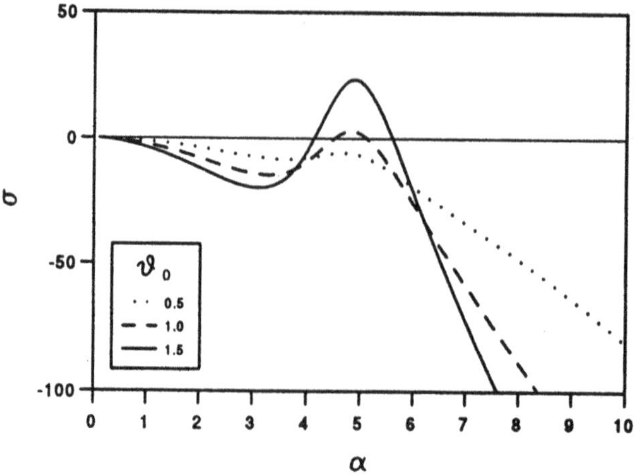

Figure 3: Growth rate of bed perturbations σ as a function of the wavenumber α for $d_{so} = 0.01$ and for different values of the the Shield stress θ_0.

It can be seen that for $\theta_0 = 0.5$ the unidirectional uniform flow is still stable. When the Shields stress is increased, unstable disturbances can develop leading to a spanwise periodic three-dimensional flow. Disturbances with a wavenumber of about 4.8 exhibit the maximum growth rate. This value describes perturbations with a wavelength 1.3 times the flow depth , a configuration that matches very well the one obtained in the experiment of Nishiya & Makino as reported by Hayashi et al. (1985). This experiment was characterized by a very large aspect ratio, a condition that matches reasonably the hypothesis formulated in the present theory. If the aspect ratio is not sufficiently large, the wavelength can be strongly influenced by the presence of the sidewalls.

6. Conclusions

Flow in an infinitely wide channel with an erodible bottom is studied, showing that, under suitable conditions, the uniform flow is unstable with respect to spanwise periodic perturbations. A mechanism of formation for sand ridges is then identified, with no need for the presence of the sidewalls as stated by previous works on the subject (Nezu & Nakagawa 1989).

A critical value of the basic Shields stress θ_{0c} that bounds the unstable region is determined, and information on the wavelength of the most unstable disturbance are found. Predictions of the characteristic wavenumber agree with the experiments if the aspect ratio is sufficiently large, while the threshold for instability appears to be overestimated.

If a spanwise variation of the bed roughness is considered, a secondary flow is generated with a sense of rotation that agrees with experimental observations. Since this behavior might be somewhat unexpected in view of the approximate way in which the bed roughness enters the problem, this result shows that the mechanics of this kind of secondary flow is fairly well grasped by the simplified theory presented herein. A natural future development of this work is the study of the formation of sand ridges with an heterogeneous sediment. This would hopefully lead to a deeper understanding of the entire phenomenon of longitudinal streaks.

References

1. M. Colombini, *J. Fluid Mech.* **254** (1993) 701.
2. J. K. Culberston, *Geological Survey Research* (US Geological Survey, Prof. Paper 575, 1967) D214.
3. A. O. Demuren and W. Rodi, *J. Fluid Mech.* **140** (1984) 189.
4. H. J. De Vriend, *J. Hydraul. Res.* **15** (1977) 37.
5. T. B. Gatsky and C. G. Speziale, *J. Fluid Mech.* **254** (1993) 59.
6. T. Hayashi, M. Ohashi and Y. Kotani, in *Recent Studies on turbulent phenomena*, ed. T. Tatsumi, H. Maruo & H. Takami (Tokio Institute of Technology, Tokio 1985) 243.
7. S. Ikeda, *J. Hydraul. Div. ASCE* **107** (1981) 389.
8. S. Ikeda, *J. Hydraul. Engng ASCE* **108** (1982) 1369.
9. S. R. McLean, *Mar. Geol.* **42** (1981) 49.
10. A. Müller and X. Studerus, *Proc. 18th IAHR Congress, Cagliari, Italy* (1979) B19.
11. D. Naot and W. Rodi, *J. Hydraul. Div. ASCE* **108** (1982) 948.
12. I. Nezu and H. Nakagawa, *J. Hydraul. Engng ASCE* **110** (1984) 173.
13. I. Nezu and H. Nakagawa, *Proc. 23rd IAHR Congress, Ottawa, Canada* (1984) B65.
14. I. Nezu, H. Nakagawa and A. Tominaga, in *Turbulent Shear Flows IV* (Springer, 1985) 246.
15. C. G. Speziale, *J. Fluid Mech.* **178** (1987) 459.

ON THE INITIATION OF BED DEFORMATION OF A MEANDERING STREAM

M. S. Yalin
Department of Civil Engineering, Queen's University
Kingston, Ontario, Canada K7L 3N6

and

A. M. Ferreira da Silva
Department of Civil Engineering, Queen's University
Kingston, Ontario, Canada K7L 3N6

ABSTRACT

A method is presented for the computation of flow in an initial meandering channel: flow rate is constant, the bed of the rectangular cross-section is horizontal. The method rests on the radial equation of motion and the equation of continuity. The consideration of the usual longitudinal equation, which makes it impossible to achieve for large sinuosities an outgoing flow between the crossover and the apex, is replaced by that of a "shift" between the centrifugal force and the transverse pressure force. The shift progressively vanishes with the decrement of sinuosity, and for any given sinuosity it tends towards zero when the banks are approached.

1. Introduction

The current approach to the determination of bed topography of meandering streams is mainly due to H. J. De Vriend [1], [6], N. Struiksma [16], J. M. Nelson, J. D. Smith and S. R. McLean [12], [13], [15]. According to this approach, the prediction of bed deformation rests on a two-dimensional, vertically-averaged flow, given by the three field-functions $\mathbf{u}$, υ_* and h. Knowing the vector-field υ_*, one determines, with the aid of a bed-load relation, the corresponding $\mathbf{q}_s$-field which is then used (in the transport continuity equation) to calculate the associated bed variations. Starting from the flow past an initially flat bed (at $t = 0$), one computes from the transport continuity equation the bed level increment $(\delta y_b)_1$ corresponding to the time step δt_1. Knowing $(\delta y_b)_1$, one recomputes the field-functions (so as to correspond to $(\delta y_b)_1$): the procedure is repeated for δt_2, δt_3, ..., etc., until an equilibrium surface representing $\nabla \mathbf{q}_s = 0$ is achieved. The results obtained by the above mentioned authors are remarkably realistic - the measured and computed bed topographies compare very favourably (see e.g. Figs. 9 to 12 and Fig. 19 in Ref. [13]).

The present paper is essentially an extension of the aforementioned works; however, the main concern is not so much the equilibrium bed topography, but the nature and formulation of the initial meandering flow - which, eventually, generates the bed topography. Following the previous works, it is assumed that the curved channel flow is rough turbulent, the flow rate Q is constant, the transport is by bed-load, the channel centreline is a sine-generated curve [8], [9], the parallel banks are (virtually) rigid. Most of the large natural rivers have a rather large width-to-depth ratio (often a 3-digit number); considering this, the B/h-ratio is treated as "large".

2. Channel Geometry

It has been mentioned in the Introduction that the channel centreline is a sine-generated curve (see Fig. 1). Hence, the variation of θ along l_c is given by

$$\theta = \theta_0 \cos\left(2\pi \frac{l_c}{L_m}\right) ,$$

the sinuosity $\sigma = L_m / \Lambda_m$ being determined (see Ref. [17]) by the (initial) deflection angle θ_0 as

$$\sigma = \frac{1}{J_0}(\theta_0) .$$

Since σ is a function of θ_0 alone, in the following the terms "small" and "large" sinuosities are to be interpreted as "small" and "large" θ_0, respectively.

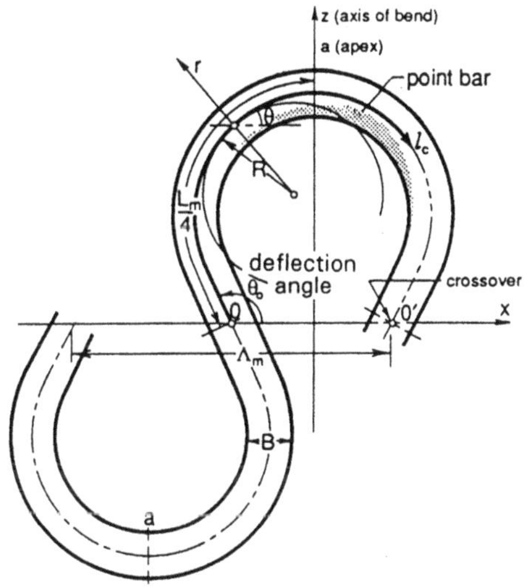

Fig. 1 (from Ref. [17])

3. Erosion-Deposition Patterns

Experiment shows (see Fig. 2) that if θ_0 is "small", the erosion-deposition zones are as illustrated in (the schematic) Fig. 3a (i.e. on the flanks of the apex). Hence, if θ_0 is "small", the deepest erosion occurs at the inner bank of a section that is between O and a. If, however, θ_0 is "large", then the erosion-deposition zones are located at (or around) the apex itself as shown in Fig. 3b: the deepest erosion occurs at the outer bank of a section that is somewhat downstream of the apex.

Clearly, the banks must be eroded mostly in those locations where the bed adjacent to them is eroded; and similar reasoning applies to deposition. Hence, the erosion-deposition pattern in Fig. 3a must lead to the migration of the channel; the

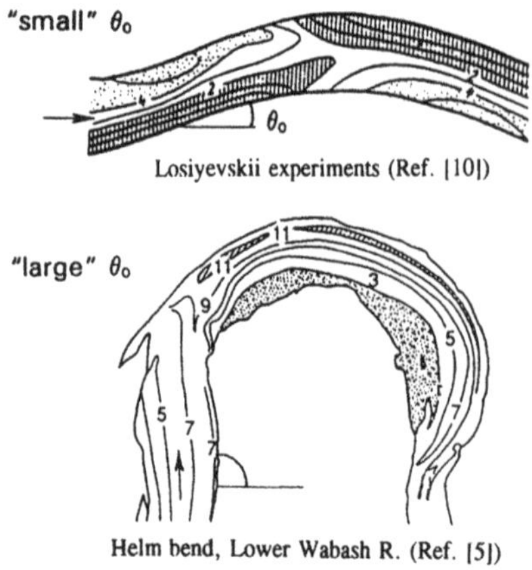

Fig. 2

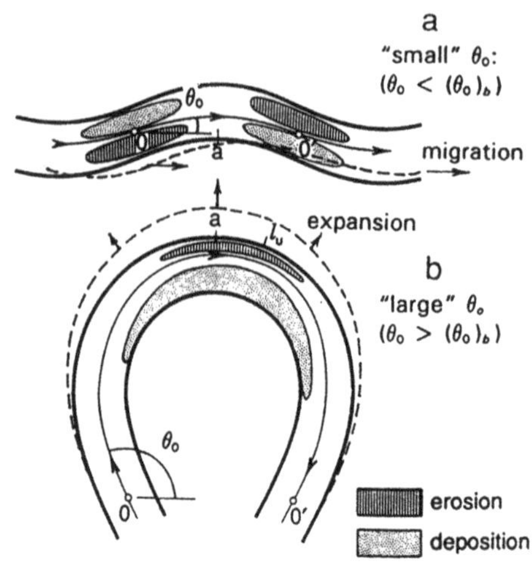

Fig. 3

pattern in Fig. 3b, to its expansion. This is in agreement with experiment. Indeed, Fig. 4 (which is produced on the basis of experimental large river data presented in Ref. [7]) indicates that for "small" θ_0 (or σ) the channel is mainly migrating, whereas for "large" θ_0 it is mainly expanding. According to this Figure, the boundary $(\theta_0)_b$ between "large" and "small" θ_0 should be between 60° and 75°. Strictly speaking, the value of $(\theta_0)_b$ must vary depending on the relative roughness (or, which is the same, depending on the friction factor c) and the width-to-depth ratio B/h. [The insert in Fig. 4 merely indicates that, theoretically, the loops of a regular meander cannot grow beyond $\theta_0 = 126°$, i.e. beyond $\sigma = 8.5$.]

4. Flow Patterns

Although to the knowledge of the authors, no special experiments involving the systematic variation of the deflection angle θ_0 were conducted, one can nonetheless assert that for "small" θ_0 the flow adjacent to the convex (inner) bank should converge along $a'a$ (see Fig. 5a), the largest angle of convergence ω_{max} being in the neighbourhood of O. But this means that the largest positive convergence of the transport rate $\nabla \mathbf{q}_s$ and, consequently, the largest negative $\partial y_b/\partial t$ should also occur in the neighbourhood of O (see equations inserted in Fig. 5). Hence, the initial flow corresponding to "small" θ_0 should initiate the erosion and deposition at the inner bank at O and O' respectively, as indicated in Fig. 3a.

If, however, θ_0 is sufficiently large, then the (anticlockwise bending (in plan)) flow entering the clockwise loop at O may not be able to follow the curvature of the inner bank, and it will diverge from it (see Fig. 5b). The largest divergence must be expected to be in the neighbourhood of the apex a where the bank curvature R_a^{-1} is the largest. Clearly, in this case the largest negative $\nabla \mathbf{q}_s$ and, consequently, the largest positive $\partial y_b/\partial t$ should also occur around a. Accordingly, the initial flow corresponding to large θ_0 must lead to the erosion-deposition zones that are located as shown in Fig. 3b.

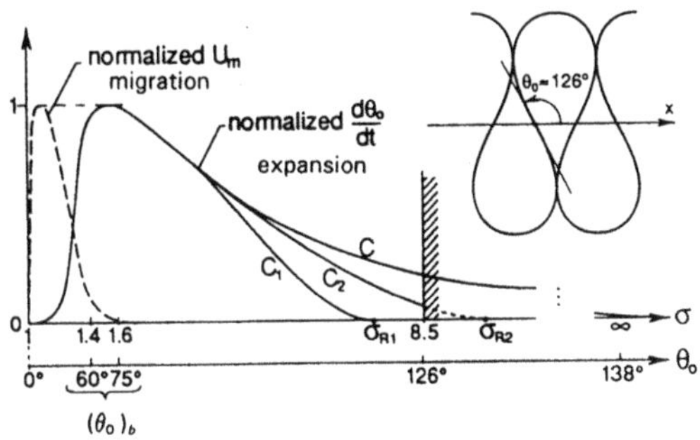

Fig. 4 (after Ref. [7])

Yet, in all of the previous computations the maximum velocity of the flow past the flat initial bed occurs at the apex of the inner bank, no matter how small or large the sinuosity, i.e. θ_0, might be (see e.g. the velocity diagrams in Fig. 6). This presents considerable difficulties in explaining the initiation of deposition at the apex of the inner bank on the basis of the initial flow alone. Consequently, in all of the previous works one had no alternative but to invoke other (than the initial flow) concepts to explain the deposition at a and, among them, the cross-circulation was mostly used.

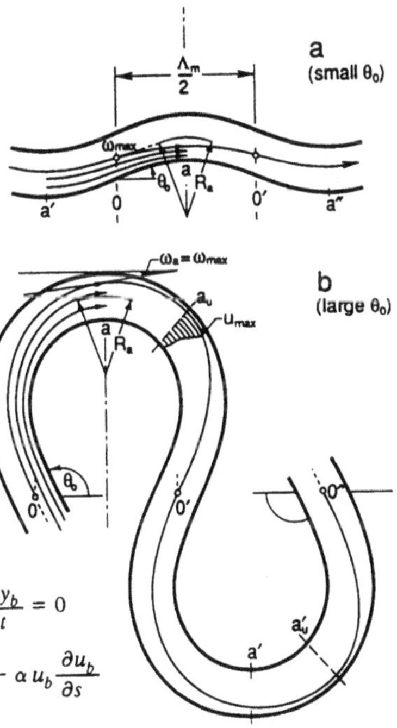

$$\nabla \mathbf{q}_s + \frac{\partial y_b}{\partial t} = 0$$

$$\frac{\partial y_b}{\partial t} = -\alpha u_b \frac{\partial u_b}{\partial s}$$

Fig. 5

It should be noted here that, as a rule, it is the large rivers which possess the large sinuosities, i.e. the large deflection angles θ_0. And it is also the large rivers which have the 3-digit values of their aspect-ratio B/h and, consequently, hardly any cross-circulation (Fig. 7). This fact has been repeatedly pointed out by the researchers dealing with large natural streams. For example, according to Matthes [11] "transverse circulation only takes place in distorted experimental models, and in such natural channels whose width is small as compared with the depth ... sediment entering the stream from the scour of the concave bank becomes deposited

Fig. 6 (from Ref. [15])

on the downstream convex bank on the same side of the river, and only a small portion of the eroded material crosses the channel" (see also e.g. [3], [4], [7]).

In summary, the present approach to the determination of bed topography basically is in line with the previous works - the authors tend to disagree only with regard to the mechanism of initiation of bed deformation. They think that the convective behaviour of the vertically-averaged flow *alone* should be sufficient to induce the erosion-deposition zones in the required locations - without the need to invoke any further concepts. Moreover, the inconsistencies between the expected and computed zones of maximum velocity suggest that the vertically-averaged equations of motion, as they stand, do not reflect adequately the behaviour of the vertically-averaged initial flow. These equations are discussed below.

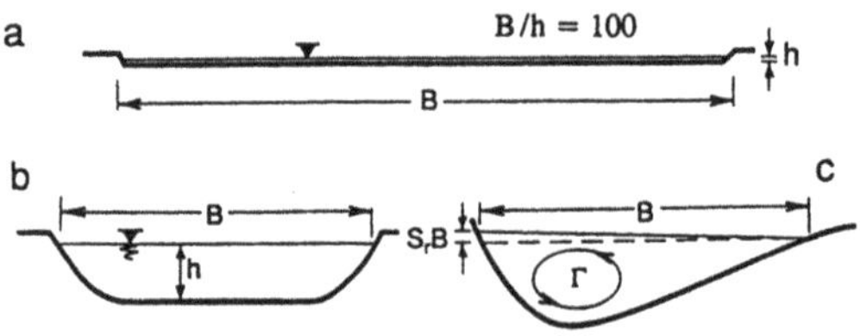

Fig. 7 (from Ref. [17])

5. Basic Equations

Following [1], [14], [15], the flow is studied (in plan) by the following (radial and longitudinal) vertically-averaged equations of motion and continuity

$$\frac{u}{N+1}\frac{\partial hv}{\partial l_c} + v\frac{\partial hv}{\partial n} = -hg\left(\frac{\partial y_b}{\partial n} + \frac{\partial h}{\partial n}\right) + \frac{hu^2}{(N+1)R} - \frac{uv}{c^2} \tag{I}$$

$$\frac{u}{N+1}\frac{\partial hu}{\partial l_c} + v\frac{\partial hu}{\partial n} = -\frac{hg}{N+1}\left(\frac{\partial y_b}{\partial l_c} + \frac{\partial h}{\partial l_c}\right) - \frac{huv}{(N+1)R} - \frac{u^2}{c^2} \tag{II}$$

and

$$\frac{1}{N+1}\frac{\partial hu}{\partial l_c} + \frac{\partial hv}{\partial n} + \frac{hv}{(N+1)R} = 0 \ . \tag{III}$$

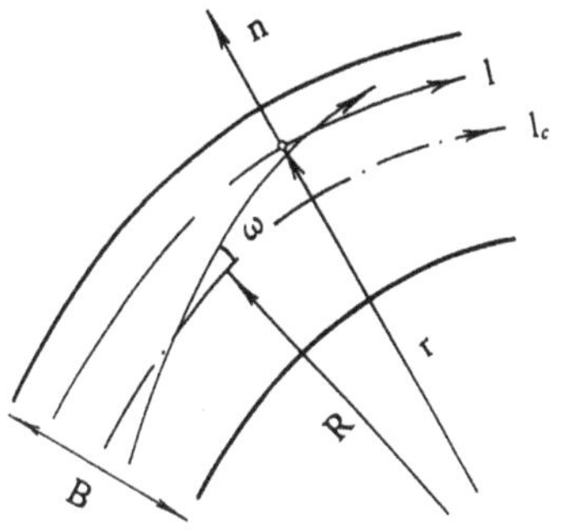

Fig. 8

Introducing $\omega = v/u$ ($\approx \tan\omega$), together with the local coordinates l and r (see Fig. 8), these Eqs. can be expressed also as

$$\frac{\partial \omega}{\partial l} = -\frac{g}{u^2}\frac{\partial h}{\partial r} + \frac{1}{r} - \frac{\omega}{hc^2} \qquad \text{(I')}$$

$$-\frac{\partial \omega}{\partial r} = \frac{g}{u^2}\left(S_b - \frac{\partial h}{\partial l}\right) - \frac{1}{hc^2} \qquad \text{(II')}$$

$$\frac{\partial hu}{\partial l} + \frac{\partial \omega hu}{\partial r} + \frac{\omega hu}{r} = 0\ . \qquad \text{(III')}$$

Consider the flow behaviour in the region Oa (from crossover to apex) in the light of the (longitudinal) equation II′ (whose validity, in the author's view, is questionable).

Fig. 9 shows schematically the flows corresponding to $\omega > 0$ and $\omega < 0$ ((i) and (ii) respectively). The right-hand side of Eq. II′ is formed by three terms (1), (2) and (3) which at the inner bank (*IB*) and the outer bank (*OB*) have some positive or negative values. This is depicted by the framed three diagrams reflecting the variation of these terms across the flow width *B*. Observe, that irrespective of whether ω being positive or negative, the nature of these diagrams (on the right of II′) remains the same. Yet, the left-hand side of II′ is not always the same. If $\omega > 0$, then $-\partial\omega/\partial r$ varies along *B* as shown in (i), whereas if $\omega < 0$ then, as in (ii) - which is the reverse of (i). Eq. II′, as it stands, yields *always* for the left-hand side the diagram shown in (ii), that is, it always yields the ingoing (along *Oa*) flow having negative ω.

It follows that the longitudinal Eq. II′ reflects adequately *only* the flow patterns corresponding to small θ_0; it is not adequate with regard to the flows corresponding to large θ_0 where the flow along *Oa* is directed towards the outer bank ($\omega > 0$). The deterioration of validity of II with the increment of θ_0 (i.e. of sinuosity) is only to be expected because (as has been repeatedly pointed out by many researchers) the nature of the flow in a loop is not determined by the conditions of that loop alone, but by the total geometry of the sequence of the loops - and the more so, the larger is sinuosity. But this means that the equations of motion in general and Eq. II′ in particular must be modified in the future so as to reflect the influence of the total geometry.

The next section is an attempt to remedy the situation, i.e. to achieve the "outgoing" flow (along *Oa*) for large θ_0.

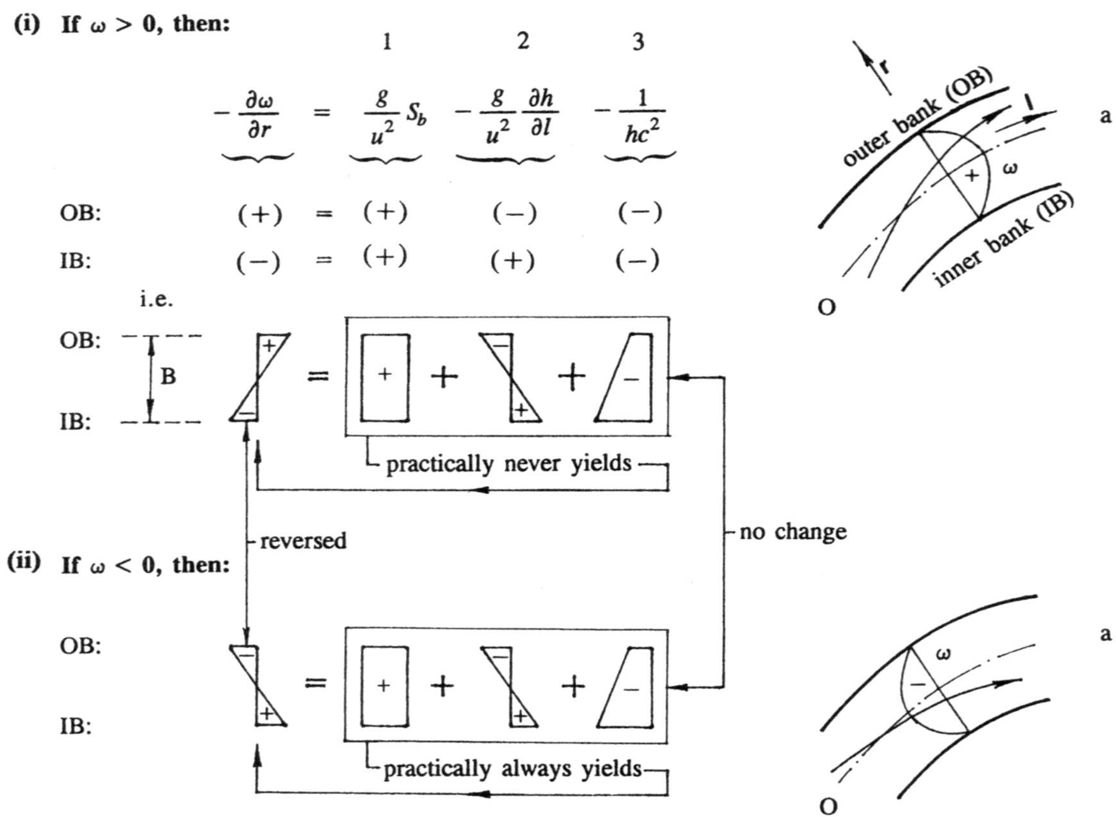
(i) If ω > 0, then:
1
2
3
$-\frac{\partial\omega}{\partial r} = \frac{g}{u^2}S_b - \frac{g}{u^2}\frac{\partial h}{\partial l} - \frac{1}{hc^2}$
OB: (+) = (+) (−) (−)
IB: (−) = (+) (+) (−)
i.e.
OB:
B
IB:
practically never yields
reversed
no change
(ii) If ω < 0, then:
OB:
IB:
practically always yields
r
outer bank (OB)
l
a
ω
inner bank (IB)
O
a
ω
O

Fig. 9

6. Shift Method

In the literature, the centrifugal force $\rho u^2/r$ is, as a rule, identified with the transverse pressure force γJ_r (see e.g. [2], [1], [15]). It is postulated here that $1/r$ and gJ_r/u^2-diagrams are basically identical, yet they are not exatly congruent: the transverse slope diagram is shifted along the abcissa to the right by a small amount k (see Fig. 10). The introduction of this shift, that is, the slight reduction of the transverse slope $\partial h/\partial r$ along Oa and its elevation along aO' is sufficient to achieve the outgoing flow along Oa, i.e $\omega > 0$, as required for large θ_0.

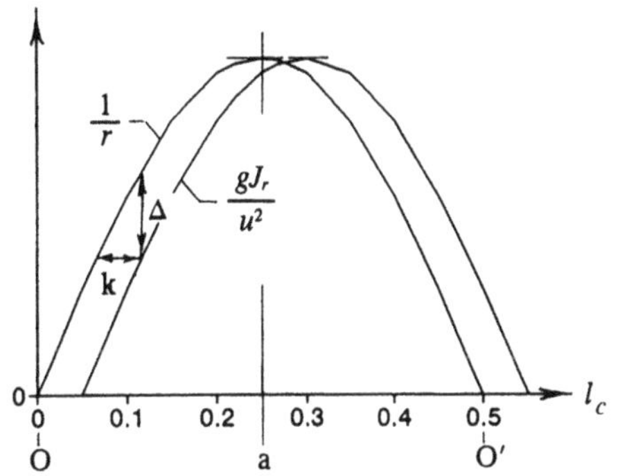

Fig. 10

Consider Eq. I', which can be expressed as

$$\frac{\partial \omega}{\partial l} + \left(\frac{1}{hc^2}\right)\omega = -\frac{g}{u^2}J_r + \frac{1}{r} \ .$$

Here, the right-hand side is the ordinate difference Δ caused by the shift k. Thus it is assumed that the value of gJ_r/u^2 at the section l_c is equal to $1/r$ at the section $l_c - k$. The introduction of the shift k and of the consequent ordinate difference Δ, which varies as a certain known function of k and l_c, makes it possible to arrive at the following three equations

$$-\frac{g}{u^2}J_r + \frac{1}{r} = \Delta$$

$$\frac{\partial \omega}{\partial l} + \left(\frac{1}{hc^2}\right)\omega = \Delta$$

$$\frac{\partial hu}{\partial l} + \frac{\partial \omega hu}{\partial r} + \frac{\omega hu}{r} = 0 \ ,$$

from which the three unknown functions u, ω and h can be solved - without involving (the inadequate for "large" θ_0) Eq. II' at all. It should be mentioned that when $\omega = 0$, then Δ and thus k must be zero, i.e. the value of k becomes zero at the banks (where $n = \pm B/2$). Hence, the symmetrical form

$$k = 4\,k_c\,[0.25 - (n/B)^2]$$

was adopted, k_c being the largest value of k at the centreline. One of the objectives of the present research is to reveal how the value of k_c varies depending on the sinuosity, i.e. θ_0.

Fig. 11 is an example of a flow computed from the present shift method; it corresponds to an "outgoing" flow in the sense of Fig. 5b.

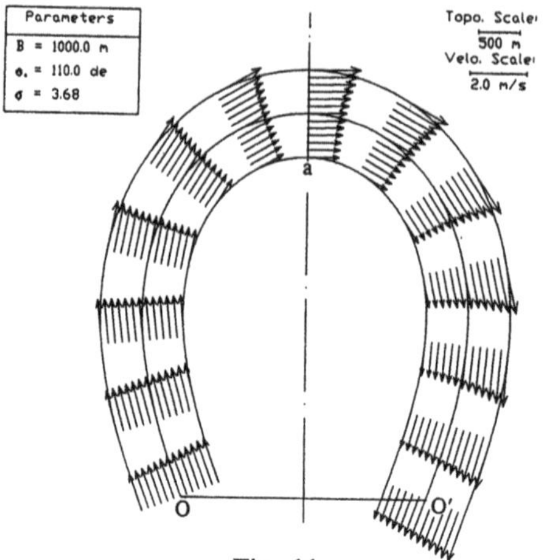

Fig. 11

7. Concluding Remarks

From the content of the present paper, it should be clear that the equations of motion and continuity (I), (II) and (III), which are undoubtedly correct, are not, at the present, interpreted and evaluated adequately.

The shift method introduced herein shows how effectively the flow patterns given by these equations can be modified as to agree with the actual flows. The present efforts of the authors are directed towards the interpretation of the terms of the aforementioned equations on the basis of laboratory measurements conducted by the junior author.

8. References

[1] De Vriend, H.J.: *A mathematical model of steady flow in curved shallow channels*. J. Hydr. Res., Vol. 15, No. 1, 1977.

[2] Engelund, F.: *Flow and bed topography in channel bends*. J. Hyd. Engrg., ASCE, Vol. 100, No. HY11, Nov. 1974.

[3] Friedkin, J.F.: *A laboratory study of the meandering of alluvial rivers*. U.S. Waterways Exp. Sta., Vicksburg, Mississippi, 1945.

[4] Hooke, R.L.: *Shear-stress distribution in stable channel bends*. Discussion, J. Hyd. Div., ASCE, Vol. 106, No. HY7, July 1980.

[5] Jackson, R.J.: *Velocity-bed-form-texture patterns of meander bends in the lower Wabash River of Illinois ans Indiana.* Geol. Soc. Am. Bull., Vol 86, 1975.

[6] Kalkwijk, J.P., De Vriend, H.J.: *Computation of the flow in shallow river bends.* J. Hydr. Res., Vol. 18, No. 4, 1980.

[7] Kondratiev, N., Popov, I., Snishchenko, B.: *Foundations of hydromorphological theory of fluvial processes.* (In Russian) Gidrometeoizdat, Leningrad, 1982.

[8] Leopold, L.B., Langbein, W.B.: *River meanders.* Sci. Am., 214, 1966.

[9] Leopold, L.B., Wolman, M.G.: *River channel patterns: braided, meandering and straight.* U.S. Geol. Survey Prof. Paper 282-B, 1957.

[10] Makaveyvev, N. I.: *River bend and erosion in its basin.* Press of the Academy of Sciences of the USSR, Moscow, 1975.

[11] Matthes, G.H.: *Basic aspects of river meanders.* Amer. Geophys. Union, 1941.

[12] Nelson, J.M., Smith, J.D.: *Flow in meandering channels with natural topography.* in "River Meandering", S. Ikeda and G. Parker eds., American Geophysical Union, Water Resources Monograph, 12, 1989.

[13] Nelson, J.M., Smith, J.D.: *Evolution and stability of erodible channel beds.* in "River Meandering", S. Ikeda and G. Parker eds., American Geophysical Union, Water Resources Monograph, 12, 1989.

[14] Shimizu, Y., Itakura, T.: *Calculation of bed variation in alluvial channels.* J. Hydr. Engrg., ASCE, Vol. 115, No. 3, March 1989.

[15] Smith, J.D., McLean, S.R.: *A model for flow in meandering streams.* Water Resources Res., Vol. 20, No. 9, 1984.

[16] Struiksma, N., Crosato, A.: *Analysis of a 2-D bed topography model for rivers.* in "River Meandering", S. Ikeda and G.Parker eds., American Geophysical Union, Water Resources Monograph, 12, 1989.

[17] Yalin, M.S.: *River mechanics.* Pergamon Press, Oxford, 1992.

List of Symbols

g	acceleration due to gravity
t	time
ρ	density of fluid
l_c	direction along channel centreline in plan; $l_c = 0$ at inflection point
n	radial direction; $n = 0$ at the centreline
l, r	longitudinal and radial coordinates as shown in Fig. 8
s	natural longitudinal coordinate of streamlines
R	magnitude of the radius of curvature of the channel centreline
N	n/R
L_m	meander length
Λ_m	meander wavelength

θ_0	deflection angle at $l_c = 0$
$J_0(\theta_0)$	Bessel function of first kind and zero-th order of θ_0
σ	channel sinuosity
Q	flow rate
B	flow width
h	flow depth
S_b	longitudinal slope
u	longitudinal component of the depth-averaged local flow velocity vector (along l_c or l)
v	radial component of the depth-averaged local flow velocity vector (along n or r)
ω	$\approx \tan \omega = v/u$
c	dimensionless Chézy friction factor
$\mathbf{q}_s$	specific volumetric bed-load rate vector
y_b	bed elevation
u_b	local flow velocity at the bed (no vertical averaging)

Site Investigations

A FIELD STUDY OF WAVE-CURRENT INTERACTIONS OVER A RIPPLED SAND BED

A HANNAY[1], J J WILLIAMS[2], J R WEST[1] and L E COATES[1]
[1]*School of Civil Engineering, The University of Birmingham, Edgbaston, Birmingham, B15 2TT, England*
[2]*Proudman Oceanographic Laboratory, Bidston Observatory, Bidston, Birkenhead, Merseyside, L43 7RA, England*

ABSTRACT

The results from a field experiment in approximately $30\,m$ of water of wave-current (w-c) interactions over a mobile rippled sand bed are presented. Open head toroidal EMCMs were deployed to measure tidal and wave-induced velocities at $z = 31$ and $70\,cm$ above the bed. Mean tidal speeds within the bottom boundary layer ranged from approximately $20\,cm/s$ to $90\,cm/s$ and root-mean square (RMS) orbital velocities ranged from approximately $10\,cm/s$ to $24\,cm/s$ with a mean wave period of $10\,s$. Spectral analysis of the burst velocity signals was applied and the horizontal velocity spectra were split into wave and turbulent energy components. Shear stress estimation using logarithmic profile, Reynolds stress and turbulent kinetic energy (tke) methods gave varying results. Shear stresses from the logarithmic profile method showed a large overprediction and the eddy correlation method is shown to be unreliable when significant horizontal wave velocities and large low frequency oscillations are present in the turbulent velocity signals. The tke method, based on an empirical relationship between tke and bed shear stress, is shown to give the most reliable results and the closest agreement with empirical predictions of shear stress for flow over a rippled sand bed. RMS orbital velocity results show significant deviation from predictions based on linear wave theory and exhibit an increase with height above the bed between $z = 31$ and $70\,cm$. Low frequency energy components have been quantified using a cumulative energy spectrum splitting technique and exhibit a high correlation with mean speed.

1. Introduction

An essential precursor to accurate prediction of marine sediment transport is the detailed understanding of wave-current (w-c) interaction and shear stress distribution in the bottom boundary layer. The estimation of bed shear stress in w-c flows is far from simple, due to the complex flow processes in the near-bed region. Wave and steady current situations have been modelled successfully and a number of theoretical models of combined w-c flows exist (eg. Grant and Madsen, 1979; Davies et al., 1988). However, a lack of good quality field data has resulted in inadequate testing of w-c bottom boundary layer theories. Dyer and Soulsby (1988) showed that, qualitatively the models generally predict enhancement of bed shear stress and apparent roughness attributable to waves, but quantitatively the model results varied significantly. The present study reports analyses of field data obtained under w-c conditions, where tidal speeds ranged from approximately $20\,cm/s$ to $90\,cm/s$ and root-mean squared (RMS) orbital velocities ranged from approximately $10\,cm/s$ to

24 cm/s, and aims to improve present understanding of the physical processes within the near-bed region.

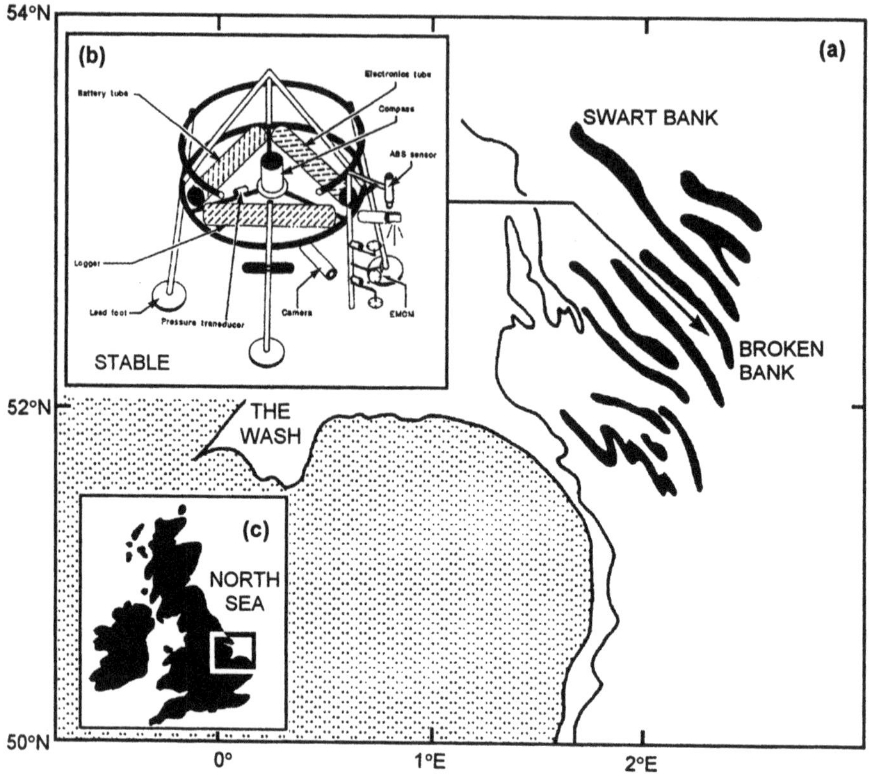

Fig. 1 (a) and (c): Location and arrangement of the Norfolk Sandbanks, (b): the STABLE rig deployed SW of Broken Bank.

2. Site Description and Instrumentation

Hydrodynamic data were collected using STABLE in the vicinity of well-developed sandbanks in approximately 30 m of water off the Norfolk coast, UK, during November 1988, Fig. 1. STABLE (Sediment Transport And Boundary Layer Equipment, Humphery, 1987) is an autonomous multi-sensor instrument capable of measuring simultaneous hydrodynamic conditions due to waves and currents and the resulting response of sediment. The position of the STABLE bed frame is shown in Fig. 1(a) and was deployed on the SW side of Broken Bank. Orthogonal pairs of 10 cm toroidal electromagnetic current meters (EMCMs) at approximate heights $z = 31$ and 70 cm above the bed were used to record near-bed mean and orthogonal turbulent flow components. A pressure sensor was used to give mean and fluctuating water depths. The instruments were sampled at $4\,Hz$ for $512\,s$ in 'burst' mode. Sequential bottom photographs recorded the local bed topography and were used to estimate the physical bottom roughness of the rippled sand bed. During the experiment bursts were recorded at $2\,hr$ intervals. The following analysis includes 18 bursts where the resultant current speed at both heights was greater than $20\,cm/s$.

Table I. Summary of flow variables: B is the burst number, d is the water depth, T_z is the zero up-crossing wave period, $\overline{S}_z$ is the mean tidal speed and σ_{wz} is the RMS orbital velocity at $z\,cm$ above the bed, ϕ is the acute angle between the waves and current. The roughness length z_0 is estimated using a logarithmic profile and bed shear stress values are obtained using profile (τ_p), eddy correlation (τ_r) and tke (τ_e) methods. The average z_0, estimated from the bed photographs is $0.26\,cm$.

B	d m	T_z s	$\overline{S}_{31}$ cm/s	$\overline{S}_{70}$ cm/s	σ_{w31} cm/s	σ_{w70} cm/s	ϕ deg	z_0 cm	τ_p N/m^2	τ_{r31} N/m^2	τ_{r70} N/m^2	τ_{e31} N/m^2	τ_{e70} N/m^2
5	29.4	10	45.0	57.6	12.6	14.9	28	1.7	3.8	0.3	3.5	1.6	1.9
6	28.7	10	24.4	31.9	12.9	14.7	17	2.2	1.4	0.1	1.7	0.7	0.9
7	28.0	9	21.1	29.4	10.4	12.4	34	4.0	1.7	0.5	1.3	0.4	0.4
8	28.2	10	48.0	65.7	13.9	16.1	25	3.4	7.5	1.9	2.8	1.4	1.5
9	28.9	10	22.0	30.5	12.2	14.3	5	3.7	1.7	0.8	0.2	0.5	0.6
10	29.7	10	32.1	43.3	12.5	14.6	40	3.0	3.0	0.1	1.2	0.7	0.7
11	29.8	10	61.3	80.5	11.3	12.7	25	2.3	8.9	0.3	1.7	2.1	2.5
12	29.4	10	51.7	68.3	13.6	16.2	18	2.5	6.7	1.0	1.6	2.1	2.8
14	28.3	8	60.0	81.8	16.2	20.2	23	3.3	11.6	1.9	1.6	1.8	2.0
15	28.6	8	55.7	78.6	15.9	18.7	15	4.3	12.7	2.0	2.8	1.9	2.7
17	29.6	10	41.1	55.8	15.6	18.1	35	3.2	5.2	0.7	0.8	1.0	1.0
18	29.1	0	43.8	59.0	16.5	19.5	22	3.0	5.6	0.6	0.4	1.3	1.6
20	27.9	9	59.9	82.1	17.4	21.6	24	3.4	11.9	2.2	2.3	1.8	2.2
21	28.4	9	45.8	64.1	19.3	23.5	16	4.1	8.1	2.6	1.2	1.1	1.4
23	30.0	10	47.2	63.4	16.1	18.4	30	2.9	6.4	0.8	1.1	1.3	1.5
24	29.9	10	66.0	88.8	14.0	16.2	18	2.9	12.5	0.7	0.7	2.2	2.9
26	28.2	11	40.5	52.4	16.3	19.9	46	2.0	3.4	0.6	0.5	0.7	0.7
27	28.5	11	66.4	90.9	16.1	19.3	25	3.4	14.4	2.7	3.1	1.7	2.1

3. Analyses and Results

Mean tidal velocity components, $\overline{U}$ and $\overline{V}$ were measured at heights $z = 31\,cm$ and $70\,cm$ above the seabed. Here, U is the fore-aft velocity (positive into STABLE) and $\overline{V}$ is the transverse velocity (positive from port to starboard of STABLE). The transverse velocity was consistently the dominant flow component and the velocities throughout the experiment were largely unobstructed by STABLE. The mean tidal speed $\overline{S}$ at each height was calculated by

$$\overline{S} = \left(\overline{U}^2 + \overline{V}^2\right)^{0.5} \tag{1}$$

and the main tidal direction relative to the axes of STABLE, was given by

$$\psi = \tan^{-1}\left(\frac{\overline{V}}{\overline{U}}\right) \tag{2}$$

The tidal ellipse is relatively narrow and the current around the Norfolk Sandbanks is essentially co-directional with the main tidal axis approximately 110° relative to the fore-aft axis of STABLE. The flood direction is approximately 340° relative to N, and approximately parallel with the axis of Broken Bank, Fig. 1(a). The tidal velocity components at $z = 31$ and $70\,cm$ are given in Table I.

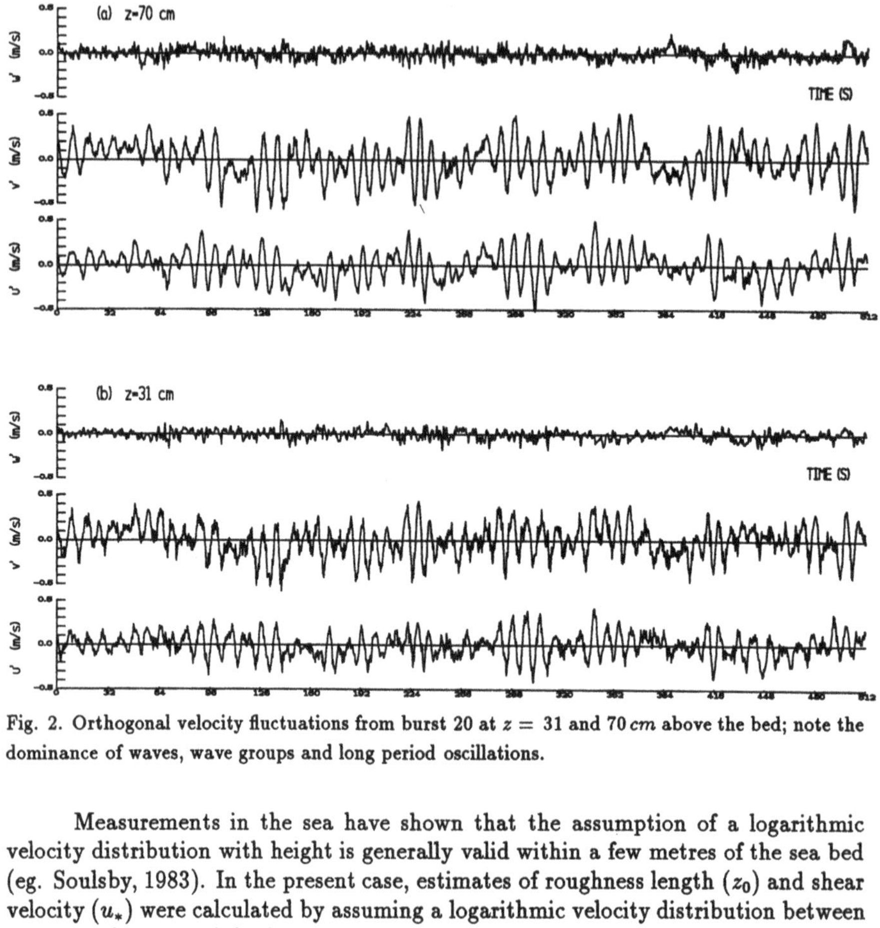

Fig. 2. Orthogonal velocity fluctuations from burst 20 at $z =$ 31 and $70\,cm$ above the bed; note the dominance of waves, wave groups and long period oscillations.

Measurements in the sea have shown that the assumption of a logarithmic velocity distribution with height is generally valid within a few metres of the sea bed (eg. Soulsby, 1983). In the present case, estimates of roughness length (z_0) and shear velocity (u_*) were calculated by assuming a logarithmic velocity distribution between $z = 31$ and $70\,cm$, of the form

$$\overline{S} = \frac{u_*}{\kappa} \ln\left(\frac{z}{z_0}\right) \tag{3}$$

Bed shear stress τ_p, defined as ρu_*^2, and z_0 estimates obtained using Eq. 3 are given in Table I.

Fig. 2 shows a typical burst (burst 20) of orthogonal (u', v', w') fluctuating velocity components at $z = 31$ and $70\,cm$. The fore-aft and transverse Reynolds stress components were calculated from the velocity signals and combined to give a resultant shear stress τ_r

$$\tau_r = \rho\left(\overline{u'w'}^2 + \overline{v'w'}^2\right)^{0.5} \tag{4}$$

Estimates of τ_r are shown in Table I for $z = 31$ and $70\,cm$ above the bed.

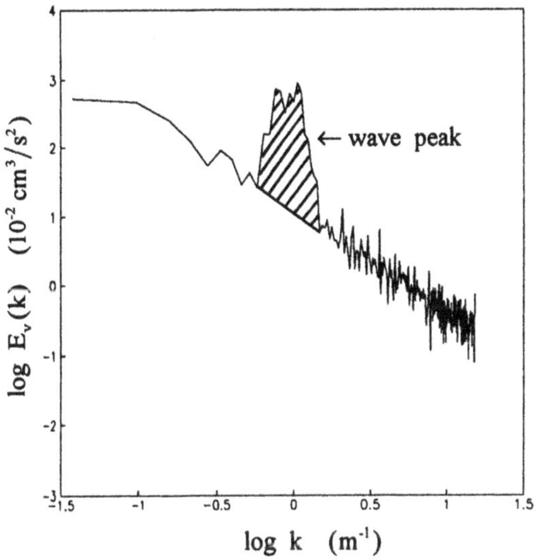

Fig. 3. Typical logarithmic energy density wavenumber spectrum, (burst 20, transverse velocity at $z = 70\,cm$ above the bed).

All horizontal velocity spectra obtained using a Fast Fourier Transform routine comprise of a characteristic turbulence spectrum and a superimposed wave spectrum, with a peak at the dominant wave frequency. The turbulent kinetic energy (tke) was estimated using a method described by Soulsby and Humphery (1990) which involves splitting a spectrum into wave and turbulent energy components. A typical velocity spectrum is shown in Fig. 3, for the transverse flow component v, at $z = 70\,cm$ for burst 20. The shaded area associated with the wave peak, $\overline{v_w'^2}$ (Fig. 3) was removed to leave the turbulent variance, $\overline{v_t'^2}$. This splitting technique was applied to horizontal velocity spectra only as the vertical velocity spectra were found to contain negligible amounts of wave energy. The tke (E) was calculated using the relationship

$$E = \frac{1}{2}\left(\overline{u_t'^2} + \overline{v_t'^2} + \overline{w_t'^2}\right) \tag{5}$$

where $\overline{w_t'^2}$ is the total energy beneath the vertical velocity spectrum. The empirical relationship between E and shear stress, quoted by Soulsby (1983) is used here where the shear stress τ_e is given by

$$\tau_e = 0.19\rho E \tag{6}$$

Values of τ_e obtained using the tke method are given in Table I at $z = 31$ and $70\,cm$.

RMS wave orbital velocities were calculated at each height by combining the horizontal wave variances derived from splitting the spectrum (Soulsby and Humphery, 1990).

$$\sigma_w = \left(\overline{u_w'^2} + \overline{v_w'^2}\right)^{0.5} \tag{7}$$

The wave direction (θ_w) relative to the axes of STABLE was calculated using

$$\theta_w = \tan^{-1}\left(\frac{\overline{v_w'^2}}{\overline{u_w'^2}}\right)^{0.5} \tag{8}$$

The mean wave direction calculated from θ_w at $z = 31$ and $70\,cm$ was subtracted from the mean current direction, where the mean current direction was calculated as the average between the two heights. Estimates of σ_w and the mean acute angle (ϕ) between the mean current and wave direction, which is often required as an input parameter in w-c interaction models are given in Table I. The zero up-crossing wave periods (T_z) were calculated from the zero-mean burst pressure signal and were found to agree consistently with the dominant peak in the horizontal velocity spectra.

4. Discussion of Results

4.1. Wave Activity

Table I shows σ_w values at $z = 31$ range from $10.4\,cm/s$ to $19.3\,cm/s$ and from $12.4\,cm/s$ to $23.5\,cm/s$ at $z = 70\,cm$. Fig. 4 shows σ_w at $z = 31$ against σ_w at $z = 70$. The regression line has a slope of 0.75 and the orbital velocities exhibit a correlation coefficient, $r = 0.99$. Linear wave theory suggests that the difference between orbital velocities so close to the bed should be very small, hence the slope of the line should be almost unity. Fig. 4 shows that the orbital velocities at $z = 70\,cm$ are consistently larger than the orbital velocities at $z = 31$ cm, the deviation from unity increasing with increasing wave activity. Fig. 4 also shows the results presented by Soulsby and Humphery (1990) for σ_w at $z = 10\,cm$ and $z = 40\,cm$. The data correlate well ($r = 0.96$) and exhibit a slope of 0.82. In common with the present data, the deviation from the unity line increases with increasing wave velocity.

The interpretation of these results must give consideration to the limitations of the present measuring system and analysis techniques. The high correlations seen in Fig. 4 gives confidence in the spectral splitting technique. Whilst at first sight the consistency of the results suggests a possible calibration error of the EMCMs, tests on single EMCMs in a re-circulating flume resulted in identical calibration curves. Each horizontal velocity signal (eg. Fig. 2) is a combination of the velocity signals recorded by two heads - the signals were combined to give flow directions relative to the axes of STABLE. A calibration error would need therefore to be very large to account for the differences in orbital velocities reported here and is considered to be unlikely.

The validity of the spectrum splitting technique in estimating correct magnitudes of orbital velocities must also be considered. Other studies in combined w-c flows have calculated orbital velocities directly from velocity signals. In the present study, when $\overline{S}$ was greater than approximately $20\,cm/s$, low frequencies and turbulent fluctuations contaminated the wave velocity signal. This resulted in inadequate and inconsistent estimation of the zero up-crossing wave period and orbital velocity. In the case of burst 19 (not shown in Table I), $\overline{S} = 15.1$ and $18.2\,cm/s$ at $z = 31$ and $70\,cm$ respectively, and the velocity time series did not exhibit significant low frequency oscillations. The horizontal velocity signals were smoothed using a 4-point running average technique to remove the data spikes due to turbulent events.

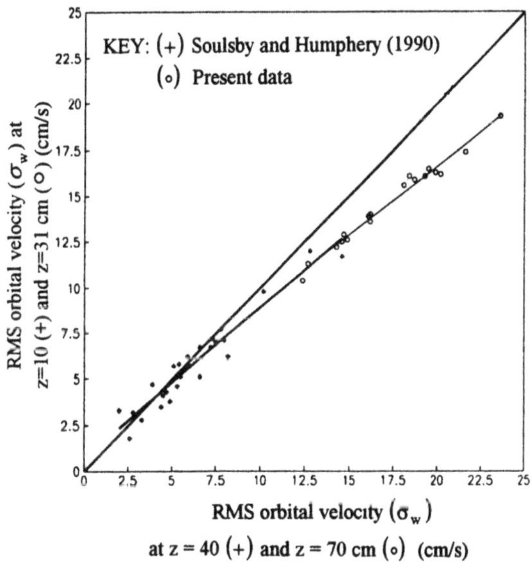

Fig. 4. Ratio of σ_w at each sensor height; + - Soulsby and Humphery (1990) at $z = 10$ and $40\,cm$; o - present data at $z = 31$ and $70\,cm$.

The four estimates of zero up-crossing wave period from the velocity signals at each height agreed to within $\pm 2\,s$. The significant wave velocity components for u and v were calculated as the mean wave velocity of the highest third waves. The horizontal components were combined to give the resultant significant wave velocity U_s, the value of which equals $30.5\,cm/s$ and $36.7\,cm/s$ at $z = 31$ and $70\,cm$, respectively. The results from splitting the spectrum, where $2\sigma_w$ is analogous to U_s, (Myrhaug and Slaattelid, 1989), agree to within $\pm 0.5\,cm/s$, helping to substantiate the spectral splitting technique. Since the differences in orbital wave velocity with height have been observed using independent methods, these results therefore suggest a real phenomenon. The wave direction calculated from the significant velocity components in burst 19 agreed to within $\pm 2^\circ$ of the mean directions calculated using Eq. 8. The zero up-crossing wave period estimated from the velocity signals agreed with T_z from the pressure signal, and the peak in the wave velocity spectrum to within $0.5\,s$. Investigation of the horizontal velocity signals given in Fig. 2 shows that the magnitude of the wave velocities is indeed larger at $z = 70$ than at $z = 31$ cm. This is consistent with the results presented in Fig. 4 and shows that the difference in orbital velocity is not a consequence of the spectral splitting analysis.

To explain the difference in orbital velocity with height, further field data is required to resolve the oscillatory flow closer to the bed. Possible explanations of increasing orbital velocity with height must come from theoretical considerations of w-c interactions, although no mathematical model, to the authors knowledge, has simulated the results shown in Fig. 4. Davies et al., (1988) show numerically that the addition of a current ($\overline{U}_{100} \approx 100\,cm/s$) on a wave ($U_s = 100\,cm/s$) will result in the formation of a w-c oscillatory boundary layer. The thickness of this layer will extend up to approximately $10\,cm$ above the bed. Fig. 2(b) from Davies et al. shows that the near-bed orbital velocity increases with height up to approximately $10\,cm$. This could be an underprediction given the complicated processes in reality and might explain the difference in orbital velocities with height.

4.2. Low Frequency Oscillations

Low frequency oscillations have been observed in a wide variety of flows and are apparent in the velocity signal measured by Bowden and Ferguson (1980). Their data show high frequency turbulent fluctuations superimposed on long period $(1 - 10\,min)$ oscillations in the horizontal and vertical velocity signals, recorded at $100\,cm$ above the bed. The long period oscillations are clearly seen in the $u'w'$ product signal. Similarly, although shorter period oscillations are apparent in Fig. 2, low frequency oscillations seen in the velocity signals discussed here have typical periods between $30\,s$ and $5\,min$. In a study conducted in the Frazer River, Canada, a contribution to the total momentum flux of approximately 30 % was attributed to multi-minute, low frequency oscillations (Lapointe, 1993). Whilst many researchers have observed low frequency oscillations, the generating mechanisms remain unresolved, and require further investigation.

To quantify the relative amounts of low frequency energy present in each velocity record, a method was developed to separate the high and low frequency and wave energy components by using cumulative velocity spectra. All the horizontal cumulative energy spectra showed three distinct frequency bands. Essentially these three bands correspond to energy at low and high frequencies and to wave energy. The two points of maximum curvature in each spectrum were chosen to separate the energy into relative frequency contributions. The waves occur within a very narrow frequency band and were consistently the dominant source of energy in the burst records considered here. The magnitudes of σ_w calculated for the two sensor heights using this splitting technique exhibit a correlation coefficient $r = 0.99$ at each height and slopes approximately equal to 1.0 with the σ_w results shown in Table I. The wave energy calculated from the cumulative method includes all the energy between the two limiting frequencies. Despite this, the high correlation between the two spectral splitting techniques suggests that the cumulative method adequately splits the energy into wave, low and high frequency components.

To compare the relative magnitudes of low frequency contributions at $z = 31$ and $70\,cm$, the resultant component was calculated by combining the u and v low frequency components. The low frequency variance gives $r = 0.93$ between $z = 31$ and $70\,cm$, with consistently larger (upto 30% for high variances) values recorded by the higher sensors. The low frequency variances give $r = 0.91$ and 0.84 with $\overline{S}^2$ at $z = 31$ and $70\,cm$, respectively. Thus, there is some evidence to show that the generation of these low frequency oscillations, is related in some way to the mean tidal flow. The length of record used in the present study could not adequately describe the distribution of low frequency energy, since the spectra suggest that the energy is spread over a wide frequency range. Turbulence is essentially a three-dimensional process and in a flow of approximately $1\,m/s$, a turbulent eddy would have to be of the order of $300 - 350\,m$ long to produce fluctuations of approximately $5\,min$ period. Eddies with these dimensions would be elongated two-dimensional features and the generation mechanism of such events is far from obvious.

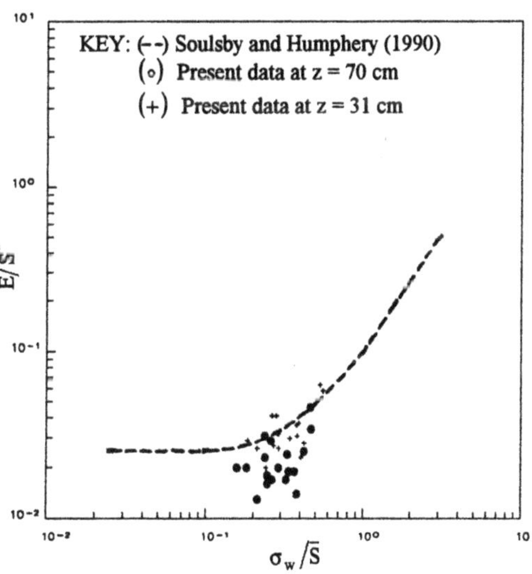

Fig. 5. Normalised tke with w-c ratio showing data from Soulsby and Humphery (1990) and from the present study at $z = 31$ and $70\,cm$.

4.3. Turbulent Kinetic Energy

The turbulent kinetic energy E, calculated using Eq. 5 was non-dimensionalised with $\overline{S}^2$ and plotted against w-c ratio ($\sigma_w/\overline{S}$) for each burst velocity signal at $z = 31$ and $70\,cm$; the results are presented in Fig. 5. Within the bottom boundary layer the shear stress is proportional to $\overline{S}^2$, and E is proportional to shear stress (Soulsby, 1983). Therefore, the value of $E/\overline{S}^2$ may be assumed constant for a fixed bed roughness, providing the flow is steady. Fig. 5 also shows the results presented by Soulsby and Humphery (1990) at z = 40 cm above the bed. They concluded that $E/\overline{S}^2$ was constant and approximately equal to 0.025 for values of $\sigma_w/\overline{S} \leq 0.2$. For values of $\sigma_w/\overline{S} > 0.2$, $E/\overline{S}^2$ increased rapidly up to a value 10 times larger than the current only value for $\sigma_w/\overline{S} \approx 2$. The authors concluded that the increase in $E/\overline{S}^2$ for increasing w-c ratio was attributable to wave-generated turbulence close to the bed, diffusing up in to the flow from the wave boundary layer.

The results from the present study agree quantitatively with those presented by Soulsby and Humphery for $0.1 < \sigma_w/\overline{S} < 0.6$. Further work is needed for a larger range of w-c conditions as the limited variation in the value of $\sigma_w/\overline{S}$ restricts any firm conclusion that the value of $E/\overline{S}^2$ increases rapidly with increasing w-c ratio. The data presented in Fig. 5 show also that the value of non-dimensional tke is generally larger at $z = 31$ than at $z = 70\,cm$ for a constant w-c ratio. $E/\overline{S}^2$ at the two heights give $r = 0.99$, with the values of $E/\overline{S}^2$ at $z = 31$ approximately 50% larger than the values at $z = 70\,cm$. These results suggest that the assumption of a constant stress layer (eg. Huntley, 1988) may for the present conditions be inappropriate and measurements closer to the bed are required to clarify the concept of a constant stress layer when significant horizontal wave velocities are present.

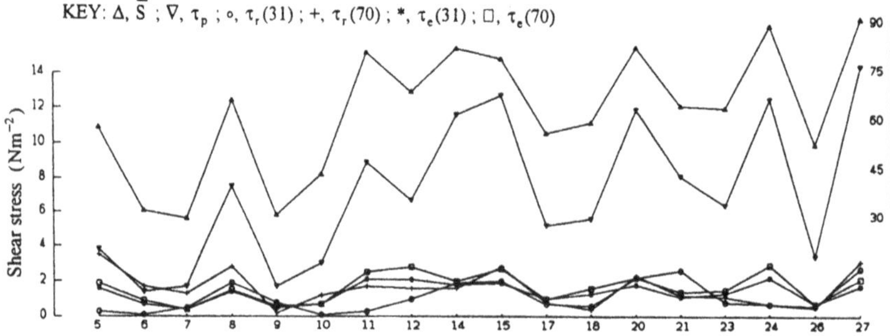

Fig. 6. Tidal variation in $\overline{S}$ at $z = 70\,cm$ and five estimates of shear stress using logarithmic profile, tke and Reynolds stress methods.

4.4. Shear Stress Estimation

The shear stresses presented in Table I show large differences between the three methods used. The calculation of bed shear stress contains many uncertainties but accurate estimation is essential for the prediction of sediment transport rates.

In steady flow, close to the sea bed, the decrease with height of the shear stress has been reported to be small. Bowden and Ferguson (1980) found no significant variation in Reynolds stress with height within the bottom $2\,m$. Soulsby (1983) stated, in addition to the shear stress, the ratios of orthogonal variances, $\overline{u_t'^2}$: $\overline{v_t'^2}$: $\overline{w_t'^2}$ are also constant with height and proportional to u_*^2 within the surface layer. This region of near-bed flow is often referred to as the constant stress layer (eg. Huntley, 1988). The tke results presented in Fig. 5 showed that a constant stress layer was not observed in the present study. Consequently, using tke and Reynolds stress techniques to estimate shear stress should result in a difference in the stress estimate with height above the bed.

The quadratic friction law, given by

$$\tau_0 = \rho C_D \overline{S}^2 \tag{9}$$

can be used to calculate the drag coefficient C_D, from estimates of bed shear stress and mean velocity measurements, usually at $100\,cm$ above the bed (eg. Soulsby, 1983). The drag coefficient often quoted (eg. Drake and Cacchione, 1992) for steady flow at $z = 100\,cm$ over a rough bed (eg. ripples and gravel) is 3×10^{-3}. Eq. 9 has been used here to estimate a mean drag coefficient using velocity measurements at $z = 70\,cm$ and shear stresses estimated from the logarithmic profile, tke and Reynolds stress methods, where the slope of a line assumed to pass through the centroid of the data and the origin of the co-ordinate system equals ρC_D.

The Reynolds stress method has been investigated and the results have been presented in Figs. 6 and 7. Poor correlations of the stresses at each height with $\overline{S}^2$, the effects of waves and low frequency oscillations, and the sensitivity to slight sensor misalignment in w-c flows suggest that use of the Reynolds stress method to estimate shear stress and fluid drag is inadequate for the present conditions.

The shear stresses τ_p do not appear to adequately estimate the bed shear stress for the present conditions given the large magnitude of the estimates given in Table I and Figs. 7 and 8. Correlation r between τ_p and $\overline{S}^2$ at $z = 70\,cm$ is 0.97, this is not unexpected given that the dominant parameter in Eq. 3 is $\overline{S}$. Eq. 9 leads to a drag coefficient of 17.3×10^{-3}, approximately five times the steady flow drag coefficient at $z = 100\,cm$. Values of τ_p will include the influence of the wave boundary layer, seen as an increased roughness felt by the flow, as observed by Drake and Cacchione (1992). It must be stressed that using only two sensor positions and assuming a logarithmic velocity profile is notoriously prone to error, and estimates of τ_p are very sensitive to uncertainties in sensor position. This experiment did not adequately measure the height of the sensors above the bed during deployment, and the sensor positions are given assuming a slight rig settlement of $10\,cm$, determined from flume tests with mobile sand, Humphery (personal communication). Given all the uncertainties described above, the shear stress estimates from the logarithmic profile method must be viewed with considerable caution.

The tke method utilises the velocity spectra and is based on the ratio of bed shear stress to tke, the value of which is often assumed constant under widely varying conditions. An alternative method of shear stress estimation is the inertial dissipation method which considers turbulent velocity spectra within the inertial subrange (eg. Huntley, 1988). The measurement system used in the present study did not adequately resolve the range of frequencies necessary to observe an inertial subrange, hence the inertial dissipation method which is closely related to the tke method was not considered here. The tke method is robust since it does not require assumptions of the values of κ and the Kolmogorov constant. It can be applied easily in w-c flows although it is sensitive to tke variations with height. The tidal variation in the estimates of τ_e calculated using the tke method are shown in Fig. 6. Values of τ_e at $z = 31$ and $70\,cm$ give $r = 0.98$ and the slope of the regression line showed that the magnitude of τ_e was larger at $z = 70\,cm$ than at $z = 31$ cm. The largest estimates of τ_e showed approximately 20% increase with height, whilst the smallest estimates agreed at each height. The values of τ_e gave correlation coefficients $r = 0.90$ and 0.85 with $\overline{S}^2$ at $z = 31$ and $70\,cm$ respectively, and were averaged to give a single estimate of stress $\overline{\tau_e}$. The $\overline{\tau_e}$ values shown in Fig. 7 give $r = 0.87$ with $\overline{S}^2$ at $z = 70\,cm$ and a drag coefficient at $z = 70\,cm$ of 2.6×10^{-3}, which is in close agreement with the estimated drag coefficient calculated from the Reynolds stress method. Wave energy components have been successfully removed from the horizontal velocity spectra which still contain low frequency energy components of periods greater than approximately 15 s.

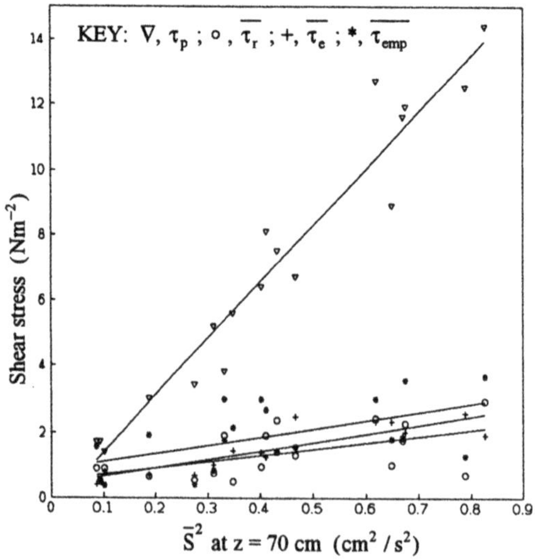

Fig. 7. Shear stress estimates calculated from the logarithmic velocity profile ($\overline{\tau_p}$), tke ($\overline{\tau_e}$), and Reynolds stress ($\overline{\tau_r}$) method and shear stress ($\overline{\tau_{emp}}$) assuming a constant roughness, $z_0 = 0.26\,cm$, against $\overline{S}^2$ at $z = 70\,cm$.

For a mean roughness length of $z_0 = 0.26\,cm$ estimated from the bed photographs (described in the following section), C_D calculated using Eqs. 3 and 9 is 4.5×10^{-3} at $z = 70\,cm$. Given the uncertainties associated with assuming a logarithmic velocity profile and z_0 calculations (Eqs. 10 and 11), this value is within the expected limits. Using a constant roughness equal to $0.26\,cm$ in Eq. 3 results in an alternative semi-empirical prediction of bed shear stress τ_{emp}, for the given mean tidal velocities in Table I. The values of τ_{emp} for each burst give $r = 0.53$ and 0.56 with $\overline{S}^2$, at $z = 31$ and $70\,cm$ respectively. Fig. 7 shows the average semi-empirical shear stress ($\overline{\tau_{emp}}$) that exhibits a correlation coefficient $r = 0.55$ with $\overline{S}^2$ at $z = 70\,cm$, and gives $C_D = 3.9 \times 10^{-3}$, which is surprisingly within close agreement with the drag coefficient calculated using the relationship between $\overline{\tau_e}$ and $\overline{S}^2$. The poor correlations of τ_{emp} and $\overline{\tau_{emp}}$ with $\overline{S}^2$ gives strong evidence to suggest that the bed roughness changes throughout the study and further evidence that the logarithmic profile is not valid for the present conditions. The estimated drag coefficients from the tke and Reynolds stress methods closely agree and are predicted to within an order of magnitude by assuming steady flow theory (Eqs. 3 and 9) with an observed mean roughness length from the bed photographs.

4.5. Bed Roughness Length

In steady flow, the sea-bed roughness length z_0 can be estimated relatively accurately using mean velocities at several heights and Eq. 3. When bedforms are present in current dominated combined flow, the logarithmic velocity profile will estimate the roughness (eg. Drake and Cacchione, 1992) due to form drag and skin friction over the bedforms, and will also include a roughness due to the effects of the wave boundary layer. Essentially, the wave boundary layer will be seen by the current as an apparent or increased bed roughness. If significant horizontal wave velocities are present, the logarithmic velocity profile is unlikely to adequately predict the bed roughness. An indirect method of estimating the bed roughness utilises bottom photographs to calculate the height and wavelength of bedforms via formulae based on known bedform geometry. Probably the most common is the relationship proposed by Grant and Madsen (1982).

$$k_s = 28\frac{h^2}{L} \tag{10}$$

where $z_0 = k_s/30$ and k_s is the Nikuradse roughness length. The bed photographs recorded during the experiment yielded estimates of mean ripple wavelength L. The mean wavelength of ripples in a given photograph was calculated using appropriate scaling corrections, and ripple height h was estimated using the relationship proposed by Allen (1970).

$$h = 0.074L^{1.19} \tag{11}$$

The photographs gave wavelengths between 10 and $26\,cm$, hence heights between approximately 1 and $4\,cm$. The mean ripple wavelength and height for the whole experiment was estimated at $18\,cm$ and $2.3\,cm$ respectively, giving a roughness $z_0 = 0.26\,cm$.

Given the scatter in the results presented by Allen, this result must be treated as an order of magnitude estimate only. Further, some bursts contained significant amounts of suspended sediment and those photographs could not be used to estimate bedform lengths. Soulsby (1983) suggested that the roughness of a typical rippled sand bed was of the order of $0.6\,cm$ in steady tidal currents, which agrees to within an order of magnitude of z_0 from the present observations. The mean z_0 calculated from assuming a logarithmic velocity distribution was $3.1\,cm$. These results show an order of magnitude overprediction compared to the bottom photograph results, a disparity that is not unexpected due to the uncertainties associated with the logarithmic profile method, doubts about the exact magnitude of z_0 and the unsteady nature of the flow.

5. Conclusions

The present study, examining w-c interactions with a mobile rippled sand bed, has yielded the following conclusions.

The use of the logarithmic profile method to estimate z_0 results in a large overprediction compared to the observed physical bed roughness from bottom photographs. Similarly, estimates of bed shear stress were thought to be overpredicted due to the combined influences of the wave boundary layer, flow over mobile sand ripples and the limited data points in the mean velocity profile.

Estimates of bed shear stress derived from Reynolds stress measurements do not adequately describe the bottom shear when significant oscillatory horizontal velocities are present. The sensitivity of the results to slight sensor misalignment must be investigated with respect to field studies of the type described here.

Values of τ_e derived from the empirical relationship between tke and shear stress have been shown to give credible estimates of mean shear stress and the drag coefficients agree to a first approximation with the drag coefficients assuming constant bed roughness and steady flow formulae.

Normalised turbulent kinetic energy results suggest an increase in near-bed tke with increasing w-c ratio. This conclusion is in qualitative agreement with the results presented by Davies et al. (1988) but requires further field measurement verification for a wider range of conditions.

RMS wave orbital velocities calculated from splitting the spectrum agreed well with significant wave velocities calculated from the horizontal velocity signals when mean current speeds were less than $20\,cm/s$. The RMS orbital velocities at $z = 70\,cm$ were consistently larger than RMS orbital velocities at $z = 31$, the difference in orbital velocities increased with increasing wave activity. This result could explain the existence of a w-c oscillatory boundary layer extending beyond the height of the highest measuring point, but needs further investigation.

Low frequency energy components apparent in the velocity signals were quantified using a cumulative spectra splitting technique, and were shown to exhibit a high correlation with $\overline{S}^2$. Generating mechanisms for these low frequency oscillations needs further investigation.

6. Acknowledgment

This work was undertaken as part of the CSTAB research program. It was funded by the Commission of the European Communities, Directorate General for Science, Research and Development under MAST contract No. MAS2-CT92-0024C. This research was supported by Proudman Oceanographic Laboratory and the Science and Engineering Research Council. Access to seabed photographs provided by the Department of Oceanography, University of Southampton, was also appreciated.

7. References

1. J.R.L. Allen, *Physical Processes and Sedimentation* (Unwin Books, London, 1970).
2. K.F. Bowden and S.R. Ferguson, Variations with Height of the Turbulence in a Tidally-Induced Bottom Boundary Layer, in *Marine Turbulence*, ed. J.C.J. Nihoul (Elsevier, Amsterdam, 1980) 259 - 286.
3. A.G. Davies, R.L. Soulsby and H.L. King, A Numerical Model of the Combined Wave and Current Bottom Boundary Layer, *J. Geophys. Res.* **93(C1)** (1988) 491 - 508.
4. D.E. Drake and D.A. Cacchione, Wave-Current Interaction in the Bottom Boundary Layer during storm and non-storm conditions- Observations and Model Predictions, *Continental Shelf Res.* **12(12)** (1992) 1331 - 1352.
5. K.R. Dyer and R.L. Soulsby, Sand Transport on the Continental Shelf, *Ann. Rev. Fluid Mech.* **20** (1988) 295 - 324.
6. W.D. Grant and O.S. Madsen, Combined Wave and Current Interaction with a Rough Bottom, *J. Geophys. Res.* **84** (1979) 1797 -1808.
7. W.D. Grant and O.S. Madsen, Moveable Bed Roughness in Unsteady Oscillatory Flow, *J. Geophys. Res.* **87** (1982) 469 - 481.
8. J.D. Humphery, STABLE- An Instrument for studying Current Structure and Sediment Transport in the Benthic Boundary Layer, *Proc. Electronics for Ocean Tech. Conf.* (Heriot-Watt Univ., Edinburgh, IERE, London, 1987) **27** 57 - 62.
9. D.A. Huntley, A Modified Inertial Dissipation Method for Estimating Seabed Stresses at low Reynolds Numbers, with application to Wave/Current Boundary Layer Measurements, *J. Phys. Oceanogr.* **18** (1988) 339 - 346.
10. M. Lapointe, Monitoring Alluvial and Suspension by Eddy Correlation, *Earth Processes and Landforms* **18** (1993) 157 - 175.
11. D. Myrhaug and O.H. Slaattelid, Combined Wave and Current Boundary Layer Model for fixed Rough Seabeds, *Ocean Eng.* **16(2)** (1989) 119 - 142.
12. R.L. Soulsby, The Bottom Boundary Layer of Shelf Seas, in *Physical Oceanography of Coastal and Shelf Seas*, ed. B. Johns (Elsevier Science Pub., Amsterdam, 1983) 189 - 266.
13. R.L. Soulsby and J.D. Humphery, Field Observations of Wave-Current Interaction at the Sea Bed, *Proc. NATO Adv. Res. Workshop on Water Wave Kinematics*, ed. A. Torum (Molde, Norway, 1989).

SAND TRANSPORT ON MACRO-TIDAL BEACHES

YOLANDA FOOTE, DAVID A. HUNTLEY and TIM O'HARE
Institute of Marine Studies, University of Plymouth, Plymouth, Devon, PL4 8AA, UK.

ABSTRACT

Field measurements of cross-shore velocities from a macro-tidal beach are used to compute velocity moments and, following the energetics modelling approach, infer patterns of sediment transport through a tidal cycle. These predictions are compared with sediment fluxes obtained by combining the velocity data with measured suspended sand concentrations. The measured velocity moments are relatively insensitive to the incident wave conditions and exhibit consistent "shape functions" over the nearshore, with dominant contributions due to incident waves and mean flows. The sediment flux comparisons indicate that the moments generally provide good predictions except in the case of transport due to incident waves where the direction of transport is often a result of the presence of bedforms. A tentative model for sediment transport on a macro-tidal beach, based on a shape function approach, is also presented.

1. Introduction

Macro-tidal beaches, with tidal range in excess of 5m, are found in many locations, most notably in Europe on the shores of the Northwest European shelf seas. The geomorphological features of macro-tidal beaches differ in significant ways from micro- or meso-tidal beaches, notably in tending to have a mild intertidal beach slope separating steeper high tide and low tide beach slopes. However, despite these qualitative as well as quantitative differences, there have been few studies of the detailed dynamics of sand movement on macro-tidal beaches, and understanding of the processes responsible for the observed morphologies is therefore very limited (Short, 1991).

The purpose of this paper is to discuss some of the processes which can operate on macro-tidal beaches, to present some preliminary results from a recent macro-tidal beach study, the British Beach and Nearshore Dynamics (B-BAND) programme (Davidson et al., 1992), and to outline work currently in progress on this problem. Specifically the paper concentrates on two main topics. The first is an investigation of the applicability of 'energetics' modelling of sand transport, based on the water velocity field, to transport rates observed at one of the B-BAND sites. The second is a brief discussion of the special effects present on macro-tidal beaches, and an assessment of some potential influences on beach profiles.

1.1 Energetics modelling of beach sand transport

The basic concept of equating the energy needed to move sediment to a fixed proportion of the available energy supplied by the water movement originated with Bagnold (1966), and was applied to beach sand movement by Bowen (1980) and Bailard (1981, 1987). The resulting equations predict that sand transport rates, both as bedload and as suspended load, are related to moments of the local velocity field, including the skewness and more complex moments involving moduli of the velocity. Following the theoretical work of Bailard, Guza and Thornton (1985) used field measurements of velocities on a micro-tidal beach to study the relative importance of the Bailard moment terms. They showed that the most significant terms were:

Bedload transport: $\overline{u^3}/\left(\overline{u^2}\right)^{3/2}$ Suspended load transport: $\overline{|u|^3 u}/\left(\overline{u^2}\right)^2$

(Note that Guza and Thornton inadvertently have the bedload term identified as the suspended load term and vice versa). These moment terms are multiplied by functions of beach slope and total velocity variance to provide a prediction of transport rate for each mode of transport.

A number of interesting questions arise from this predictive scheme. Firstly, how do these moments vary with offshore position? Secondly, does the normalisation in these terms make them insensitive to incident wave heights and therefore in some sense 'universal' parameters? Thirdly, what are the relative contributions of incident waves, long period motion (surf beat) and mean flows to the total moments? Finally, and perhaps most importantly, how do these moment terms relate, if at all, to the true sand transport rates? In the following sections we will investigate these questions using some of the B-BAND data sets.

1.2 Contributions to velocity moments

We begin by addressing the third question posed above, namely the relative contributions of incident waves, long period motion and mean flows to the total moments. We assume that the total velocity field is made up of a sum of contributions:

$$u = \bar{u} + u_S + u_L \tag{1}$$

where $\bar{u}$ is the mean flow, u_S is the incident wave flow and u_L is the long period flow.

For bedload transport, the skewness is the normalised time-average of the cube of the velocity. Substituting the form of u given above and expanding the cube results in eight potentially non-zero time-averaged terms (Foote et al., 1992):

1. $\overline{u}^3$ The cube of the mean flow
2. $\overline{u_S^{\,3}}$ The skewness of the incident waves
3. $\overline{u_L^{\,3}}$ The skewness of the long period motion
4. $\overline{3u_S^{\,2}u}$ Movement by the mean flow of sand stirred by incident waves
5. $\overline{3u_L^{\,2}u}$ As 4, but stirring by long period motion

6. $\overline{6uu_S u_L}$ A three way correlation, expected to be zero

7. $\overline{3u_L^2 u_S}$ Correlation between long period variance and incident wave velocity

8. $\overline{3u_S^2 u_L}$ Correlation between incident wave variance and long period velocity.

For the moments, each of these terms should be normalised by $(u^2)^{3/2}$.

For suspended load transport, the analysis is more difficult since the total velocity appears as its modulus, which does not readily separate into different terms. For the present analysis, we make the reasonable assumption that u_S is very much larger than either the mean, $\bar{u}$, or the long period motion, u_L. This is indeed the case except very close to the shoreline, where incident motion is much reduced and long period motion increases. With this assumption, the suspended load moment terms expand into three terms:

1. $\overline{\left(u_S^2\right)^{3/2} u_S}$ Incident wave term

2. $\overline{\left(u_S^2\right)^{3/2} u_L}$ Long period term

3. $\overline{\left(u_S^2\right)^{3/2} u}$ Mean flow term

with each of these terms normalised by $(u^2)^2$. In the following sections we will investigate the relative magnitudes of these terms in the B-BAND field data.

2. The B-BAND field data

Details of the B-BAND measurement programme and data are given in Davidson et al. (1992) so it is only necessary to give a brief summary of the relevant points here.

The data were collected during the deployment at Spurn Point on the north-east coast of the UK (figure 1) in April 1991. Measurements were made at a station consisting of three electromagnetic current meters, at 0.10, 0.25 and 0.63 m above the bed, with three optical backscatter (OBS) sensors for suspended sediment concentration, at heights of 0.04, 0.10 and 0.25 m. The station, one of several at the site, was deployed on the inter-tidal low gradient portion of the beach (figure 2), and variations of velocity field (positive onshore) and sand response were measured as the tidal rose and fell over the station. Data was logged at the top of the beach through long cables. The logging regime was a 17 minute run, at a sampling rate of 2 Hz., every 20 minutes over the tidal cycle from just above one low tide to just before the following low tide.

Three specific tides are singled out for analysis here, representing a variety of incident wave conditions. Tide 164PM (from the afternoon of the 16th April 1991) experienced a violent storm, with incident swell breaking at a height in excess of 3m.

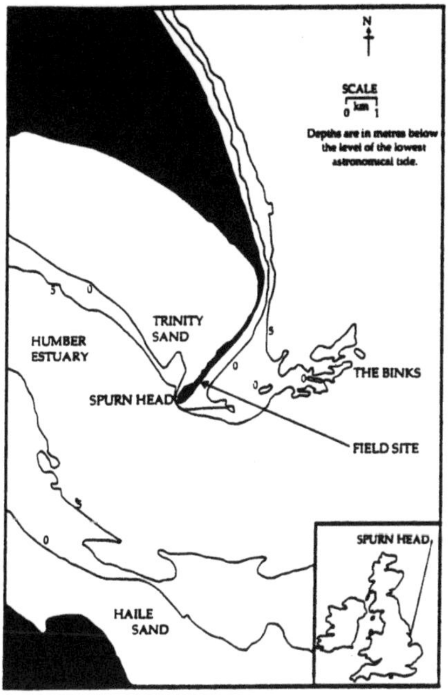

Figure 1 Location of the field site at Spurn Head.

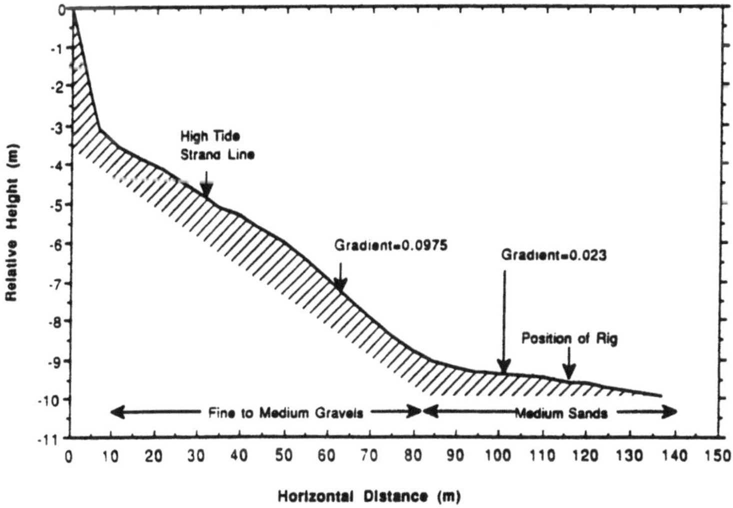

Figure 2 Beach profile at Spurn Head, 23 April, 1991.

Unfortunately this data set is of limited duration due to instrument failure. During Tide 184PM incident waves were mixed swell and local seas with breaker heights of approximately 1.6m. Tide 234PM was during a calm period, with swell breaking with a height varying from 0.7m on the flood to 1.0m on the ebb.

3. Results of the moment analysis

It was expected that the relative magnitudes of the moment terms would be different inside the surf zone from outside the surf zone. An initial analysis was therefore undertaken in which the average moments from these two regions were computed separately.

The results for the dominant bedload moment, the skewness, are shown in figure 3, where the magnitudes of the terms are represented by bars. Outside the surf zone (figure 3a) the largest contribution comes from the skewness of the incident waves themselves, but terms 4 and 8 also contribute significantly. Term 4 represents the movement, by the mean flow, of sand stirred by the incident waves, and term 8 is caused by correlation between the incident wave variance, essentially the envelope of the incident waves, and the long period flow. The significance of the latter term has been pointed out by Foote et al. (1992). Inside the surf zone figure 3b reveals a rather different pattern. The incident wave skewness continues to contribute significantly, but the largest term, in the opposite direction to the incident wave skewness, is term 4, reflecting the larger, offshore-directed mean flows inside the surf zone. Here also term 5 becomes significant, showing that the stirring effect of the long period motion becomes significant as the shoreline is approached.

It is particularly striking in figures 3 that the magnitudes of the terms is similar for each of the tides, despite the differences in incident wave conditions. We will investigate this further below.

Figure 4 shows the equivalent results for the suspended load transport moments. Outside the surf zone all three terms are significant, with the long period term, representing correlation between long period velocity and incident wave variance, predicting offshore transport. Inside the surf zone the long period terms appear on average to be small. Again the relative magnitudes of the terms are similar for each tide.

In order to investigate the variation of these moments with offshore position we have plotted the values for each tide against mean water depth in figures 5 and 6. Figure 5a shows the incident wave skewness term (bedload term 2). As expected from the spatially averaged values (figure 3a), the values for each tide appear to fall on the same general curve, despite the differences in incident wave conditions. Without the normalisation of the moment values by $(u^2)^{3/2}$, the magnitudes of u_s^3 at 4m mean depth differ by an order of magnitude between tides 164PM and 184PM, and incident wave heights differ by a factor of two. The similarity of the normalised values of the moments suggest that the normalisation results in values which are relatively insensitive to wave height variations. There is also some evidence that the scaling with mean depth is more appropriate than the more intuitive scaling with the ratio of depth to breaker depth (e.g. Roelvink and Stive, 1989). For example, the breaker depths for 184PM, and the somewhat different breaker depths on flood and ebb for 234PM, are marked along the depth axis, and there is no clear evidence that the respective moment values are displaced

Figure 3a Averaged values of the largest bedload moments outside the surf zone

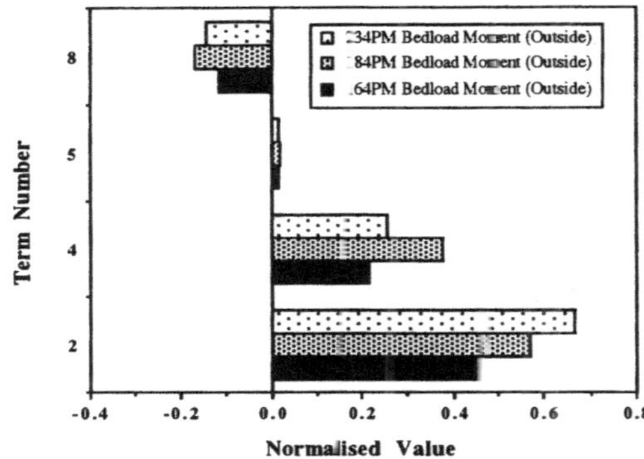

Figure 3b Averaged values of the largest bedload moments inside the surf zone

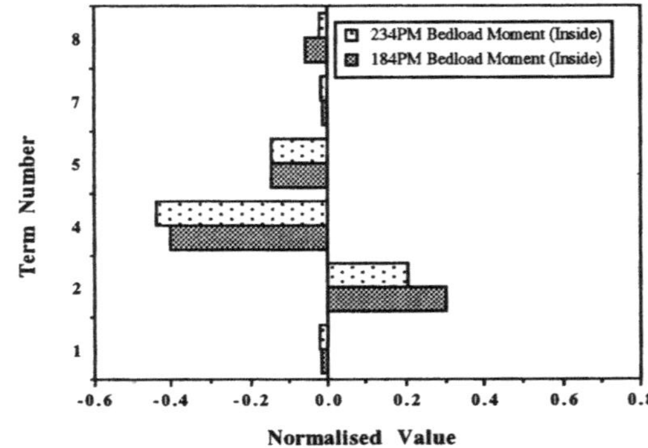

Figure 4a Averaged values of the suspended load moments outside the surf zone

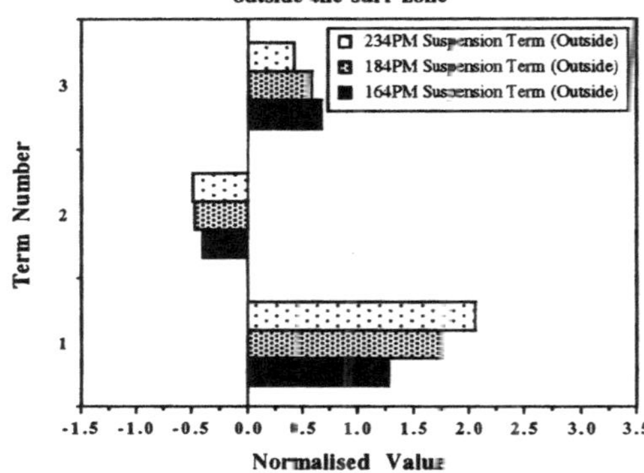

Figure 4b Averaged values of the suspended load moments inside the surf zone

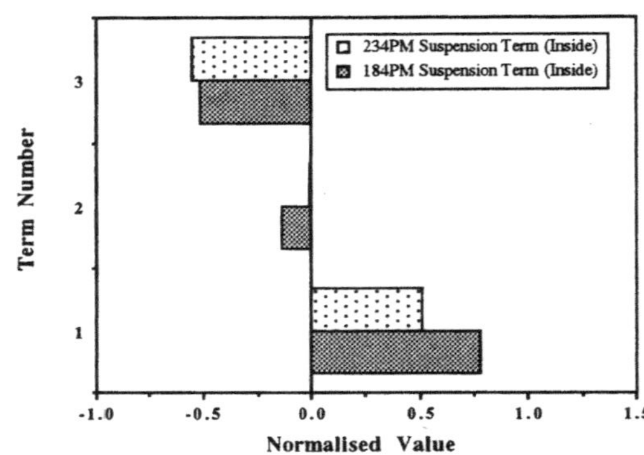

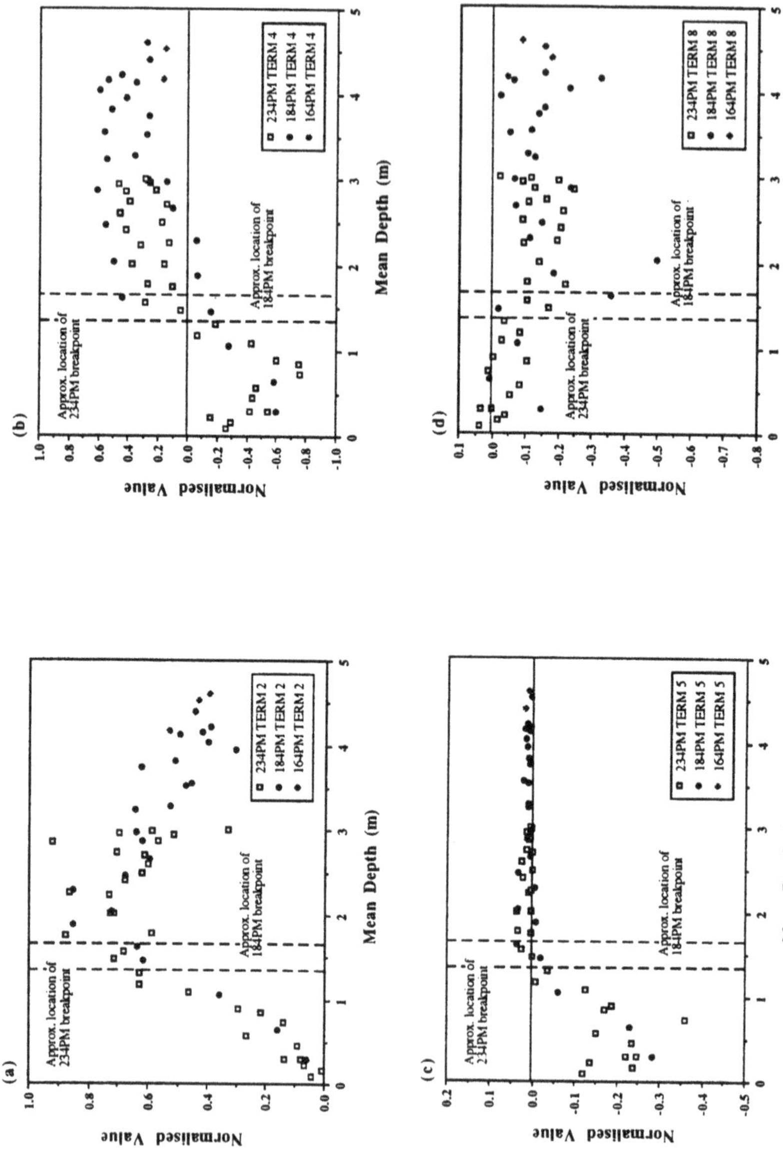

Figure 5 'Shape functions' (variations with mean water depth) for the largest components of the bedload moment term (a) Term 2 (short wave skewness) (b) Term 4 (short wave stirring and mean flow transport) (c) Term 5 (long wave stirring and mean flow transport) (d) Term 8 (transport due to correlation between short and long waves).

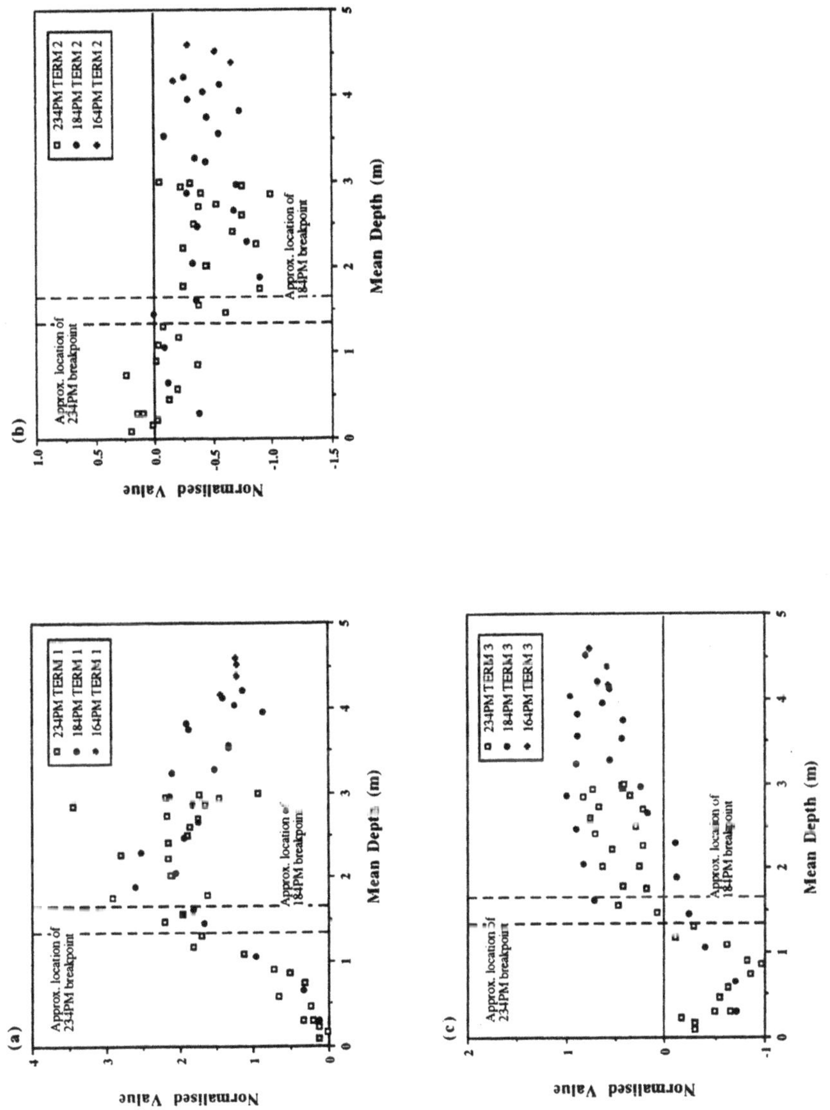

Figure 6 'Shape Function' for components of the suspended load moment term (a) short waves (b) long waves and (c) mean flows.

relative to each other by these differences; in particular the 234PM values show no such displacement between flood and ebb values. These comments on the scaling of the moments apply equally to the other moments in figures 5 and 6, and suggest that, at least for this beach, some quasi-universal 'shape functions' exist for sand transport predictors. Scaling by mean water depth is an unexpected result which needs further confirmation and study.

The general shape of the incident wave skewness (figure 5a) shows that the predicted transport direction is always onshore, with a peak value at a depth of about 2.5m declining to zero at the shoreline and diminishing more slowly seawards, in general agreement with the wave height variations. Bedload term 4 (figure 5b) shows the variation expected from the mean flow, with offshore flow inshore and onshore flow in deeper water. The influence of long wave stirring (figure 5c) is only significant in shallower water, where the mean flow is offshore. Bedload term 8 (figure 5d) shows the expected negative correlation between incident wave envelope and long period motion in the offshore region, consistent with the expectation of forced long period motion. In shallower water this correlation reduces in magnitude and shows some indication of becoming positive in the inner surf zone, as expected in the region where long period modulation of water depth, and hence breaker height, should be more important than the forcing (Foote et al., 1992).

The three suspended load moments show qualitatively very similar 'shape functions' to their equivalent bedload terms. Thus the short wave term (figure 6a) matches the skewness term (figure 5a), the long period term (figure 6b) reflects the same processes as the correlation term (figure 5d) and the mean term (figure 6c) shows the same pattern as figure 5b.

The <u>total</u> 'shape functions' for each tide are shown in figures 7 and 8. Qualitatively these shapes are similar to the mean flow terms, with predicted offshore transport in the inner region and onshore transport further offshore, and this reflects the fact that the other terms, though large, tend to oppose each other. Thus the observed moments in a general way support the classic break-point bar scenario, with mean flow-dominated offshore transport within the surf zone and onshore transport seaward of the break point.

4. Comparisons between moment predictions and observed suspended sand fluxes

It is clearly of interest to test the energetics moment predictors of sand transport rates with direct observations. With the present data sets this is only possible for suspended sand transport, using the OBS sensors. In principle we should also be comparing the moment predictors with depth-integrated transport rates, but in the present preliminary analysis we consider only the sand flux measured at a single height of 0.10m above the bed. The assumption therefore is that this flux is representative of the depth integrated transport. The work of Russell (1993) and others suggests that this assumption is at least qualitatively reasonable in the energetic nearshore zone where sediment suspension concentrations are high.

Using concurrent velocity, u, and suspended sediment concentration, c, measurements, the local sand flux can be calculated as:

$$\overline{uc} = \overline{u_S c_S} + \overline{u_L c_L} + \bar{u}\bar{c} \qquad (2)$$

Figure 7 'Shape Function' for the total bedload moment

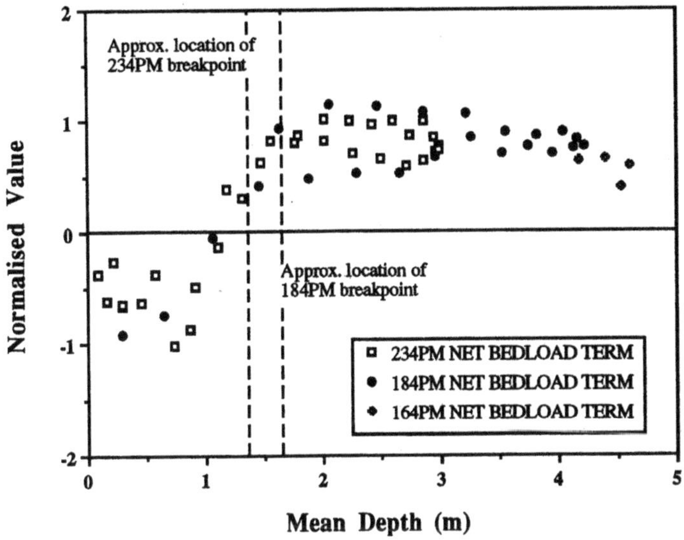

Figure 8 'Shape Function' for the total suspended load moment

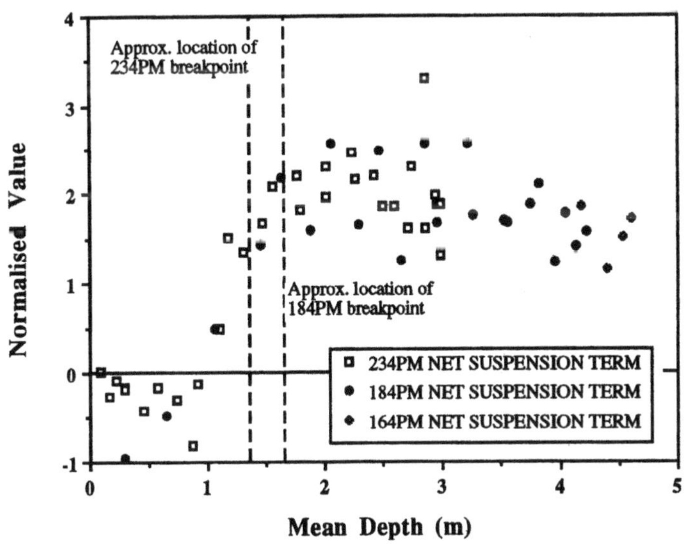

The first second and third terms on the right-hand side of this equation then relate to the incident wave, long period and mean flow suspended transport moment terms respectively.

Figure 9 compares these moment and flux measurements, for values averaged across the surf zone and seaward of the surf zone. The moment values are un-normalised, as required for absolute transport predictions. However, no attempt has been made at this stage to compare absolute values of moment and flux estimates: the purpose of figure 9 is to compare relative values for the different tides and different modes of suspended sand transport.

A number of conclusions can be drawn from figure 9. Outside the surf zone, the moment predictions and fluxes compare very well for both mean flow and long period flow components. In fact a simple constant factor applied to the moment predictors would predict these transport modes for each tide to within about a factor of two. However, the moments are seen to be very poor predictors of the transport due to incident waves. In each case the sign of the predictions is opposite to the observed fluxes, and magnitudes are in relatively poor agreement also. It is possible that this lack of agreement reflects the complex behaviour of sand transport due to waves over bedforms, particularly rippled beds which are likely to occur outside the surf zone. Davidson et al. (1993) discuss the possible influence of bedforms in the incident wave sand transport at Spurn Point. They find a distinct asymmetry of sand suspension concentration on the flood and ebb stages of tide 234PM which they attribute to the presence of degrading ripples on the ebb. Clearly the moment predictors cannot provide satisfactory results under such conditions.

Results from inside the surf zone are less clear, but again qualitative agreement between moments and fluxes is found for mean flow and long period modes of transport but poor agreement is found for incident wave transport.

5. Discussion of transport processes specific to macro-tidal beaches

The moment transport functions discussed above suggest that it is possible to identify 'shape functions' for transport predictors which have a consistent shape as functions of mean water depth. On a macro-tidal beach, the simplest process which might be occurring is passive advection of such shape functions over the inter-tidal range.

To visualise the influence of such advection, consider a transport shape function which reflects the observed onshore transport outside the surf zone, the offshore transport within the surf zone, and includes a narrow swash zone region of onshore transport. If this transport shape is advected back and forth over a beach, a number of distinct regions arise, differing in the range of transport conditions they experience. Figure 10 shows schematically what can arise for a particular choice of tidal excursion and surf-zone width.

Integrating the transport function over a tidal cycle at each offshore location produces an overall transport function for this macro-tidal beach which takes the form shown in figure 11a. As might be expected, this is an elongated version of the initial transport shape function. Its influence on an initially plane beach profile is suggested in the lower part of figure 11b, where regions of transport divergence display erosion and regions of convergence display accretion. Interestingly, the resulting profile shape is similar to the observed macro-tidal beach profile, with a wide, low-slope inter-tidal

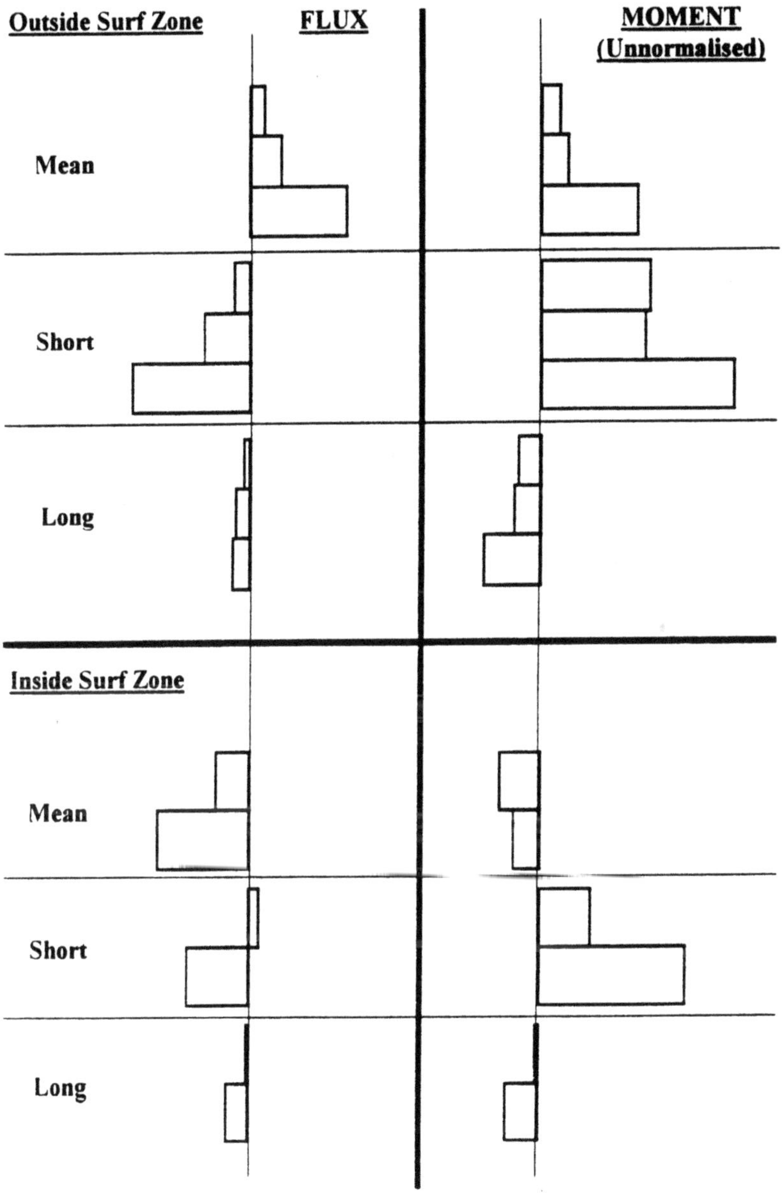

Figure 9 Comparison between averaged suspended load moments and measured suspended load fluxes. A single arbitrary multiplier has been used to scale the moment terms relative to the flux terms.

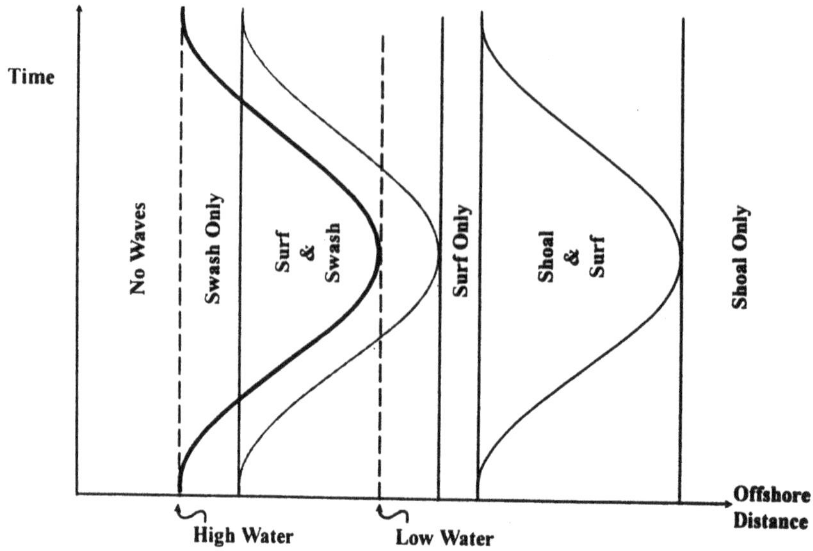

Figure 10 Cross-shore transport zones resulting from tidal excursion of the shape function.

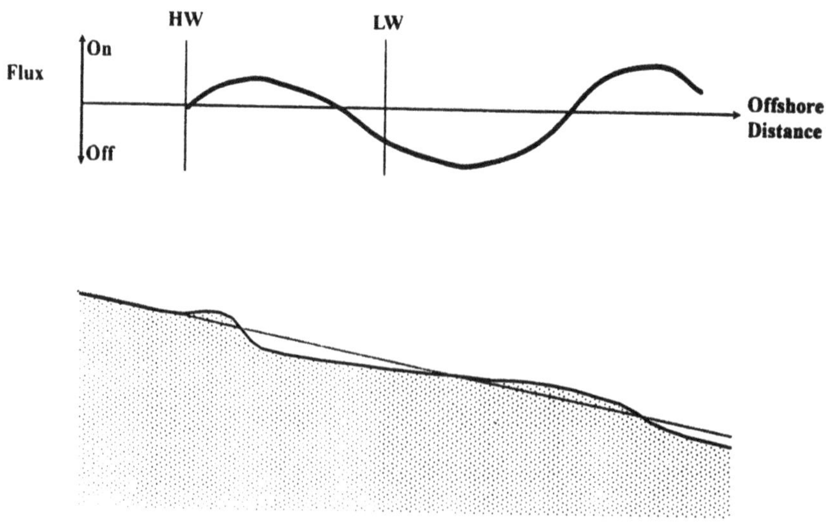

Figure 11 The net transport over a tidal cycle, and the possible resulting bed profile.

region and a steeper high tide profile and a steeper profile below the low tide level. In this model therefore the macro-tidal beach profile arises largely as a stretched 'break-point bar'. This simple schematic model clearly shows some promise in providing a possible explanation for the characteristic profiles of macro-tidal beaches and further work developing more quantitative models is in progress.

Other effects on macro-tidal beaches are more subtle. In phase with the tide level is the effect of changing effective beach profile as tide level changes. It is known that long period waves are particularly sensitive to beach slope at the shoreline (e.g. Holman and Bowen 1982) so it is likely that transport functions will be modified by this process. There is also a number of possible effects which are in quadrature with the tide level, i.e. they differ between flood and ebb conditions. Davidson et al. (1993) discuss the possible influence of bedforms which lag behind the changing hydrodynamic conditions, and cite observations of suspended sediment concentration as evidence for this effect. The B-BAND data sets also show marked interaction between longshore wave-induced currents and the ebb and flood tidal flows, and further work is in progress to quantify the importance of this effect. One other effect which requires investigation is the influence of changing groundwater levels within the beach as tide levels change. Turner (1994) has conducted some important field and theoretical studies on this topic and shown that it may be an important process.

6. Conclusions

1. Moments of the velocity field relevant to energetics modelling of sand transport rates have been found to vary consistently with mean water depth and are relatively insensitive to incident wave conditions. The dependence on mean depth rather than depth scaled by breakpoint depth is unexpected and requires further investigation. Nevertheless the suggestion of consistent 'shape functions' for these transport-predicting moments should simplify modelling of sand transport on beaches.

2. The influence of incident wave velocities is the strongest factor in the moment predictors, but incident waves combined with mean flows and with long period flows also provide significant contributions. Near the shoreline, interaction between long period motion and mean flows also becomes significant, at least for bedload transport. However the overall moment shape functions are qualitatively similar to the variation of mean flow with mean depth, with the other contributions tending to oppose each other. Thus the total transport predictions suggest offshore transport inside the surf zone and onshore transport outside the surf zone.

3. Comparison of the suspended sand transport moment predictors with observed sand fluxes reveals that the predictors can provide transport estimates accurate to within about a factor of two for the influence of mean flows and long period flows, but that predictions of the influence of incident waves is very poor, with even the direction of transport being wrong in most cases. It is suggested that this may be the result of the flow over bedforms such as ripples, whose influence cannot be properly modelled by the energetics approach.

4. The suggestion of consistent 'shape functions' for sand transport predictions has lead to the development of a model, as yet schematic, for the development of macro-tidal beach profiles. Advection of a simple transport shape function by the tidal excursion has been shown to lead to a beach profile similar to the characteristic macro-tidal profile. Work is continuing on this model. However, it is recognised that a simple advection of a constant transport shape function is unlikely to explain all features of macro-tidal beaches. A full model will need to include the effects of changing effective beach profile, non-equilibrium bedforms, interactions between wave-induced and tidal mean flows and perhaps asymmetry in the behaviour of ground water in the beach.

7. References

Bagnold, R.A., 1966. *An Approach to the Sediment Transport Problem from General Physics.* U.S Geol. Survey, Washington, D.C., Prof. Paper **422-1**, 37pp.

Bailard, J.A., 1981. An energetics, total load sediment transport model for a plane sloping beach. *J. Geophys. Res.*, **86,** 10,938-10,954.

Bailard, J.A., 1987. Surf zone wave velocity moments. *Proc. Coastal Hydrodynamics*, ASCE, New York. 328-342.

Bowen, A.J., 1980. Simple models of nearshore sedimentation: Beach profiles and alongshore bars. in *The Coastline of Canada*, ed. S. B. McCann. Geological Society of Canada Paper 80-10, Ottawa.

Davidson, M.A., Russell, P.E., Huntley, D.A., Hardisty, J. and Cramp, A., 1992. An overview of the British Beach and Nearshore Dynamics (B-BAND) programme. *Proc. 23rd Int. Conf. on Coastal Engineering*, ASCE, New York, 1987-2000.

Davidson, M.A., Russell, P.E., Huntley, D.A. and Hardisty, J., 1993. Tidal asymmetry in suspended sand transport on a macrotidal intermediate beach. *Marine Geology*, **110**, 333-353.

Foote, Y., Huntley, D., Davidson, M., Russell, P., Hardisty, J. and Cramp, A., 1992. Incident wave groups and long waves in the nearshore zone. *Proc. 23rd Int. Conf. on Coastal Engineering*, ASCE, New York, 974-989.

Guza, R.T. and Thornton, E. B., 1985. Velocity moments in the nearshore zone. *J. Waterways Port, Coastal and Ocean Engineering*, **111,** 235-256.

Holman, R.A. and Bowen, A.J., 1982. Bars, bumps and holes: Models for the generation of complex beach topography. *J. Geophys. Res.*, **87**, 457-468.

Roelvink, J.A. and Stive, M.J.F., 1989. Bar-generating cross-shore flow mechanisms on a beach. *J. Geophys. Res.*, **94,** 4785-4800.

Russell, P.E., 1993. Mechanisms for beach erosion during storms. *Continental Shelf Research: Special Issue on Nearshore and Coastal Oceanography,* **13(11),** 1243-1266.

Short, A.D., 1991. Macro-meso-tidal beach morphodynamics - and overview. *J. Coastal Res.*, **7(2),** 417-436.

Turner, I., 1994. Water-table outcropping on macro-tidal beaches: A simulation model. *Marine Geology,* in press.

SEDIMENT SUSPENSION BY TURBULENCE IN THE SURF ZONE

G G Smith and G P Mocke

Research Engineers, CSIR, P O Box 320, Stellenbosch, 7600, South Africa

The process of suspension of sediment under broken waves in the surf zone is investigated through experimental data analysis and computational modelling. Sediment concentration and hydrodynamic measurements from field experiments are used, as well as a large database of time-averaged recordings. The sediment suspension due purely to diffusive turbulence is investigated via a vertical mixing apparatus. The ensemble of measurements show sediment suspension to occur in the form of intermittent events. In addition, a tendency for increased suspension toward the end of wave groups is found, indicating the progressive stirring effect of successive waves in a group. These results highlight the predominant role of turbulence in sediment suspension. The wave-period averaged turbulence structure is found to be well predicted by a two equation (k,ε) turbulence model. A parametrization for the bottom reference concentration using appropriate wave breaker magnitudes provides an effective bottom boundary condition for modelling the suspended sediment distribution. Measured profiles were found to be well predicted through incorporation of the vertical diffusivity as determined from the turbulence model within the classical advection-diffusion equation.

INTRODUCTION

Suspended sediment concentrations in the surf zone have been measured to be up to several orders of magnitude higher than is typical for an unbroken wave regime, particularly at the higher elevations above the bed. In assessing the influence of wave breaking on the suspension process, Nielsen (1984) and Shibayama *et al.* (1986) have identified the action of the large scale vortices generated coincident with wave plunging. The convective contribution of essentially two-dimensional vortices is, however, generally restricted to a limited transition zone, shorewards of which the surf-zone is characterised by an inner region physically analogous to a hydraulic bore.

Within this inner region Nadaoka *et al.* (1989) have identified an eddy structure whereby the surface bore is dominated by a nearly two-dimensional flow structure, bounded below by strongly three-dimensional obliquely descending eddies which bring highly intermittent turbulence to the bottom. The dominant action of this turbulence for sediment suspension is verified by means of concentration measurements and visualisation studies made by Nadaoka *et al.* (1988) and Sato *et al.* (1990) in a wave flume.

A suspension dynamic that depends primarily on wave breaker turbulence will inherently be intermittent in nature. Intermittent sediment suspension in the surf zone, being only loosely correlated to the wave cycle, would lend itself to a wave-averaged predictive modelling approach, rather than a wave cycle dependent approach. In the present exercise the suspension process is investigated with reference to field and laboratory measurements as well as wave-averaged predictive modelling.

MEASUREMENT TECHNIQUES

Field measurements

During a one-day field exercise, continuous near-bed sediment concentration measurements were obtained via an array of three optical backscatter sensors (OBS). In addition, time-averaged samples from the same levels as the OBS were made with a suction sampling device operated by a vacuum pump. The exercise was conducted at Strand near Cape Town in South Africa (Figure 1). The instrumentation was attached to a light scaffold frame (Figure 2 and Plate 1), resembling that of Black and Rosenberg (1991a,1991b). This frame was manually deployed. In addition to the sediment measurements, waves were recorded via a resistance wave gauge as well as two pressure sensors (at elevations of 0.71 m and 0.28 m above the bed). Current velocity components on/offshore and longshore were measured with two electromagnetic current meters developed by the CSIR, situated at 1.34 m and 0.91 m above the bed.

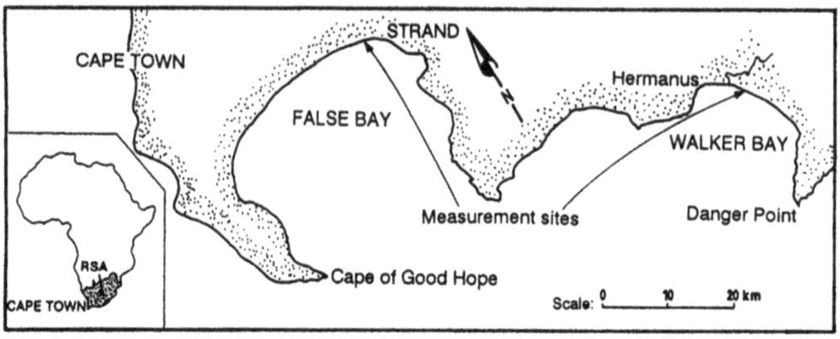

Figure 1: Locality Map

The water depth at the frame varied with the tide from 1.8 m to 1.1 m and the scaffold frame was within the surf zone except for the highest tidal conditions. Under the relatively calm conditions with minimal wind, wave breaking ranged from spilling to mildly plunging. Significant wave heights varied from 0.72 m to 1.18 m, with an average wave period of 13 seconds.

In a separate field exercise a comprehensive set of continuous measurements of suspended sediment was recorded with a vertical array of seven transmissometer sediment sensor units. These units were fixed to a frame together with a pressure transducer and an array of five time-averaged concentration suction samplers and deployed via a helicopter at Walker Bay (location shown in Figure 1) (Coppoolse *et al.*, 1992). Additional instrumentation included eight current meters, a further five pressure sensors and a wave measuring buoy, situated at fixed and variable locations, to a depth of 20 m.

Laboratory measurements

A rather unique set of measurements were obtained from a laboratory experiment in which turbulence is generated from a vertically oscillating grid within a glass-walled tank (Figure 3). The stroke and frequency of the grid can be varied according to the intensity of turbulence required. This "mixing tank" is based on similar devices used for experiments on density gradient mixing due to diffused turbulence (E. and Hopfinger, 1986). The suspension of sand from a bed of sand at the base of the tank is recorded at various elevations using an optical backscatter sensor.

During the large wave flume LIP11D beach profile study (Sanchez-Arcilla *et al*, 1994) successful measurements of both time-averaged and instantaneous sediment concentrations at a number of stations along the flume were obtained with vertical arrays of OBS sensors and suction samplers. Simultaneous recordings of pressures, currents and video images were also made. Use is made of the time averaged wave height and sediment concentration results in the present study.

Plate 1: Light scaffold measurement frame.

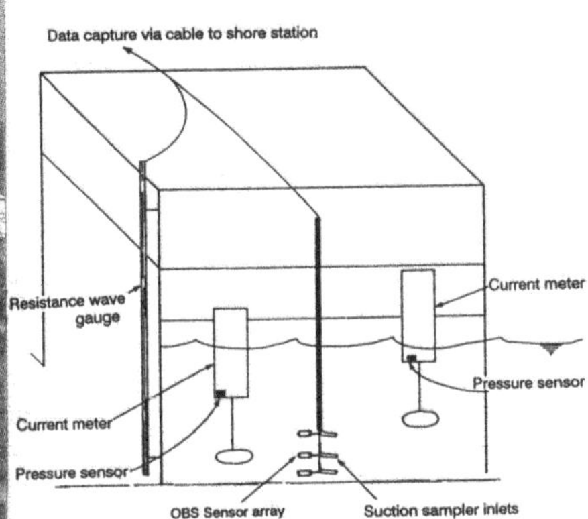

Figure 2: Schematic illustration of the frame with cable for shore-based recording

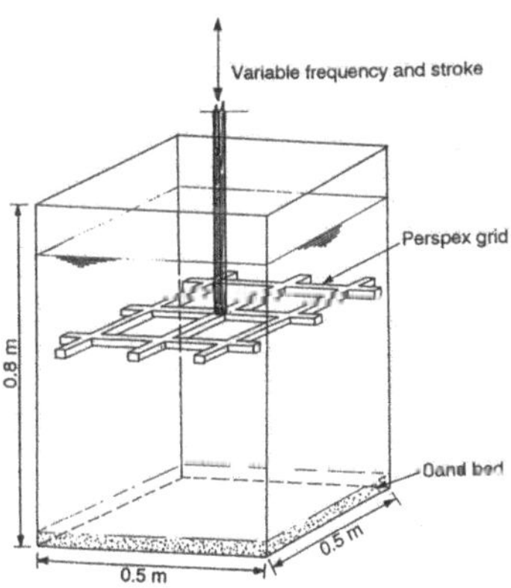

Figure 3: The mixing tank.

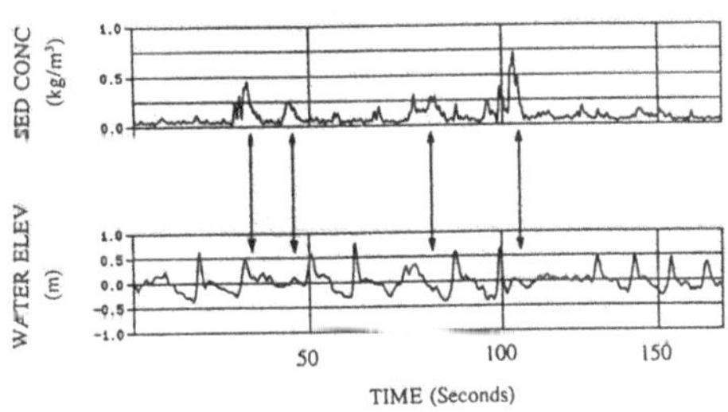

Figure 4: Time series of sediment concentration and the corresponding wave record.

ANALYSIS

Suspension relative to incident waves

Time series of sediment concentrations in the surf zone indicate intermittent periods of elevated concentrations, i.e suspension events, which are not necessarily discreetly correlated with the largest waves or their coincident orbital velocities. An example (Figure 4) is the wave occurring after some 65 seconds; although it is one of the highest waves, no corresponding suspension event is evident. In addition, no obvious link of suspension events to the wave phase is obvious, as indicated by the arrows in the figure. In exploring this aspect further, a detailed analysis of the occurrence of suspension events was carried out. This required the establishment of consistent criteria defining a suspension event. As can be seen in figure 4 the concentration record is typically made up of a quasi-steady background concentration interspersed with events. An event is defined in the present study by a magnitude of twice the average background concentration, which is determined visually, for each record of 200 seconds. Because of relatively rapidly changing conditions due to the tides, this definition was found to be suitable for the data set. Thus it was used in preference to an alternative definition based on the mean and standard deviation (Jaffe and Sallenger, 1992).

Figure 5 is a histogram of the suspension events, incorporating a total of 200 events. The figure illustrates the percentage occurrence of events occurring, at different stages after the wave peak, throughout the wave cycle. (A typical wave trace is included in the figure). If events were totally random and uncorrelated with the wave phase, then the percentage occurrence of events would be of a similar magnitude throughout the wave cycle. This is generally the case, apart from somewhat higher and lower occurrences of events at 0.125T and at 0.5T respectively, possibly due to wave orbital motions. This fairly random distribution of events throughout the wave period supports the idea of random, intermittent events occurring in the surf zone, as would be generated by intermittent, three dimensional turbulent eddies (Nadaoka *et al*, 1989).

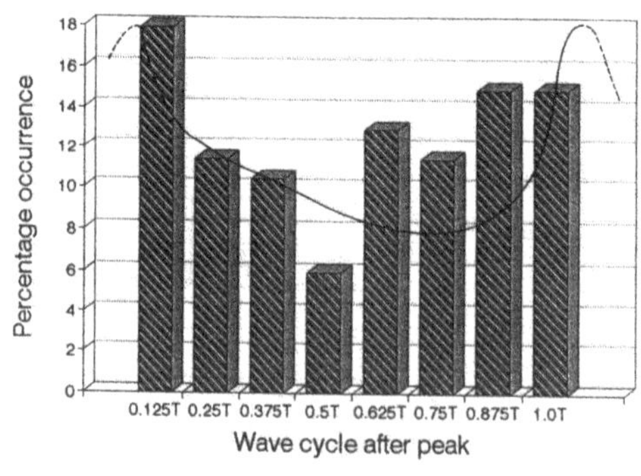

Figure 5: Histogram of sediment suspension events throughout the wave cycle

Considering that these intermittent events may be caused by turbulent eddies, it is worthwhile to investigate the effect on sediment suspension of a purely diffusive transport of turbulence, as is generated in the mixing tank. Time series of sediment concentration recorded in the tank at elevations of 5 mm and 150 mm are shown in Figure 6. In this case the mixing grid, having 25 mm square bars spaced at 125 mm, was oscillated with a frequency of 4 Hz and a stroke of 100 mm. Despite a relatively constant time averaged concentration, the occurrence of intermittent suspension events due to the turbulence can be clearly seen. It is interesting to note that the "spiky" character of this record, with distinct events occurring intermittently, is similar in nature to the concentration time series recorded in the surf zone (e.g. Figure 4 and Figure 7).

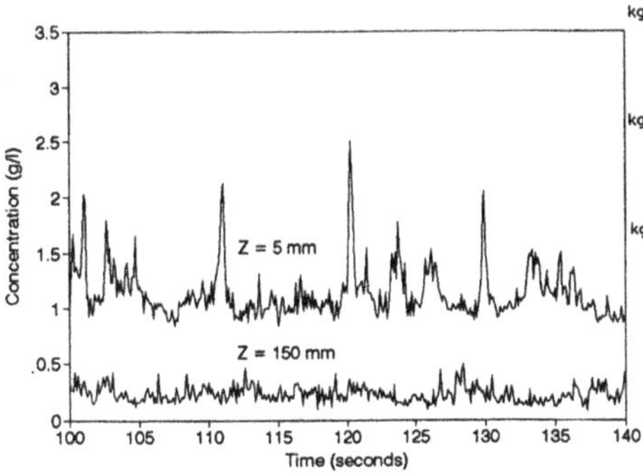

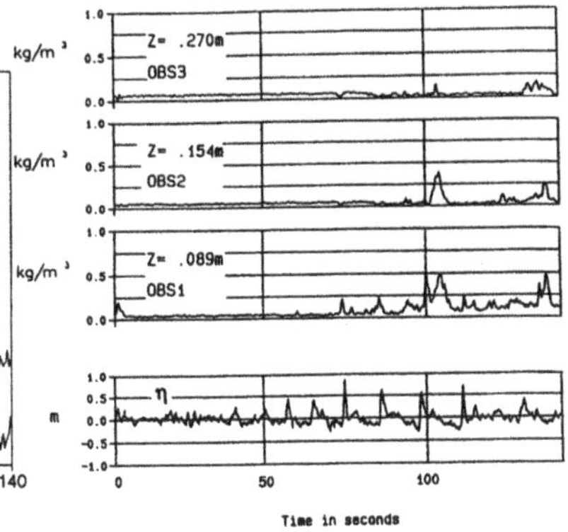

Figure 6: Sediment concentration at two elevations in the mixing tank

Figure 7: Time series of sediment concentration and the corresponding wave record.

Suspension relative to wave groups

Observations of the field data indicate that even though the intermittent nature of sediment suspension events persists, it is often enhanced during wave groups. This enhanced suspension does not occur with every wave group, but seems associated with longer wave groups and those that include larger waves. It was further observed that this suspension increases with the passage of the wave group, so that the highest concentrations are experienced toward the end of the group. This is exemplified in Figure 7, where a wave group is evident from approximately 70 seconds to 120 seconds in the record. In response to the group, the suspension events progressively increase to reach a maximum at the end of the group. This trend is also evident in Figure 4, and it is thought that this is related to a progressively increasing stirring effect of wave generated turbulence.

In investigating this further, conventional spectral-type analyses are avoided due to inherent limitations in applicability to the "spiky" non-linear nature of the concentration record. Use is rather made of the Smoothed Instantaneous Wave Energy History (SIWEH) function, together with cross-correlation. The SIWEH function allows a description of the group activity of waves in a wave record, by computing the distribution of wave energy along the time axis. The SIWEH function is defined as follows:

$$E(t) = \frac{1}{T_p} \int_{-\frac{T_p}{2}}^{\frac{T_p}{2}} \eta^2(t+\tau)\, d\tau \qquad (1)$$

where: $\eta(t)$ is the water surface displacement and
T_p is the wave period of peak spectral energy

A wave record and the corresponding SIWEH function (smoothed using the Bartlett window) are illustrated in Figures 8 (a) and (b), where the higher peaks in the function can be seen to correspond to wave groups. In a similar fashion, the SIWEH function was found to be effective in the description of periods of higher sediment concentrations, as can be seen in Figures 8(c) and (d). Although it would seem more natural to use the raw concentration record, the application of the SIWEH function facilitates both the filtering and the description of groups of suspension events. Cross correlation of the two SIWEH records (Figure 8 (e)) indicates a degree of correlation at a lag of approximately 45 seconds, i.e. the peak of the smoothed sediment concentration tends to occur at 45 seconds after the wave group peak, for this case. Since 45 seconds is approximately equivalent to three wave periods, this would seem a reasonable delay after the peak of the wave group.

When evaluating this delay time the possibility of suspension occurring a small distance from the measuring sensor, and subsequently being convected past the sensor, must be considered. This effect may add to the lag time between the suspension and wave group peaks.

An additional feature of the correlation shown in Figure 8 is a periodicity in the correlation peaks, related to the period of the wave groups (approximately 100 seconds in this case).

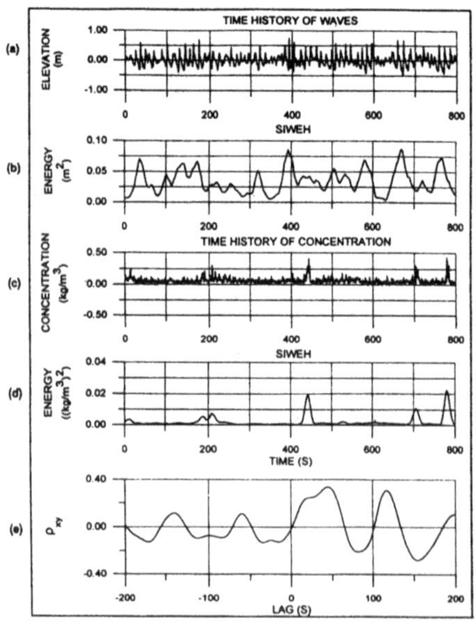

Figure 8: Sediment concentration time series and the coincident wave record, their corresponding SIWEH traces, and cross correlation between the latter.

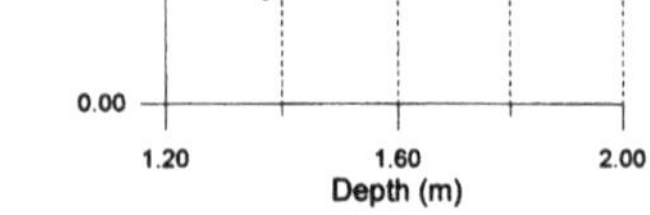

Figure 9: Lag of peak sediment suspension after the wave group peak versus water depth.

Tidal action during the measurement exercise resulted in a shift in effective water depth. The above-mentioned SIWEH and correlation analysis was repeated for times at different water depths. The resulting correlation lag between the wave group and smoothed sediment concentration was plotted against depth (Figure 9). As can be seen, the lag tends to increase with increasing depth. This is likely related to the variable distance from the surface source of turbulent stirring.

Time averaged results

Figure 10 is a plot of time-mean concentrations evaluated through averaging continuous OBS records at various elevations in the mixing tank. The elevated concentrations throughout the depth exhibit a distribution of sediment concentrations similar to what is typically measured in the surf zone, supporting the notion of turbulence driven suspension in this region.

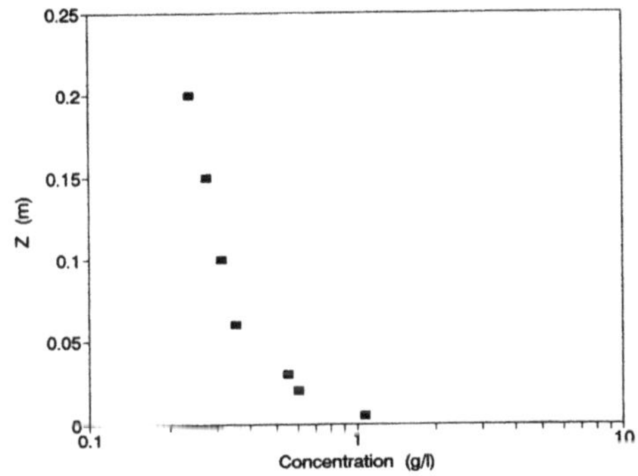

Figure 10: Time-averaged sediment concentrations versus elevation in the mixing tank.

An essential boundary condition for the mathematical modelling of suspended sediment profiles is the effective bottom reference concentration C_b. Field and laboratory breaker zone measurements extracted from the database compiled by Van Rijn (1991) as well as profiles obtained from the LIP11D exercise are examined for determining this parameter. Where measurements allowed, a fit of the predictive model discussed below was employed for determining a reference value at the effective bottom. In other cases an exponential fit of the three lower measurements was extrapolated to the bottom. This analysis provided a total of some 69 C_b values covering 5 different experimental exercises.

Previous attempts to parametrize this reference value have focused primarily on the rôle of turbulence generated by mean shear (τ) near the bottom as reflected in the Shields parameter $\theta = \tau/(\rho(s\text{-}1)gd)$. Because of the significance of near surface generated wave breaker turbulence, however, it is of fundamental importance to also incorporate this contribution in any parametrization exercise. The energy dissipation in a breaking wave may be modelled as for a periodic bore. Thus the equivalent energy available for sediment suspension is related to this dissipation, as a term of production. Modelled according to the dissipation in a hydraulic jump the breaker production term may be expressed as:

$$P = \frac{gH^3}{4h^2T} D \tag{2}$$

The dimensionless energy dissipation $D=h^2/h_t h_c$ (Svendsen, 1984), with h_t and h_c the depths below the wave trough and crest level respectively. Complementing this parameter representing the injection of TKE into the flow is the relative degree of depth penetration for this turbulence. It is proposed therefore that an appropriate scale for quantifying the breaker contribution would be P/h, where h is the mean water depth.

Combining this scale together with the Shields parameter the best fit regression curve against the database measurements is as shown in Fig. 11 and provides the following expression for predicting the bottom reference concentration:

$$C_b = 2{,}16\ \theta + 21{,}35P/h \tag{3}$$

In carrying out the exercise the relative importance of component terms such as the wave height was clearly evidenced.

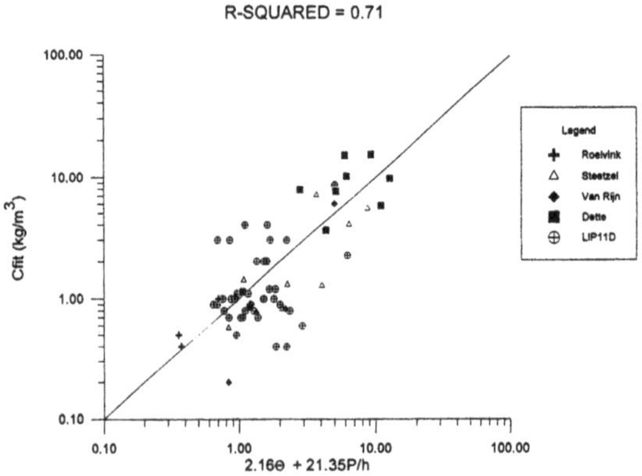

Figure 11: Plot of measured versus calculated values of bottom reference concentration C_b.

COMPUTATIONAL MODELLING

System of equations

The time-mean vertical distribution of suspended sediment is described by the time-independent form of the classical advection-diffusion equation:

$$0 = \frac{\partial}{\partial z}\left(D_c \frac{\partial C}{\partial z} + w_s C\right) \tag{4}$$

where C is the sediment concentration, w_s is the sediment fall velocity and D_c is the eddy diffusivity which is assumed equivalent to the eddy viscosity ν_t defining the turbulent momentum flux. Following the analysis of Prandtl-Kolmogorov the eddy viscosity may be related to the TKE density k ($\overline{u_i u_j} = 2k$) and its rate of dissipation ε such that $D_c = \nu_t = c_\mu k^2/\varepsilon$, where c_μ is an empirical constant. For a unidimensional flow the dimensionless form of the time-averaged transport equations for k and ε may be expressed in the following manner:

$$0 = \frac{\partial}{\partial z}\left(D_k \frac{\partial k}{\partial z}\right) + P + P_v + G - \varepsilon \tag{5}$$

$$0 = \frac{\partial}{\partial z}\left(D_\varepsilon \frac{\partial \varepsilon}{\partial z}\right) + [c_{1\varepsilon}\left(P + P_v + G\,(1 - c_{3\varepsilon})\right) - c_{2\varepsilon}\,\varepsilon]\,\frac{\varepsilon}{k} \tag{6}$$

where the above equations have been made dimensionless by characteristic flow parameters such as h and the wave celerity. The dimensionless diffusion coefficients are defined as:

$$D_k = 1/Re + \nu_t/\sigma_k\,, \quad D_\varepsilon = 1/Re + \nu_t/\sigma_\varepsilon \tag{7}$$

where the molecular viscosity term Re, the flow Reynolds number, is negligible in comparison to the turbulence contribution.

The forcing functions are the production P (Equation 2) of TKE due to wave breaker induced turbulence and $P_v = \nu_t\,(\partial u/\partial z)^2$ due to mean shear. The buoyancy term is expressed as $G = R_i D_c\,\partial C/\partial z$. This term, which includes a reference dimensionless Richardson number $R_i = g\,\Delta\rho/\rho\,h/c^2$, expresses the effect of turbulence damping due to the gradient in concentration. It may be noted that the selected length and velocity scales for defining R_i are the water depth (h) and wave celerity (c) respectively. The choice of empirical constants, discussed in Mocke (1991), are close to classical values (Rodi, 1980) except for c_μ (= 0.3).

Boundary conditions

The main forcing function for the flow is the production of turbulence in the surface roller. This production is assumed to be imposed at the upper boundary of the flow regime, which is taken at the mean water level. The surface boundary conditions for k,ε are controlled by the imposed production P_b, whilst at the bottom magnitudes computed at the limit of the vicious sublayer (Rodi, 1980) are imposed.

At the surface boundary, the vertical turbulent and gravity flux of sediment is assumed to be in equilibrium:

$$D_c \frac{\partial c}{\partial z} + w_s C = 0 \tag{8}$$

The parametrized bottom reference concentration formulation (Equation 3) is imposed at the bottom boundary, which in this case is immediately outside the viscous sublayer.

Resolution procedure

The set of equations are discretized in an implicit scheme using the finite difference method. A central differencing approach is used to construct a tri-diagonal matrix which is solved analytically by the Thomas algorithm.

COMPUTATIONAL RESULTS

Wave breaker turbulence

Initial model simulations were carried out in the absence of the sediment phase in an effort to evaluate the effectiveness of the turbulence model for predicting the observed dynamic as well as measured turbulence parameters. The experimental arrangements and description of measurements are detailed in Stive (1980) and Okayasu *et al.* (1988).

Solving the set of equations (5) and (6), predictions of time-averaged kinetic energy are found (Figure 12) to compare favourably with measurements of Stive (1980).

Due to dimensional considerations, the eddy viscosity profile displays similarities to that for k, indicating an appreciable variation over depth. Estimates of eddy viscosity as determined from the measurements of Okayasu *et al.* (1988) are likewise found to be well predicted (Figure 13).

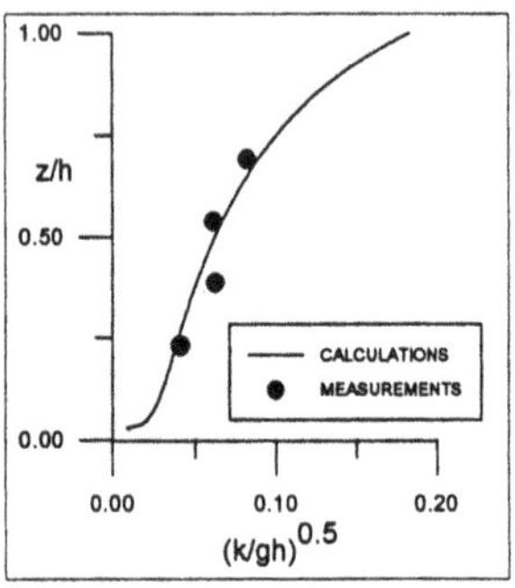

Figure 12: Intercomparison of computed TKE profile and measurements of Stive (1980).

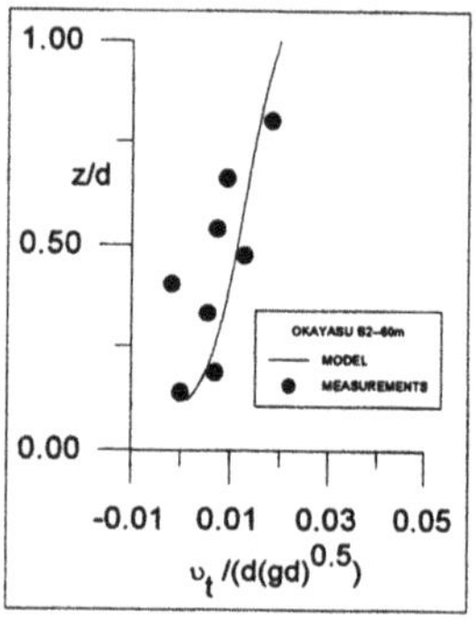

Figure 13: Intercomparison of computed (-) and estimated (•) υ_t profile of Okayasu *et al* (1988)

Suspended sediment

For the computation of the suspended sediment profiles the turbulence formulations (5) and (6) are extended by the scalar diffusion equation (4). Model comparisons are somewhat hindered however, by the fact that presently there exists no simultaneous measurements of suspended sediment and turbulence quantities. Furthermore, there is a lack of suspended sediment measurements that also include an accurate assessment of the cross-shore wave energy loss important for determining local dissipation.

Despite these limitations the model was applied to a number of cases where suspended sediment was measured under breaking waves. In these cases, which include both laboratory and field measurements, the given wave height was exploited for determination of the breaker energy production. Where the sediment fall velocity of the suspended sediment is not provided, it is estimated from the sediment size characteristics.

Although recognized as potentially significant, little attention has been given to incorporating the turbulence damping effect of a concentration gradient in previous attempts at modelling suspended sediment in the surf-zone. With the inclusion of the buoyancy term G in the equations (5) and (6) however, this contribution can be assessed. As is in fact shown in Figure 14 where a model simulation with and without the buoyancy term are compared with measurements from LIP11D the turbulence damping effect only slightly reduces the amount of sediment in suspension.

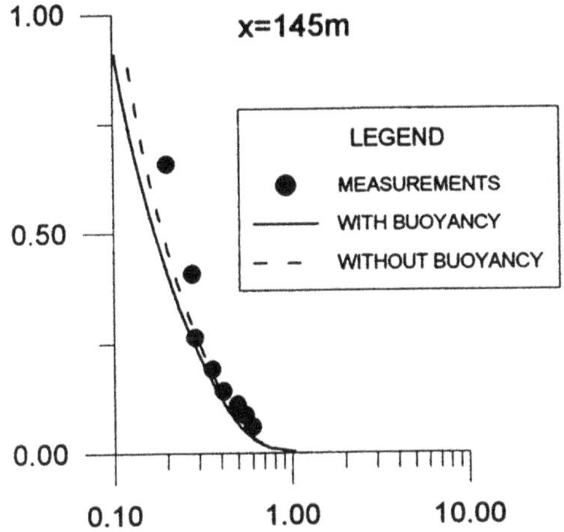

Figure 14: Predicted concentration profile with (-) and without (---) buoyancy effects (LIP11D, Test 1A, X = 145 m).

As may be observed from the comparisons with the concentration measurements presented in Figure 15(a) and (b), predictions are generally in reasonable agreement. The sensitivity of the model to the wave height and bottom friction velocity and reference concentration was found to be important,with adjustments of less than 10% resulting in improved correlations in many cases.

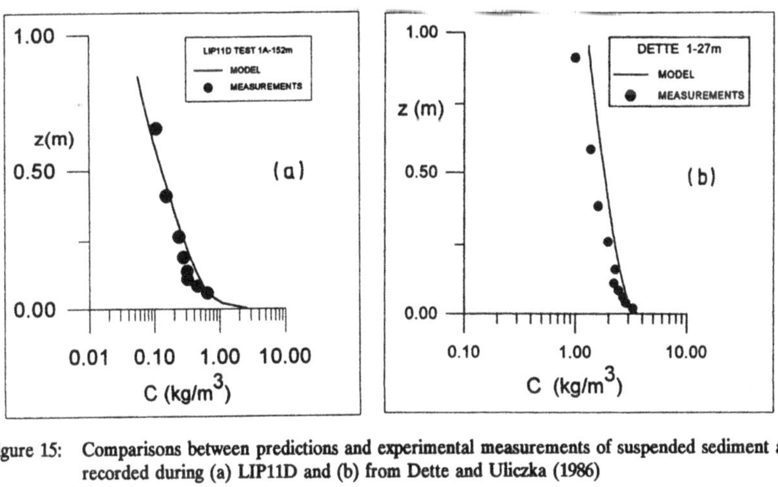

Figure 15: Comparisons between predictions and experimental measurements of suspended sediment as recorded during (a) LIP11D and (b) from Dette and Uliczka (1986)

CONCLUSIONS

Time series analyses indicate that sediment suspension in the surf zone typically occurs in the form of intermittent events. This is the nature of suspension that would be expected under the action of turbulent eddies.

The character of this sediment suspension is similar to that of laboratory-generated suspension, under a purely diffusive transport of turbulence.

A further finding, making use of the SIWEH function in the analysis, indicates a tendency for suspension toward the end, or just after the passage of larger and/or higher energy wave groups. This effect is likely the result of progressive turbulent stirring of successive waves in the group.

Parametrization of the bottom reference concentration is facilitated making use of scales relevant for describing the turbulent contribution of the breaker. In this analysis use was made, where applicable, of a predictive model for determining a reference value at the effective bottom.

The predictive k,ε turbulence model is found to adequately describe the vertical mixing due to wave breaking and hence the resultant sediment distribution.

ACKNOWLEDGEMENTS

The authors would like to sincerely thank L Engelbrecht for his contribution to statistical and modelling aspects of the study.

REFERENCES

Black, K.P. and Rosenberg, M.A. (1991a) Suspended sediment load at three time scales. *Coastal Sediments '91*, ASCE, pp 313-327.

Black, K.P. and Rosenberg, M.A. (1991b) Hydrodynamics and sediment dynamics in wave-driven environments. *Technical Report No.* **13**, Victorian Institute of Marine Sciences, Melbourne.

Coppoolse, R., Schoonees, J., Mocke, G.P., Smith, G. (1992). Hydrodynamics and Sediment Concentrations at Walker Bay. *Proc. 23rd ICCE*, ASCE, pp 3026-3039.

Dette, H and Uliczka, K (1986). Velocity and sediment concentration Fields across Surf Zones. *Twentieth Coastal Eng. Conf.*, Taiwan.

E.,Xuequan and Hopfinger, E.J. (1986). On mixing across an interface in stably stratified fluid. *J. Fluid Mech.*, vol. **166**, pp. 227-244.

Hardenberg, B. van, Hay, A.E. Sheng, Y, Bowen, A.J. (1991). Field measurements of the vertical structure of suspended sediment. *Coastal Sediments'91*, ASCE.

Jaffe, B. and Sallenger, A. (1992). The contribution of suspension events to sediment transport in the surf zone. *Proc. 23rd ICCE*, ASCE, pp 2680-2693.

Mocke, G.P. (1991). Turbulence modelling of suspended sediment in the surf zone. *Proc. Coastal Sediments '91*, ASCE, 432-446.

Mocke, G.P. and Smith, G.G. (1992). Wave breaker turbulence as a mechanism for sediment suspension. *Proc. 23rd ICCE*, ASCE, pp 2279-2292.

Nadaoka, K, Ueno, S, Igarashi, T. (1988). Sediment Suspension due to large scale eddies in the surff zone. *Proc. 21st ICCE*, ASCE.

Nadaoka, K., Hino, M. and Koyano, Y. (1989). Structure of the turbulent flow field under breaking waves in the surf zone. *J. Fluid Mech.* vol. **204**, pp 359-387.

Nielsen, P (1984). Field measurements of time averaged suspended sediment concentrations under waves. *Coastal Engineering*, **8** : 51-72.

Okayasu, A, Shibayama, T and Horikawa, K (1988). Vertical variation of undertow in the surf zone. *Proc. 21st Intl. Conf. Coastal Eng*, pp. 1420-1436.

Rodi, W. (1980). Turbulence models and their application in hydraulics - a state of the art review. IAHR, Delft.

Sanchez-Arcilla, A., O'Connor, B.A., Roelving, D.A. and Jiminez, J.A. (1994). The Delta Flume'93 Experiment. In press for Coastal Dynamics'94, Barcelona.

Sato, S., Homma, K, Shibayama, T. (1990). Laboratory study on sand suspension due to breaking waves. *Coastal Eng. in Japan*, Vol. **33**, No. 2.

Shibayama, T., Higuchi, A. and Horikawa, K. (1986). Sediment Transport due to breaking waves. *Proc. 20th Coastal Eng. Conf.*, ASCE.

Stive, M.J.F. (1980). Velocity and pressure field of spilling breakers, *Proc. 17th Coastal Eng. Conf.*

Svendsen, I.A. (1984). Wave heights and set-up in the surf-zone. *Coastal Eng.* **8**: 303-29.

Van Rijn, L. (1991). DATA BASE - Sand concentration profiles and sand transport for currents and/or waves. *Delft Hydraulics rep.* H **1148-04/05**.

STABILISING A RIVERBED BY LOCAL WIDENING - A CASE STUDY OF THE RIVER EMME, SWITZERLAND

BENNO ZARN

Laboratory of Hydraulics, Hydrology and Glaciology (VAW)
Swiss Federal Institute of Technology,
8092 Zurich, Switzerland

ABSTRACT

The continuous bed erosion of the River Emme can be arrested, and the bed stabilised to its 1982 level by local widening of the riverbed. Hydraulic modelling showed that the existing 30 m wide channel has to be increased to a width of between 65 and 85 m over a length of 460 m. In this reach a braided, partly meandering river will develop. This imposed change of river morphology will reduce the sediment transport capacity, resulting in an increase of slope in the widened reach. Additionally, the diverse flow patterns created are expected to be beneficial to the river ecology. The first stage of this local widening was carried out in winter 1991/92.

1. Introduction

At the turn of the century the major rivers in Switzerland were trained by straightening and narrowing the riverbeds to improve flood protection and to gain land for agricultural production. This caused an increase in sediment transport capacity that led, in several reaches of the Rhine, Thur and Emme Rivers, to continual riverbed erosion. Today, about eighty concrete sills stabilise the bed of the River Emme. In spite of these measures continual riverbed erosion has taken place between the existing sills. This progressive erosion was the reason for investigating the sediment budget of the River Emme [1]. The numerical simulation predicts a reduction of the slope from 4.2 ‰ (observed 1982) to 2.4±0.2‰ within a period of 50 years for the River Emme at Utzenstorf between the sills at 9.250 and 8.100 km [2]. The comparison of the observed riverbed profile in 1990 with the prediction for 1992 in figure 1 confirms the anticipated riverbed erosion. To prevent further erosion, Jäggi and Pellandini [3] propose a local widening of the riverbed as an alternative to conventional stabilisation using hard engineering solutions of sills or boulder ramps.

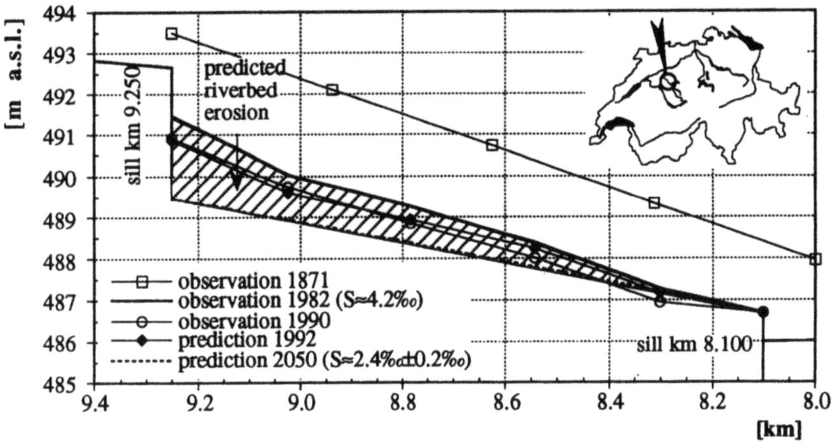

Fig. 1: Observed and predicted longitudinal profiles of the River Emme.

2. The Principle of Stabilising a Riverbed by Local Widening

The influence of riverbed width on the annual sediment load and on the slope can be calculated with one of the common sediment transport formula, by assuming a single-thread channel with a plane bed. Figure 2 shows the results for the River Emme, using the methods of Meyer-Peter and Müller [4], Keulegan [5] and Einstein [6]. The width at the maximum sediment transport capacity is often known as the optimum width (w_{opt}). At a width smaller than the optimum width the influence of the banks reduces the sediment transport capacity and at a width larger, the decrease in flow depth reduces the transport capacity. The assumption of a single-thread channel with a plane bed is only valid for widths smaller than 20 to 30 metres. With increasing width, alternate bars and finally a fully braided river morphology develops.

For stabilising the continual riverbed erosion of the River Emme the sediment transport capacity has to be reduced. This can be achieved with a widening or a narrowing of the bed width. However, an extremely narrow riverbed would decrease the discharge capacity and is therefore not practical. With a local widening, the costs of bank protection would be lower compared to widening over the whole reach between the two sills. The local widening has to be designed in such a way that the predicted slope reduction in the remaining section can be compensated for. A reduction of the sediment transport capacity leads to deposition, and hence, to a steeper slope until a new equilibrium is achieved (figure 2b).

Figures 2a) and 2b) are based on the assumption that the riverbed is still a single-thread channel with a plane bed for widths larger than 30 m, which, generally, leads to an underestimation of the flow depth and an overestimation of the slope [7]. The influence of these two variables on the calculation of the sediment transport capacity is unknown. This uncertainty is indicated with hatching in figure 2. The slope in the locally widened reach has to be determined as accurately as possible to determine the dimensions of the channel in this reach. If the slope were too low, the widening would not completely arrest the bed erosion, and if the slope were too high the degree of flood protection would be less. Because of the inaccuracy of the slope prediction it was necessary to investigate the proposed widening of the riverbed using a hydraulic model. The model used had a geometric scale ratio of 1:55 and was based on the Froude-law.

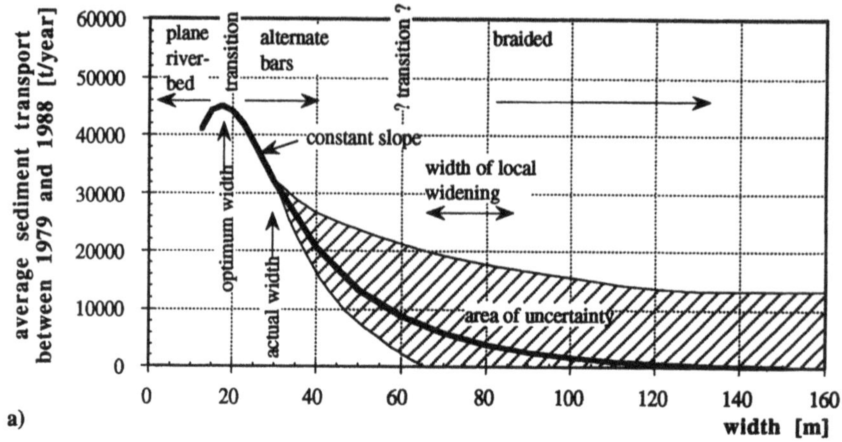

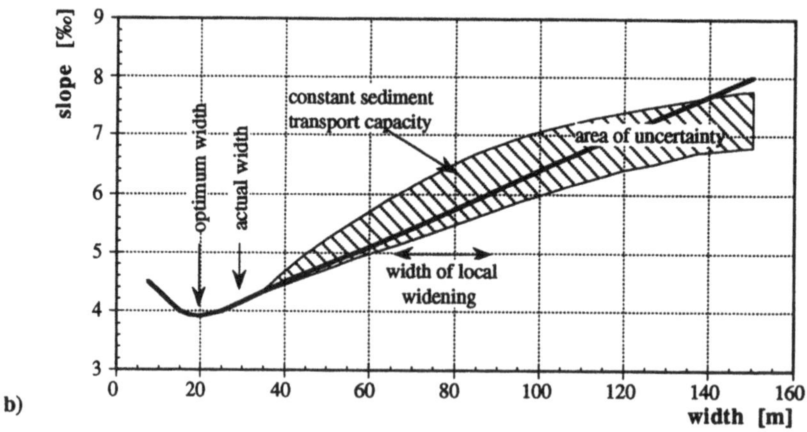

Fig. 2: Influence of riverbed width a) on the sediment transport and b) on the slope.

3. The Local Widening

The best location for the channel widening lies in the middle of the reach between the sills at 9.250 and 8.100 km. In order to stabilise the riverbed at its 1982 level the existing 30 m wide channel has to be widened to between 65 and 85 m over a length of 460 m (figure 3 and 4). If the bank protection is removed, the widening will occur naturally. This process can be sped up by excavating a 7.5 m wide strip along both flood plains then using this material to narrow the main channel. The equilibrium slope of the widened reach will be of the order of 6 ‰ (figure 5). Therefore, the predicted slope reduction of 1.8 ‰ for the remaining reach of the Utzenstorf section can be compensated for.

Four 36 m long guiding embankments in both flood plains will have to be built to prevent an uncontrolled widening, and in particular to protect the existing levees from erosion. No control measures are necessary at the expansion zone between the upstream channel and the widened reach. At the contraction zone downstream of the widening, an additional guiding embankment will have to be constructed. Relative to the 1982 riverbed level, maximum scour depths of 2.7 m were measured at the guiding embankments. Maximum scour did not necessarily occur at peak discharges. Local flow conditions were primarily responsible for scour depths.

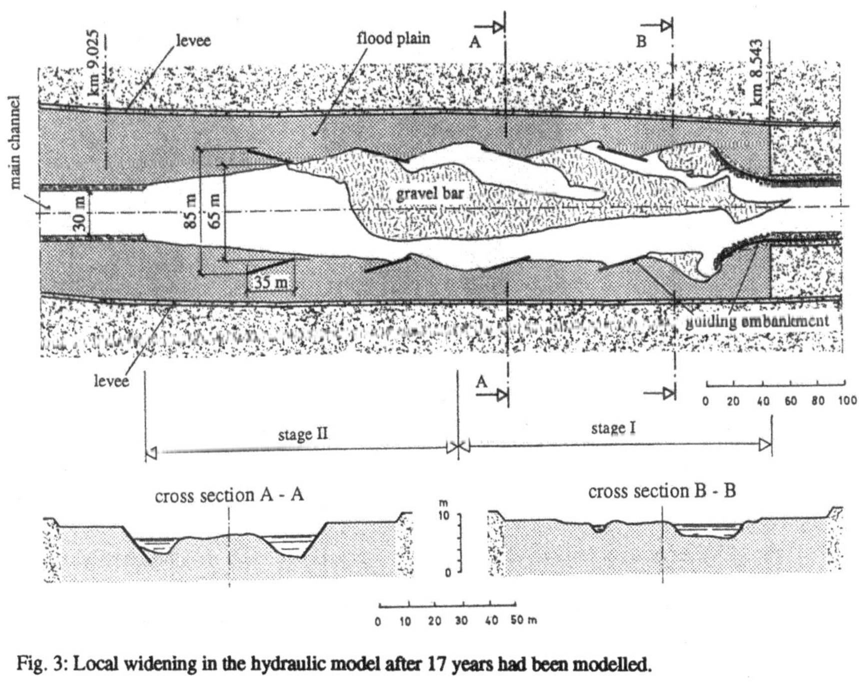

Fig. 3: Local widening in the hydraulic model after 17 years had been modelled.

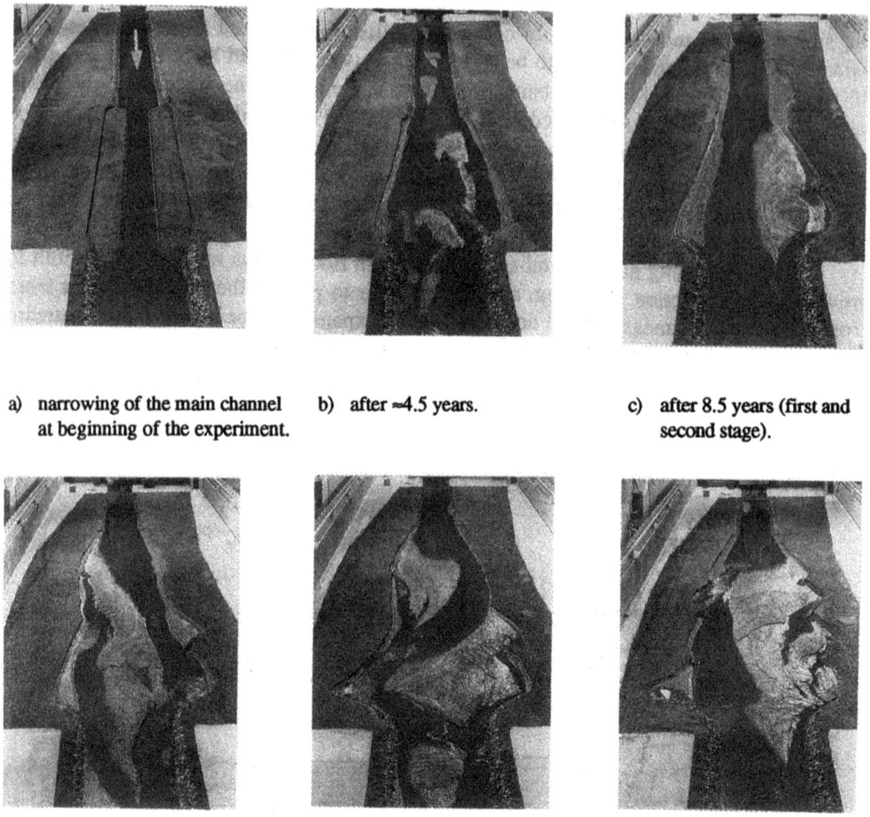

a) narrowing of the main channel at beginning of the experiment.

b) after ≈4.5 years.

c) after 8.5 years (first and second stage).

d) after ≈9.5 years.

e) after ≈10 years.

f) after ≈17 years.

Fig. 4: Development of the local widening during a long-term experiment.

Compared to 1982 conditions, the flood protection will be slightly better when the local widening is completely developed. During the formation of the local widening, the sediment input to the downstream reach will increase slightly, thus impeding the erosion there temporarily.

If the channel widening is left to nature, its development will depend largely on the magnitude, duration and sequence of floods. It is predicted that it will be at least a decade before the local widening is complete. To ensure flood protection during this period, the channel widening should be divided into two stages. The second stage commencing only when the natural development of the first stage is nearly complete.

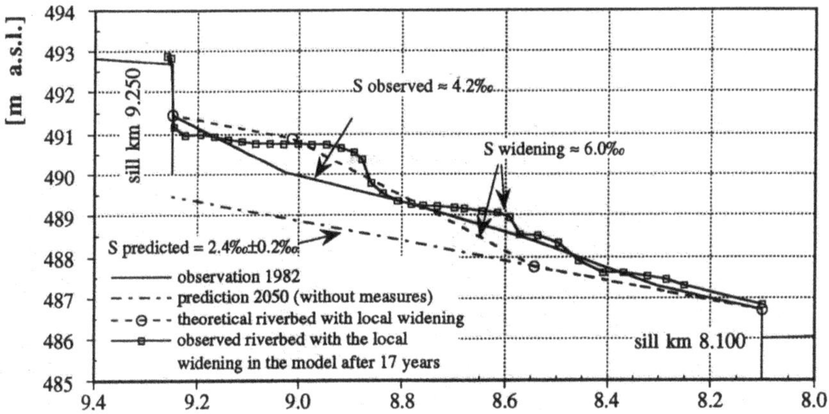

Fig. 5: Longitudinal profile observed in the model (end of second stage)[1].

4. Stage I Construction

In winter 1991/1992 the construction of the first stage was undertaken (figure 6). The proposed narrowing of the main channel to speed up the enlargement process was not necessary, because a flood with a peak discharge of about 370 m^3/s eroded 8'000 m^3 of sediment during the construction and sufficiently enlarged the riverbed. Obvious sedimentation can be observed in the widened reach (figure 7). A meandering, partly braided river with pool-riffle systems and variable flow velocities has developed due to the increased width. An inventory of algae, invertebrate, fish population, game, reptile and birds was done before the construction. It is planned to repeat the inventory after completion of the widening. The first results of the 1992 inventory show a considerable increase in the fish population. It is also expected that the ecological value will be further increased as a result of the changing river morphology [8]. This is a strong contrast compared to conventional riverbed stabilisation methods.

1 The longitudinal profiles observed in nature are average bed levels based on cross section surveys (figures 1, 5, and 7). The longitudinal profiles observed in the model are water levels measured at a discharge of 0.4 l/s (≈10 m^3/s in nature, figure 5).

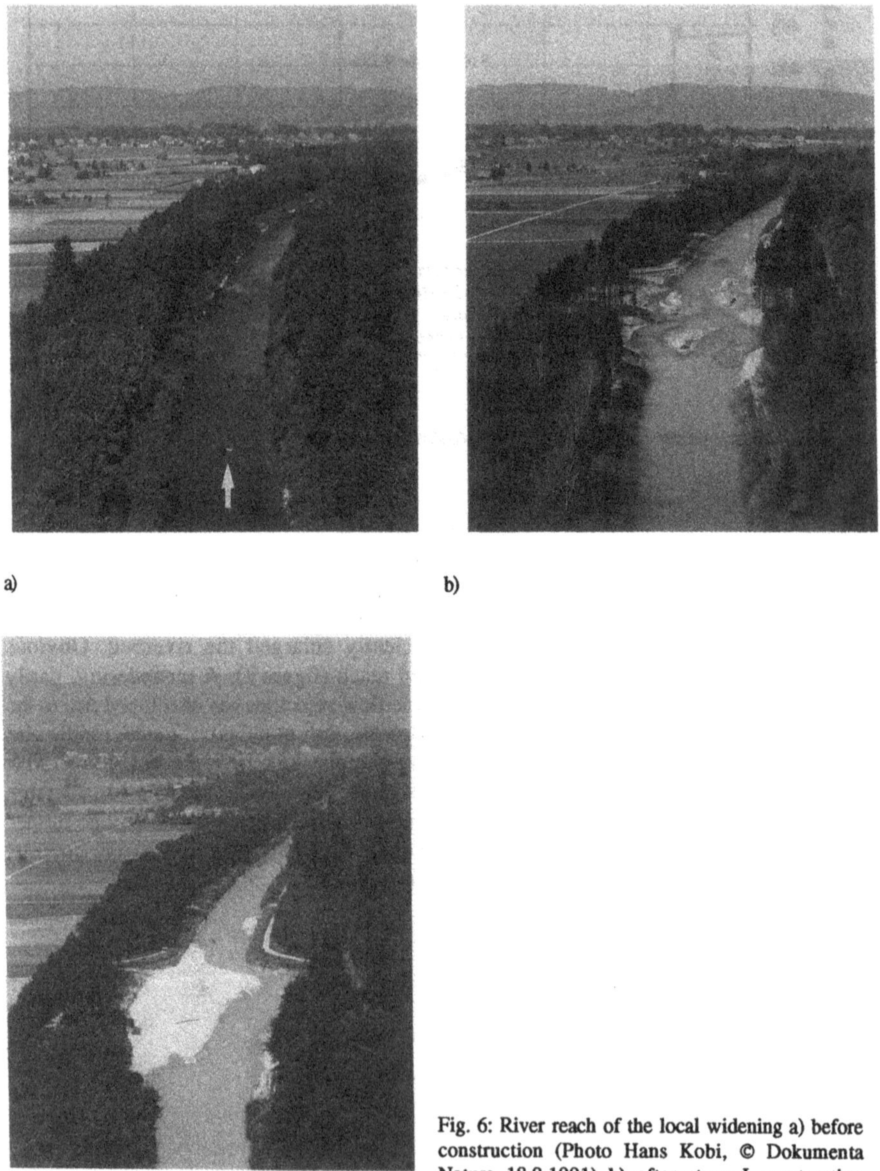

a) b) c)

Fig. 6: River reach of the local widening a) before construction (Photo Hans Kobi, © Dokumenta Natura 18.9.1991) b) after stage I construction (Photo Hans Kobi, © Dokumenta Natura 21.4.1992) and c) one year after stage I construction (9.6.1993).

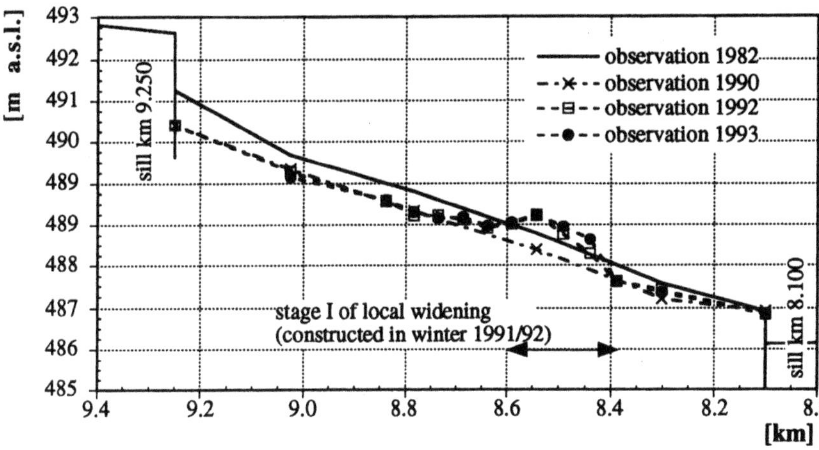

Fig. 7: Longitudinal profile observed in nature (begin of first stage development).

5. Conclusion

When designing a local river channel bed widening it is very important to know the dynamic equilibrium slope for the riverbed width, or rather, for the river morphology. The results of this investigation can be applied to other sections of the River Emme, but to develop general design specifications would require additional investigations to improve our understanding of the dependency of sediment transport on bed width, or rather, on river morphology. Local widening requires more land than conventional measures as sills or boulder ramps. However, the costs of the local widening of the River Emme were of the same magnitude as those for the more conventional methods.

6. References

1. Tiefbauamt des Kantons Bern, *Kurzfassung der Studie "Emme 2050" Tiefbauamt des Kantons Bern*, (Oberingenieurkreis IV, zweite, aktualisierte Auflage; Burgdorf, Februar 1992).
2. R. Hunziker und M. Jäggi, *Numerische Simulation des Geschiebehaushaltes der Emme*, Interpraevent 1988, Graz, **Band 3**, pp. 7-16.
3. M. Jäggi,und St. Pellandini, *Naturnahe verzweigte Flussabschnitte als Mittel gegen Sohleneintiefung*; Interpraevent 1988, Graz, **Band 4**, pp. 327-337.
4. E. Meyer-Peter und R. Müller, *Formulas for Bedload Transport*, Proc. 2nd Congr. IAHR, Stockholm, 1948.
5. G. Keulegan, *Laws of Turbulent Flow in Open Channels*, J. of Research of the

National Bureau of Standards, **Vol. 21**, Dec. 1938.
6. H.A. Einstein, *Der hydraulische oder Profilradius*, Schweizerische Bauzeitung **Nr. 8**, 1934.
7. B. Zarn, *Lokale Gerinneaufweitung: eine Massnahme zur Stabilisierung der Emme bei Utzenstorf*, Mitteilung **Nr. 118** der Versuchsanstalt für Wasserbau, Hydrologie und Glaziologie an der Eidgenössischen Technischen Hochschule Zürich, (Zürich 1992).
8. D. Shields, *Sediment and Aquatic Habitat in River Systems*, by the ASCE Task Committe, Journal of Hydraulic Engineering, **Vol. 118**, No. 5. May 1992.

SUSPENDED SEDIMENT CONCENTRATIONS NEAR THE SEABED IN NORTH SEA STORMS

D. MYRHAUG
Department of Marine Hydrodynamics, Norwegian Institute of Technology, N-7034 Trondheim, Norway

O.H. SLAATTELID
Norwegian Marine Technology Research Institute, P.O.Box 4125 Valentinlyst, N-7002 Trondheim, Norway

and

K.F. LAMBRAKOS
Exxon Production Research Company, P.O.Box 2189, Houston, Texas 77001, USA

ABSTRACT

Suspended sediment concentrations over a sandy bottom were measured for the duration (including build up and decay) of two storms in the central North Sea. The water depth was 70 meters and the seafloor consisted of fine sand of 0.20 mm average diameter. The measurements were made with two light beam transmittance meters at 0.25 and 1.00 m from the seafloor. The suspended sediment concentration data given here represent 12 minute averages. The tests included measurements of sediment concentrations, flow velocity in the boundary layer on the seafloor, steady current values over the water column, wave heights, and photographs of the seafloor. The paper presents the variation of the sediment concentrations, waves and currents during the storms. It also gives results from a study to correlate the sediment concentrations and the free stream values for the waves and currents.

1. Introduction

Suspended sediment concentrations over sandy seabeds in shallow and intermediate water depths, i.e. in coastal zones and on continental shelves, occur predominantly as a result of the combined action of waves and currents. The waves are the principal cause of the entrainment of the sediments, which are diffused into the flow by turbulent processes, and subsequently transported by the current. The combination of waves and currents, and sediment transport affects many engineering applications. Wave-current-sediment interaction considerations are made in scour and erosion studies for seabed pipelines and other near seabed structures. It is also important in developing models for predicting the movement of sediments on the seabed in combined action of waves and current, and the consequent coastal evolution.

The wave-current-sediment interaction has been investigated theoretically and experimentally by many researchers. The subject is treated in the recent books by Fredsøe and Deigaard (1992) and Nielsen (1992), where the latter also covers a review of the topic. The effect of adding current to waves is usually to increase the suspended sediment concentration at a given elevation as compared with waves alone.

This paper presents the results from the data analysis of near seabed suspended sediment concentration measurements as well as near bottom velocity measurements during two storms at the EDDA location in the North Sea. The variation of the suspended sediment concentrations, waves and currents during the storms are presented, together with results from correlation studies between the suspended sediment concentrations and the free stream value for the waves and currents. Additional analysis results of the data collected in the same experiment are given in Slaattelid et al. (1990) and Myrhaug et al. (1991, 1992). The two first references (1990, 1991) represent studies of the combined wave-current boundary layer, and the latter reference (1992) is a study of the wave boundary layer observed in the measurements.

2. Measurements

The measurements were made over the winter of 1984-85 at the EDDA (56°28.2'N 03°6.3'E) location in the North Sea in 70 m water depth. They included flow velocity close to the seabed, seabed configuration (ripples), suspended sediment concentrations, steady current over the water column, and waves by using a WAVEC directional buoy. Orthogonal horizontal components of wave plus tidal current velocity were measured at six nominal elevations 0.10, 0.25, 0.50, 1.00, 3.00 and 5.00 m from the seabed using acoustic current meters mounted on a tripod frame. A stereo-photo system, from which the seabed configuration was obtained, was also mounted in this tripod. The suspended sediment concentrations were measured at two elevations 0.25 and 1.00 m from the seafloor using light beam transmittance meters mounted on another tripod fixture. Sand traps were also mounted in this tripod. The tripod for the sediment concentration measurements was within 100 m from the flow meter tripod.

The current meters were calibrated through extensive testing in a towing tank both individually and mounted in the current meter frame. More details of the calibration results are given in Myrhaug et al. (1992). The light beam transmittance meters were also calibrated in the laboratory using bottom sediment samples from the experimental site in the North Sea (Tryggestad et al., 1987). The data were continuously sampled with a rate of 1 Hz. No measurements were made of the turbulence in the flow. The seafloor was relatively flat and uniform consisting mainly of fine sand with a mean grain size diameter (d_{50}) of 0.20 mm. More details on the measurements and instrumentation are given in Tryggestad et al. (1987).

3. Analysis

The steady current boundary layer data is based on analysis of 12 minute records covering a range of combined current and wave conditions. The data refer to coordinates where the positive x- and y-axes are taken towards east and magnetic north, respectively. The mean current velocity and direction at each current meter is calculated from the mean values of the x- and y-components of the 12 minute velocity records. The average (V_r) and standard deviation (σ) values of the flow velocity at the 5 m elevation are taken to be measures of the free stream current and the wave activity, respectively. Situations with waves plus following current and waves with opposing current have been identified by using WAVEC directional buoy data.

The logarithmic boundary layer flow model $V(z) = (u^*/\kappa)\ln((z+z_0)/z_0)$, (Schlichting, 1979), was assumed applicable for the current component in the flow and was fitted to the projected current velocities on the mean-over-depth current velocity direction calculated by vector averaging. Here u^* and z_0 are the friction velocity and roughness parameter, respectively, and $\kappa = 0.40$ is the von Karman constant. The least-squares-method was used to determine u^* and z_0 with $z + z_0$ taken as the actual elevation of the current meters above the seafloor. The actual elevations have been obtained from analysis of the stereo-photos taken each 8 hours. The current meter no. 1 elevation varied from 10.0 to 11.0 cm. These elevations are given with an accuracy of 1-3 cm. Calculated estimates of u^* and z_0 for each 12 minute record were accepted if: (1) a goodness of fit coefficient was less than 0.10; (2) the difference between the observed and predicted velocity value at each elevation was less than 15%; and (3) the standard deviation of the current's direction was less than 10 degrees. More details on these criteria, data selection and quality of the data are given in Slaattelid et al. (1990).

The suspended sediment concentration data are represented by 12 minute average, standard deviation, maximum and minimum values and are obtained by transformation of light transmission meter data to volume concentrations (C) of the suspended sediments (Tryggestad et al., 1987).

4. Results and discussion

Two storm periods, one from each of two large storms (denoted as E11 and E12) at the same site at the EDDA location in the North Sea, will be discussed here. The E11 and E12 storms lasted for about three and one day, respectively. The current velocities during the storms were approximately the same, but the tidal variation was more evident in E11 than in E12. The wave activity was about the same for the two storms, and the average wave period was the same, but with a larger variation in E12. The relative angle between waves and current varied more in E12 than in E11. The largest average concentrations measured at the 0.25 m elevation were about 100 and 120 mg/l in E11 and E12, respectively.

4.1. E11 storm results

Figure 1 shows the variation of the steady current (V_r), the wave velocity (σ), the relative angle between waves and current (θ), and the average suspended sediment concentration (C) at the 0.25 and 1.00 m elevations versus time for 67 hours during the E11 storm. The mean wave velocity period for each 12 minute record was fairly constant, ranging from 8.1 to 11.8 seconds, with an average value of about 11 seconds for the whole storm. Intervals for which the log-fit is acceptable, based on the criteria given above, are also shown in the figure. Five peaks in $C_{0.25}$ (at the 0.25 m elevation) are identified numbered 1 to 5, as well as the regions associated with each of the peaks, identified by the horizontal arrows between the dashed vertical lines.

The immediate observations which can be made from Fig. 1 are that:

there is 12 hours between the peaks in $C_{0.25}$, i.e. the variation in $C_{0.25}$ follows the semi-diurnal tidal period, but the peaks in $C_{0.25}$ occur at the minimum of V_r, which is

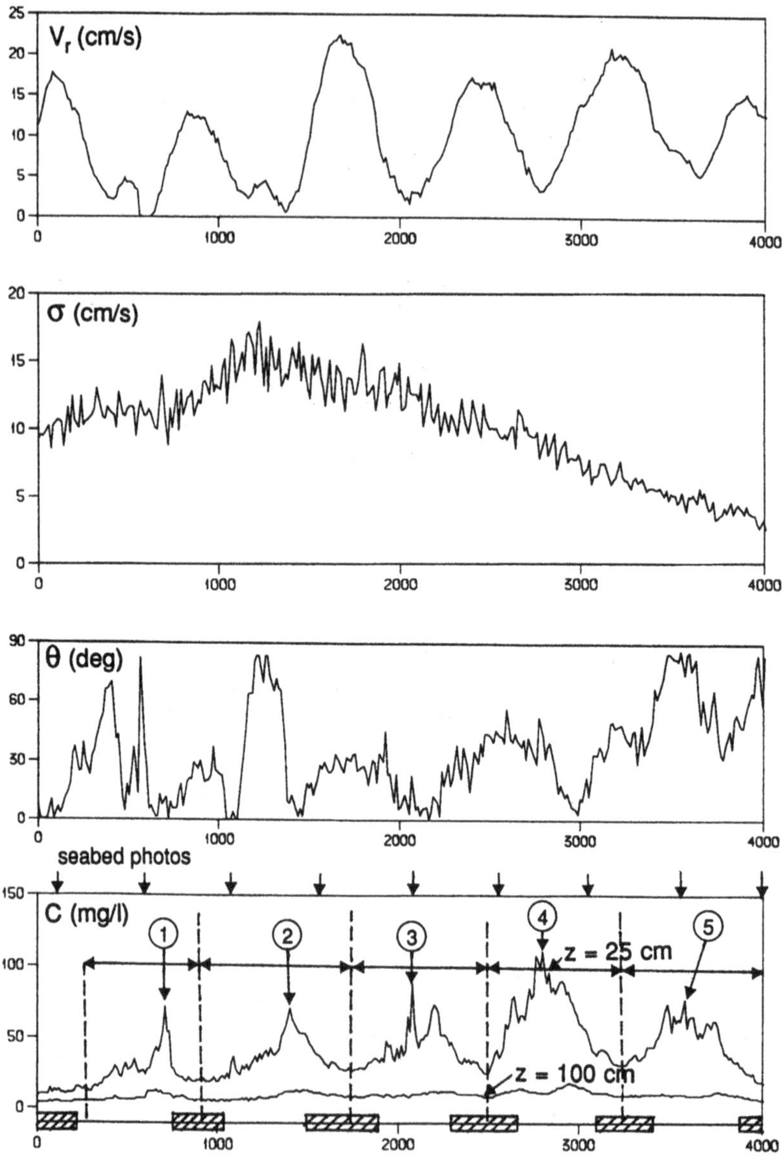

Fig. 1 Main measured flow variables versus time for the E11 storm at EDDA; current velocity (V_r); wave velocity (σ); relative angle between waves and current (θ); average suspended sediment concentration (C) at elevations 0.25 and 1.00 m above the bottom; ▨ intervals of acceptable log-fit. The horizontal arrows between the dashed vertical lines identify regions associated with each peak. Time 0 corresponds to 9.00 pm GMT November 15, 1984.

surprising compared to what is expected. The effect of adding current to waves is usually to increase the concentration, see e.g. Fredsøe and Deigaard (1992), Fig. 8.11. Possible reasons for this will be discussed subsequently;

- there appears to be little correlation between the $C_{0.25}$ peaks and the wave activity;
- the higher concentrations (peaks 3 and 4) occur after the wave activity peak in the storm;
- the log-fit is good for relatively low concentrations and poor for relatively high concentrations. However, since the concentrations are probably not high enough to affect the log-fit, that is, to give stratification effects, the reason for the poor log-fit is likely to be the low steady current velocities.

Additional information is obtained by inspection of photos of the seabed, although only mono-pictures (i.e. only one of the cameras worked) are available. The points of time at which the photos were taken, are indicated in Fig. 1. Both before and after the storm a weak ripple pattern can be observed, but the orientation of the ripples is different. During the storm the seabed is not visible in the photos due to sediment suspension. This suggests that the original ripples were washed out, or gradually changed, during the storm. These observations are also supported by the analysis results of the wave-current boundary layer interaction effects given in Fig. 2, which shows the friction velocity (u*) and roughness (z_0) parameter, respectively, versus time for three different intervals during the E11 storm. The solid curves represent the parameters obtained from the log-fit to the data. The dashed curves represent the Myrhaug and Slaattelid (1990) (hereafter denoted as MS90) fixed bed model predictions with a constant Nikuradse bed roughness of k_b = 2.5 d_{50}, and using the significant wave velocity amplitude $u_0 = 2\sigma$ to represent the waves. Overall it appears that the fixed bed model predictions agree well with the measurements for all three intervals. For a moveable bed the total bed roughness is given by $k_{btotal} = k_{bgrain} + k_{bripple} + k_{btransport}$; $k_{bgrain} = 2.5\ d_{50}$.

For the peak storm interval moveable bed effects have been taken into account by combining the MS90 model with three different moveable bed models. These models are those of Grant and Madsen (1982), Nielsen (1983) (field) and Raudkivi (1988) (hereafter denoted as GM, N and R, respectively). The peak storm results in Fig. 2 show that the moveable bed effect overpredicts u* and z_0. Thus the results in Fig. 2 suggest that ripples are not present, or at least very weak, before, during and after the E11 storm. It should be noted that the before storm period given in Fig. 2 is not included in Fig. 1, while the peak and after storm periods correspond to the third and sixth "acceptable log-fit" periods identified in Fig. 1. More details about the flow conditions for the before storm period in Fig. 2 are given in Slaattelid et al. (1990).

In the following possible correlations between the suspended concentration and the flow parameters will be investigated. First, the whole storm period will be considered. Then, the conditions associated with each of the five peaks identified in Fig. 1 will be considered.

Figure 3 shows $C_{0.25}$ versus V_r for different classes of σ and θ. Although there is large scatter, the overall trend is as already observed in Fig. 1; $C_{0.25}$ decreases as V_r increases. The scatter in the data can not be attributed to the relative direction between waves and current, since it appears that the results for all directions (0-90 degrees) are

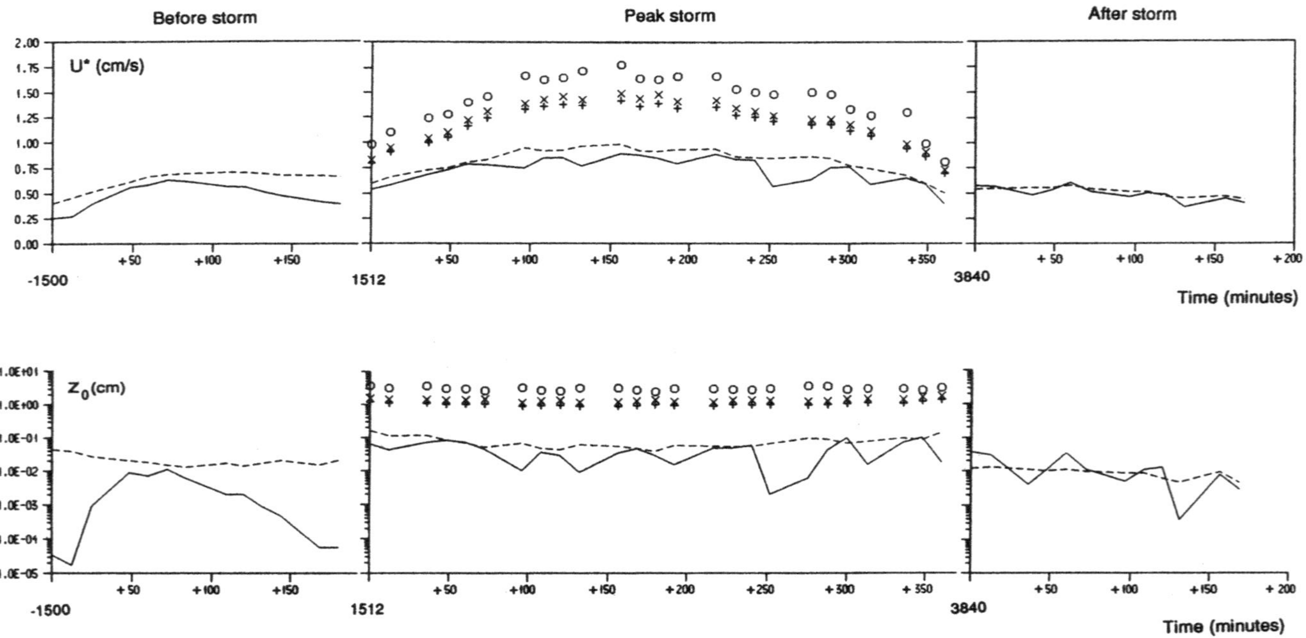

Fig. 2 u* and z_0 versus time for three different intervals during the E11 storm:— measured; - - - MS90 model with $k_b = 2.5d_{50}$; ○(MS90+GM) model; ×(MS90+R) model; +(MS90+N) model, all with $u_0 = 2\sigma$. Time refers to the time in Fig. 1.

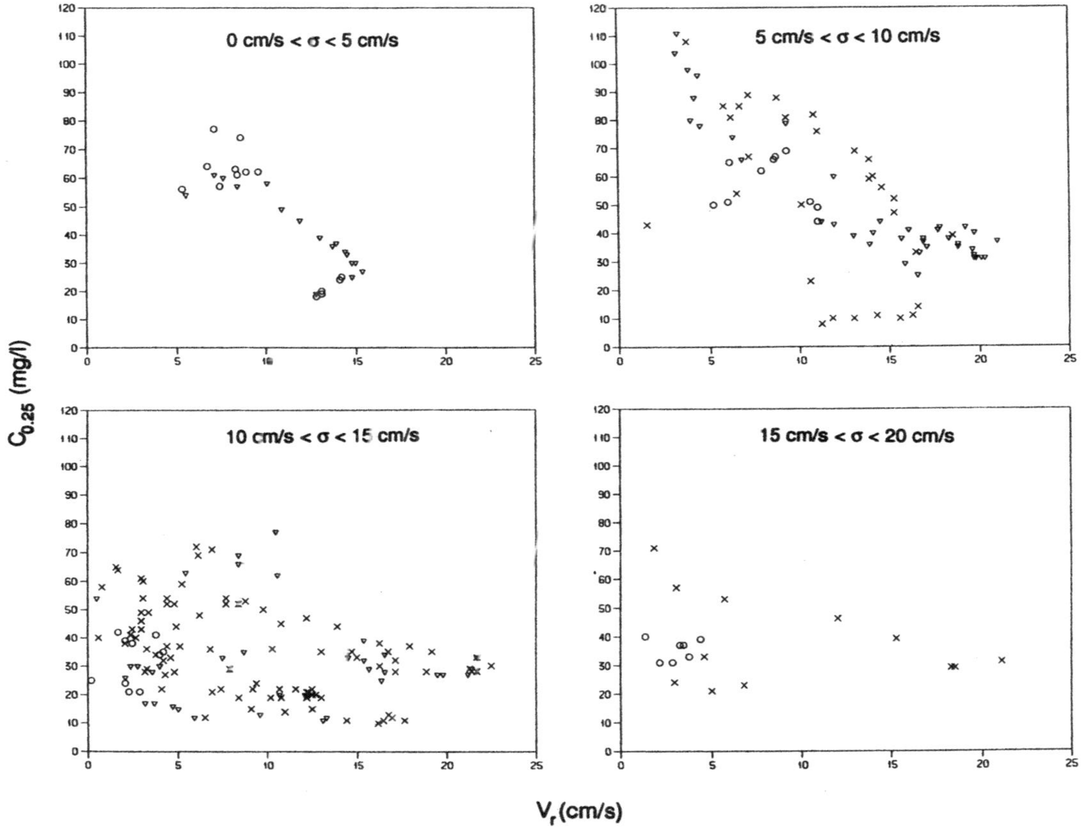

Fig. 3 $C_{0.25}$ versus V_r for different classes of σ and θ: × ($0 < \theta < 30$ deg); Δ ($30 < \theta < 60$ deg); ○ ($60 < \theta < 90$ deg).

mixed together.

Figure 4 shows $C_{0.25}$ versus V_r/σ before and after each of the five peaks identified in Fig. 1. The before peak region is the part from the first dashed vertical line to the peak, and the after peak region is the part from the peak to the following dashed vertical line. From Fig. 1 it is observed that $C_{0.25}$ values before and after the peaks mainly correspond to decreasing and increasing V_r values, respectively. It is also noted that the changes in V_r/σ are mainly due to the variation in V_r, since σ is fairly constant in the regions associated with each of the peaks (Fig. 1). Situations with waves plus following current (W+C) and waves plus opposing current (W-C) have also been identified. The overall behaviour is again that the concentration is lower for higher V_r, but it appears to be a clear distinction between the data whether they are referred to instants before or after the peaks in $C_{0.25}$. Fig. 4 shows that for the same V_r/σ, the before peak values are lower than the after peak values except for small V_r/σ. Further, it appears that as V_r/σ decreases with time (Fig. 1), $C_{0.25}$ increases along the lower part of the data until the point of maximum concentration. Then as V_r/σ increases with time, $C_{0.25}$ decreases along the upper part of the data. This effect is clear for all peaks except for peak 5. Although there are few data points for W-C, the results in Fig. 4 suggest that the conditions of W+C or W-C are unimportant.

Figure 5 shows $C_{0.25}$ versus the maximum value of the Shield's skin friction parameter $\psi' = \tau'_{bm}/\rho(s-1)gd_{50}$, where τ'_{bm} is the maximum seabed shear stress in combined wave-current boundary layer flow for sand grain size roughness based on $k_b = 2.5\ d_{50}$, ρ is the density of the fluid, s is the sediment density to fluid density ratio (taken as 2.65 for quartz sand), and g is the acceleration of gravity. These ψ' values are calculated by the MS90 model for the wave-current interaction, assuming a fixed bottom roughness and sinusoidal waves, represented by $u_0 = 2\sigma$. Fig. 5 does not show any clear trend in the data. The ψ' values also appear to be small, that is, the ψ' values associated with peaks 1 to 3 are just above the threshold value 0.05 at which the bottom sediments start to move (Nielsen, 1983). For peaks 4 and 5 the ψ' values are even lower. However, it should be noted that these ψ' values are based on predictions. In the field the bed shear stress is obviously high enough to bring sediments in suspension, and thus the ψ' field values are higher.

4.2. E12 storm results

Figure 6 shows the variation of V_r, σ, θ and C at the 0.25 and 1.00 m elevations versus time for 33 hours during the E12 storm. Intervals with acceptable log-fit are also shown in the figure. The mean wave velocity period for each 12 minute record was fairly constant, ranging from 8.8 to 13.6 seconds, with an average value of about 11 seconds for the whole storm. Two peaks in C are identified numbered 1 and 2. A third peak following peak 2 is also observed. This peak, which is unusually sharp, is attributed to either meter error or perhaps advection effects (see subsequent discussion).

The immediate observations which can be made from Fig. 6 are that:

- there is no clear correlation between the peaks in $C_{0.25}$ and the semi-diurnal tidal

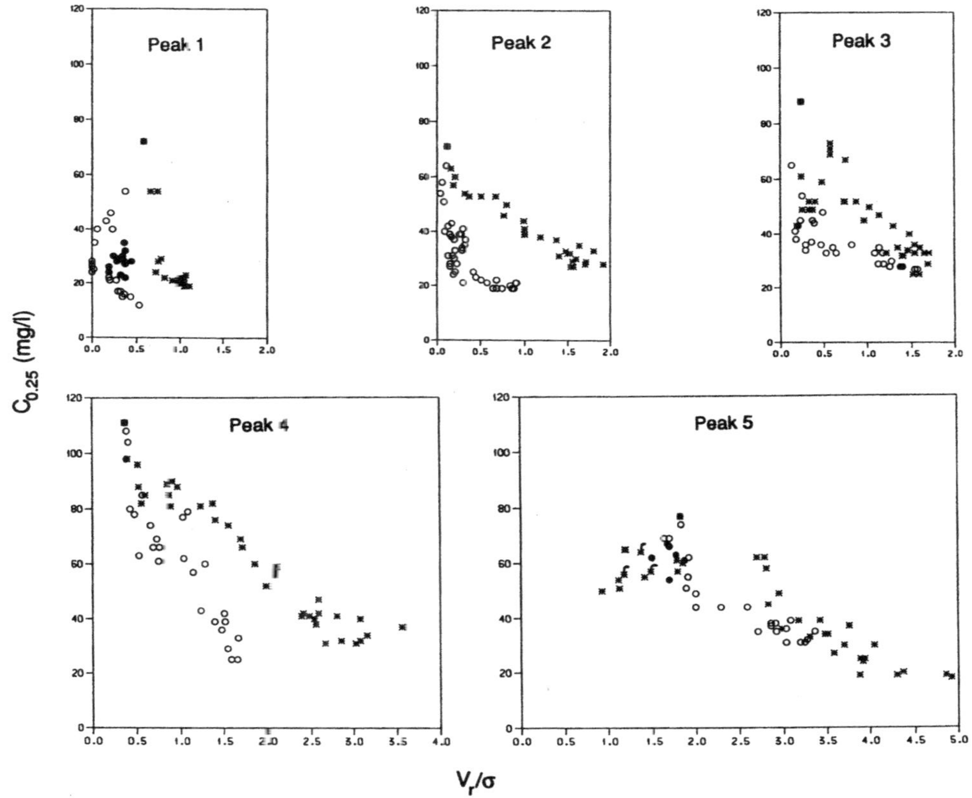

Fig. 4 $C_{0.25}$ versus V_r/σ for each region of the peaks in $C_{0.25}$ numbered 1 to 5 as identified in Fig. 1. Before peak values: ○ W+C; ● W-C. After peak values: ∗ W+C; ⁎ W-C.

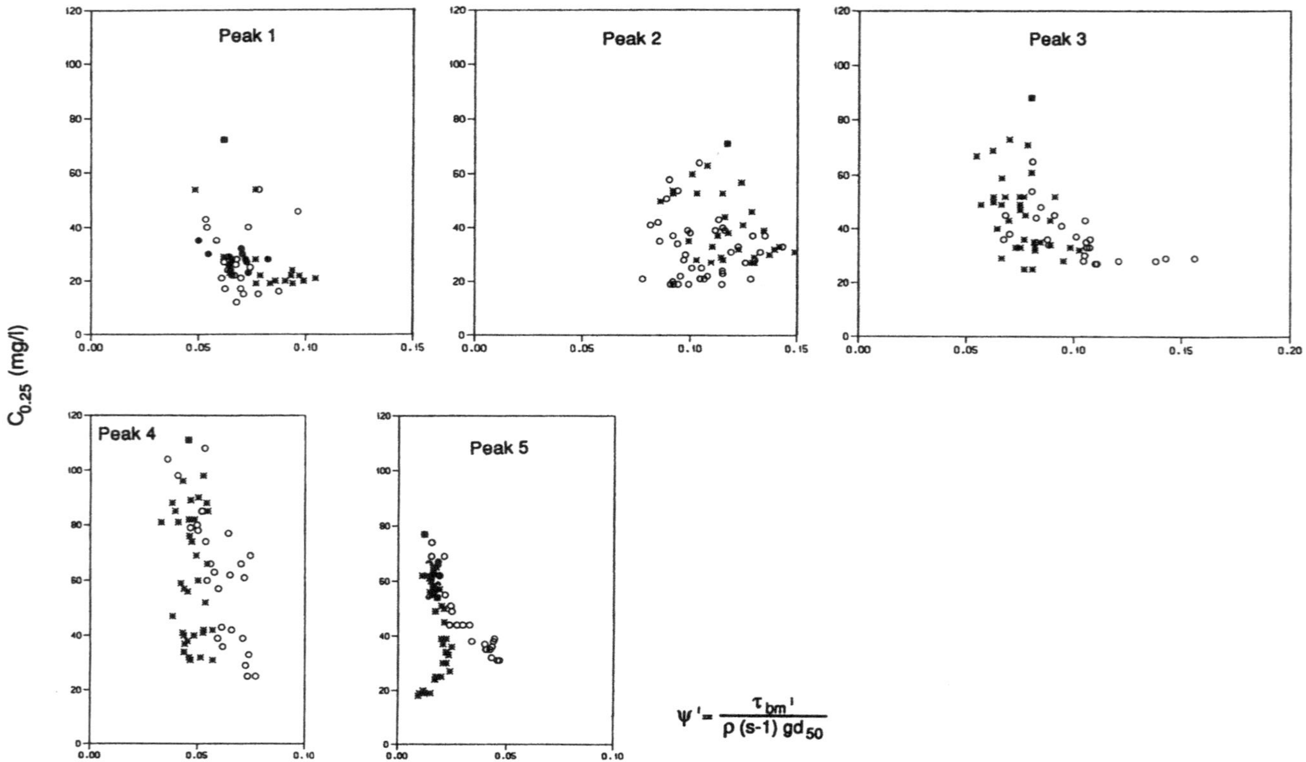

Fig. 5 $C_{0.25}$ versus Shield's parameter ψ'; all else same as in Fig. 4.

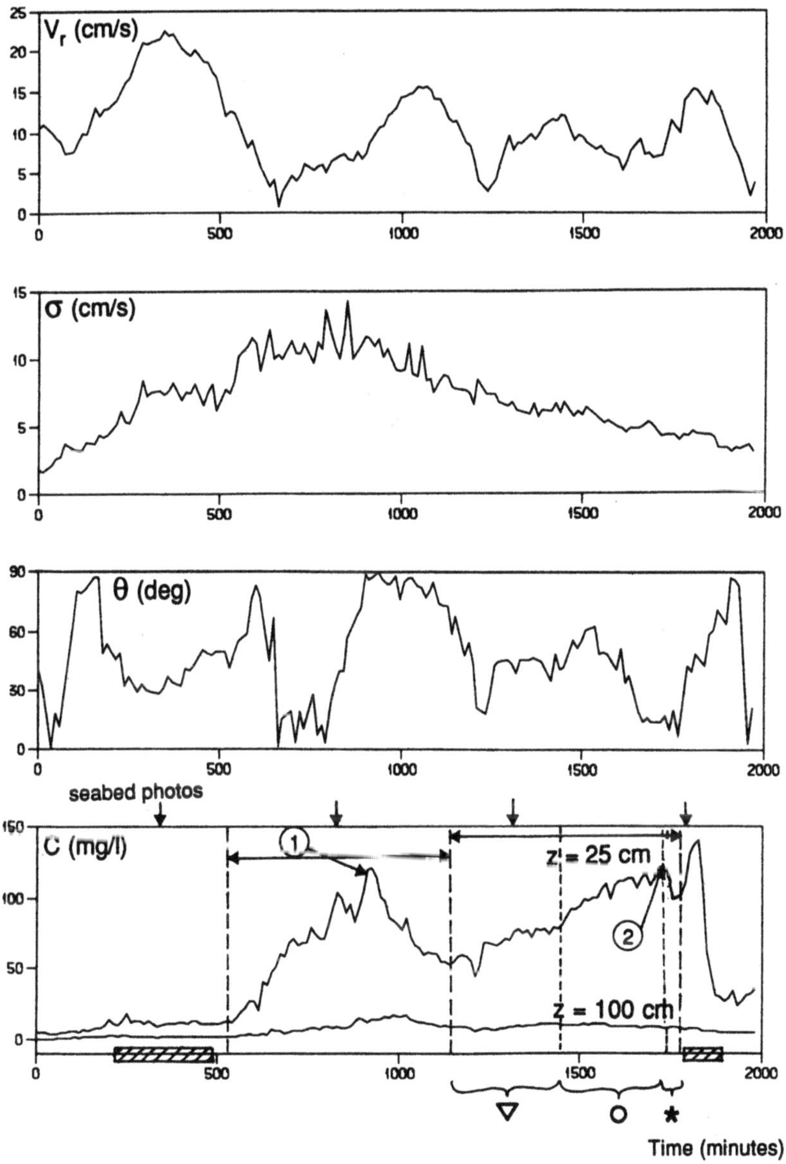

Fig. 6 Main measured flow variables versus time for the E12 storm at EDDA; current velocity (V_r); wave velocity (σ); relative angle between waves and current (θ); and average suspended sediment concentration (C) at elevations 0.25 and 1.00 m above the bottom; ▨ intervals of acceptable log-fit. The horizontal arrows between the dashed vertical lines identify regions associated with each peak. Time 0 corresponds to 09.00 am GMT November 27, 1984.

period;

- the higher concentrations occur after the wave activity peak in the storm;
- the log-fit is good early in the storm when $C_{0.25}$ is low, as well as late in the storm when the questionable peak in $C_{0.25}$ occurs, and poor at other instants.

Also for the E12 storm only mono-pictures of the seabed are available, and the points of time at which they were taken are indicated in Fig. 6. Both before and after the storm a clear ripple pattern can be observed, but the orientation of the ripples has been changed during the storm, see Fig. 2 in Slaattelid et al. (1990). It can be seen that the ripples are wave-generated before the storm, while they are current generated, or at least dominated by current, after the storm. During the storm the seabed is not visible in the photos due to sediment suspension. This demonstrates that the ripples were washed out, or gradually changed, during the storm.

By considering the whole E12 storm period the overall trend is that $C_{0.25}$ decreases as V_r increases, although this trend appears to be weaker and with a larger scatter than for the E11 storm. However, the emphasis here will be on the peak analysis results as shown in Figs. 7 and 8.

Figure 7 shows $C_{0.25}$ versus V_r/σ in the regions associated with peak 1 and 2. These regions are identified in Fig.6 together with the symbols used to distinguish before and after peak values, as well as conditions with W+C and W-C. It should be noted that changes in V_r/σ mainly reflect the variation in V_r, since σ is fairly constant in the regions associated with each of the peaks (see Fig. 6).

For peak 1 it is observed that when $C_{0.25}$ starts to increase (from about time 500 minutes), V_r/σ decreases until V_r/σ has a minimum, representing W+C conditions. After this minimum in V_r/σ, $C_{0.25}$ continues to increase to its maximum value, and these events coincide with an increase in V_r/σ, representing W-C conditions. These latter conditions illustrate what usually is expected when current is added to waves, i.e. an increase in the concentration as compared to waves alone. Then the following decrease in $C_{0.25}$ is mainly associated with a further increase in V_r/σ, representing W+C conditions except for three W-C events.

For peak 2, $C_{0.25}$ decreases as V_r/σ increases for all the data except for those in the early stage in the build-up of this peak. All these data represent W+C conditions. Although there is only three data points representing the after peak region, the data suggest a similar trend as observed for peak 1 as well as for the E11 storm.

Comparison of Figs. 4 and 7 shows that the details in the data are somewhat different, but the overall behaviour suggests that for the same V_r/σ , the before peak values are lower than the after peak values except for small V_r/σ.

Figure 8 shows $C_{0.25}$ versus the Shield's parameter defined in Fig. 5. As for the E11 storm these ψ' values are small, but since sediments are in suspension the ψ' values in the field must obviously be higher.

4.3. Discussion

The most surprising result from the analysis is that the concentration decreases when current is added to waves, which is opposite of what usually is expected. Some

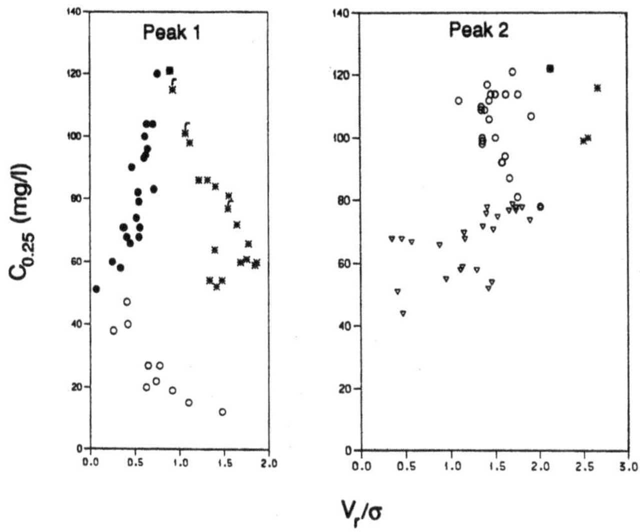

Fig. 7 $C_{0.25}$ versus V_r/σ for each region of the peaks in $C_{0.25}$ numbered 1 and 2 as identified in Fig. 6. Before peak values: ○ W+C; ● W-C. After peak values: ∗ W+C; ∗̂ W-C. Other symbols; see Fig. 6.

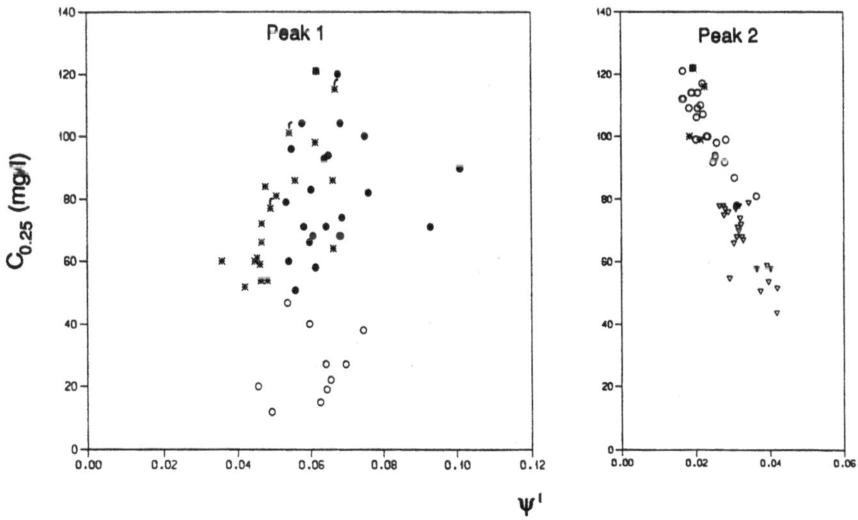

Fig. 8 $C_{0.25}$ versus ψ'; all else same as in Fig. 7.

possible explanations will be given and discussed in the following.

The high suspended sediment concentrations are associated with flow reversal in the combined wave-current boundary layer. When the current becomes dominant there is no flow reversal, and the suspended sediment concentration decreases. If free stream values are considered, flow reversal will occur when $V_r/2\sigma < 1$. For the E11 storm (see Fig. 4) the high concentrations occur at $V_r/\sigma < 2$, i.e. when flow reversal is present. The same is true for the E12 storm (see Fig. 7). The early period of the E12 storm is also consistent with this interpretation, since in that period the concentration is negligible and $V_r/2\sigma >> 1$.

Observations recently reported by Villaret and Perrier (1992) are represented in Fig. 9. These results are from laboratory measurements in combined wave-current boundary layer flow over rippled beds, where mean concentration profiles were measured over the ripple crest. The bed orbital velocity was about 30 cm/s combined with a reference current velocity of 17 cm/s. An estimate of the wave boundary layer thickness for these tests gives that the maximum overshoot of the bed orbital velocity amplitude is in the range 1 to 2 cm above the bed, while in the E11 storm the corresponding thickness is about 25 cm (Myrhaug et al., 1992), that is, corresponding to the elevation of suspended concentration measurements. Thus the region where the reduction in concentration is observed should be comparable in the laboratory and field measurements. It appears that for W-C the concentration is reduced compared to waves alone. For W+C the concentration is reduced close to the bed, while the concentration increases away from the bed compared to waves alone. This is consistent with the flow reversal explanation.

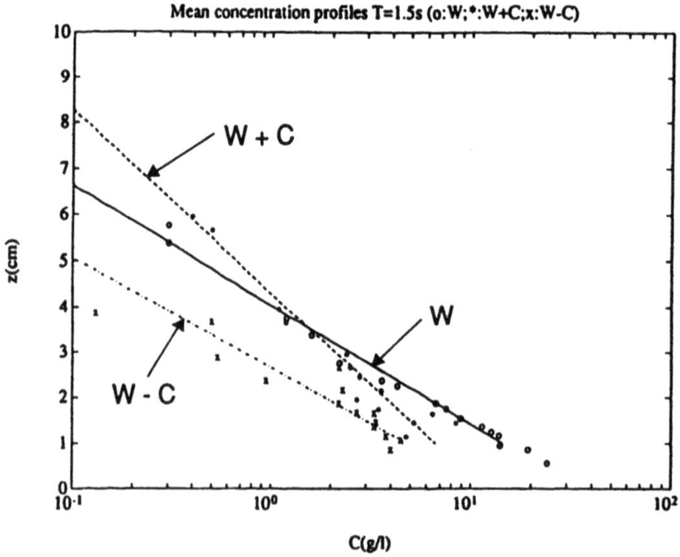

Fig. 9 Mean concentration profiles for waves (W), W+C and W-C in laboratory measurements. The symbols are data and the curves represent best fit to data (from Villaret and Perrier, 1992).

Advection from other areas with other types of bottom sediments is another possibility. However, this is unlikely because the site for the measurements were carefully chosen to be located within a larger area which was relatively flat and with uniform seabed conditions. Furthermore, advection would probably occur as a single event. Therefore this explanation should be excluded for the E11 storm, because the conditions associated with each of the peaks are very similar. However, advection might have caused the third $C_{0.25}$ peak in the E12 storm, which was excluded from further considerations in the data analysis.

Clouds of sediments associated with the flow around the tripod fixture of the light beam transmittance meters might have been stirred up by the wave action as the current decreases giving higher concentrations. It should be noted, however, that prior to the final design, the tripod was tested in steady currents in a laboratory flume where field site sediment samples were used. Thus the final design of the tripod was made to minimize the possibility of such effects. However, this explanation can not be completely excluded.

From the interpretations given above, the flow reversal explanation seems to be the most plausible. It is conceivable that the flow reversal is mainly responsible for the sediment suspension when the current component in the combined wave-current boundary layer flow conditions is relatively low. However, when the current becomes significantly large, the concentration could increase with the current.

5. Conclusions

The main conclusions which can be drawn from these data are:

1. Generally the average concentration varies with the tidal variation regardless of the wave activity. The peaks of the concentration occur at the low values of the current, and at the peaks of the current the concentration is relatively low.
2. It appears that the high concentrations occur when there is flow reversal ($V_r/\sigma < 2$) in the combined wave - current boundary layer flow conditions.
3. The analysis associated with the peaks in the average concentration at the 0.25 m elevation shows that some correlation is found between the concentration and the current to wave velocity ratio: it generally appears that (1) $C_{0.25}$ decreases as V_r/σ increases, and (2) before peak values of $C_{0.25}$ are lower than after peak values for the same values of V_r/σ, except for small V_r/σ .

6. Acknowledgements

This work was carried out as part of the MAST G8 Coastal Morphodynamics programme under contract no. MAS2-CT-92-0027. It was funded jointly by Anders Jahres Foundation for the Advancement of Science and The Research Council of Norway. The field measurements were carried out as part of the PIPESTAB project funded by Esso Norge a.s. and STATOIL.

7. References

J. Fredsøe and R. Deigaard (1992) *Mechanics of Coastal Sediment Transport* (World Scientific Publishing Co., Singapore).

W.D. Grant and O.S. Madsen (1982) *Moveable bed roughness in unsteady oscillatory flow*, J. Geophys. Res. **87** (C1), 469-481.

D. Myrhaug and O.H. Slaattelid (1990) *A rational approach to wave-current friction coefficients for rough, smooth and transitional turbulent flow,* Coastal Engineering **14**, 265-293.

D. Myrhaug, O.H. Slaattelid and K.F. Lambrakos (1991) *Time history of current boundary layer parameters in a North Sea storm*, In: R.L. Soulsby and R. Bettess (Editors), Euromech 262 - Sand Transport in Rivers, Estuaries and the Sea. (Balkema, Rotterdam), pp. 79-86.

D. Myrhaug, K.F. Lambrakos and O.H. Slaattelid (1992) *Wave boundary layer in flow measurements near the seabed*, Coastal Engineering **18**, 153-181.

P. Nielsen (1983) *Analytical determination of nearshore wave height variation due to refraction, shoaling and friction*, Coastal Engineering **7**, 233-251.

P. Nielsen (1992) *Coastal bottom boundary layers and sediment transport* (World Scientific Publishing Co., Singapore).

A.J. Raudkivi (1988) *The roughness height under waves*, J. Hydraulic Res. **26**, 569-584.

H. Schlichting (1979) *Boundary-Layer Theory* (7th Edition, McGraw-Hill, New York).

O.H. Slaattelid, D. Myrhaug and K.F. Lambrakos (1990) *North Sea bottom steady boundary layer measurements*, J. Waterway, Port, Coastal and Ocean Engineering, ASCE **116**, 614-633.

S. Tryggestad, T.E. Kollstad, K.A. Selanger and J.H. Nielsen (1987) *Measuring technique and field data for pipeline stability*, Proc. 19th Offshore Technology Conf., Paper No. 5506, Houston, Texas.

C. Villaret and G. Perrier (1992) T*ransport of fine sand by combined waves and current: an experimental study*, Report No. HE-42/92.68, Electricite de France, Direction des Etudes et Recherches, Département Laboratoire National d'Hydraulique, Chatou, France.

ON NUCLEAR GAUGING OF NATURAL STREAMS *

G.S. TAZIOLI
Dipartimento di Scienze dei Materiali e della Terra, Faculty of Engineering
University of Ancona, Via Brecce Bianche, 60131 - Ancona, Italy

ABSTRACT

The study of the phenomena which regulate bed and suspended solid transport is of great importance not only because of technical aspects but also because of economic implications. Problems relating to erosion, artificial recharge of aquifers, loss of reservoirs capacity, harbours and navigation channel maintenance, are frequent in many countries. Bed load can be quantified from measurements made by artificial tracers or by integrating samplers and suspended transport by the use of gauges based on optical, ultrasonic or nuclear systems or by the use of automatic or manual samplers. The use of nuclear gauges in permanent installations allow to measure high suspended sediment concentrations also in the presence of very severe hydrodinamic conditions. This is useful expecially in those situations in which there is a rapid variation of the suspended sediment concentration which is tipical of many mediterranean areas.

1. Introduction

In the studies concerning suspended sediment transport automatic or manual samplers and optical or ultrasonic sensors may be considered appropriate to certain environments and applications. For instance, the output of optical and acoustic gauges is sensitive not only to concentration but also to the particle size of the suspended sediment. Algal fouling, the growth of organisms, the accumulation of salts and clay on the sensor faces are all detrimental to accurate measurement [Crickmore et al., 1990].

Nuclear gauges provide an alternative to these devices and offer considerable advantages when measurements are carried out in the presence of high concentration in excess of 1000 mgl^{-1}.

Reliable concentration measurement at these levels is highly relevant to such diverse applications as evaluating the transient sediment discharge of torrential streams.

The use of gauges based on artificial radioisotopes offers many possibilities, depending on the source-detector distance and on the energy of the radiation. Quantitative measurements of suspended sediment concentration are feasible for a volume of influence which varies from a few litres to one hundred litres of turbid water.

One of the greatest advantages of the nuclear method is that measurements are independent of the diameters of sediment particles, and are practically without disturbance to the fluid. Furthermore, gauges can be made so robust that they allow continuos monitoring during very severe flow conditions.

---

* The research has been carried out with the financial support of the Italian Ministry for University and Technological research, Project on erosion and Alluvions in areas characterized by neotectonics or subsidence.

In this case the gauges are fixed to suitable structures in the stream bed or on the seabottom.

Measurements carried out using nuclear methods are suitable for concentration ranges between 1 and 200-600 gl^{-1}.

Further information on the available types of nuclear gauges, readers may consult the papers of Florkowski, 1970; Ciet & Tazioli, 1976; Tazioli & Caillot, 1983 & Crickmore et al., 1990.

2. Principles of nuclear gauges

The measurement of suspended sediment concentration with the use of nuclear gauges is based on the absorption or scattering of X or γ radiations from an artificial radioactive source, or alternatively on the measurement of natural gamma radioactivity emitted by the suspended sediment [Florkowski, 1968; Courtois et al., 1970; Tazioli & Caillot, 1983].

The use gauges equipped with artificial radioactive sources may be in transmission or back-scattering geometry. The choice of the radiations energy and the source-detector distance offers many possibilities of application as indicated in Figs. 1 and 2.

Each turbidimetric data is obtained with countrings of 10^2 - 10^3 s or longer time periods and is related to a volume of turbid water of several tens of m^3.

If a gamma-transmission gauges is considered, the intensity, I, of monoenergetic electromagnetic radiation transmitted through X cm of water containing a concentration C, by weight of sediment is related tothesourceintensity I_o, in the absence of the water and sediment, by the equation:

$$\frac{I}{I_o} = \exp\left\{-\rho_m x\left[\mu_s C + \mu_w(1-C)\right]\right\} \qquad (1)$$

Where μ_w and μ_s are the mass attenuation coefficients of water and sediment, respectively, and ρ_m is the density of the sediment-water mixture. [Florkowski, 1968]. The density ρ_m is given by:

$$\rho_m = \frac{\rho_s \rho_w}{\rho_s - C(\rho_s - \rho_w)} \qquad (2)$$

where ρ_s and ρ_w are the densities of the sediment and water.

The gamma-ray scattering gauges are based on the measurement of scattered X or γ-rays which are attenuated by the suspended sediments. The measured intensity, I, is a function of suspended sediment concentration, C. The experimental relationship, can be written as:

$$\frac{I}{I_w} = (a\,\rho_m)^n \exp(-a\,\rho_m) \qquad (3)$$

where I, is the intensity of the measured radiations in turbid water, I_w, the intensity of the radiations measured in clear water, ρ_m, the density of water-sediment mixture, a, and n, with $1 < n < 2$, are constants which characterize the gauges [Tazioli & Caillot, 1983].

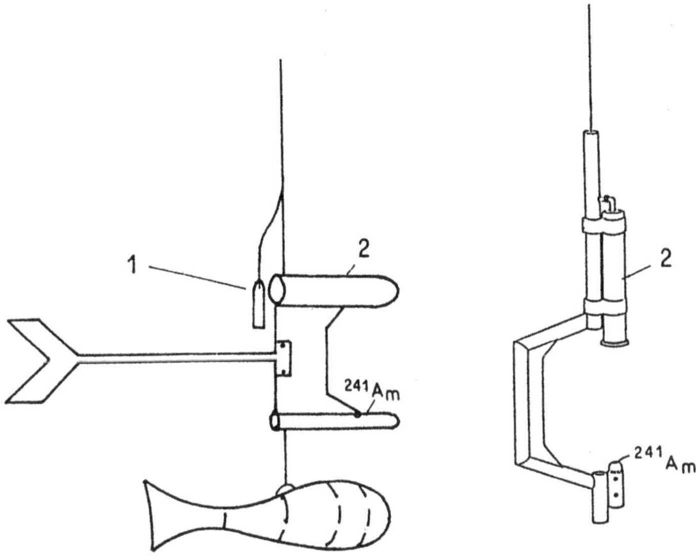

FIG. 1 - Gamma-ray transmission gauges equipped with sources of 241-Am of 100 mCi. 1: depth probe; 2: NaI(Tl) gamma scintillation detector.

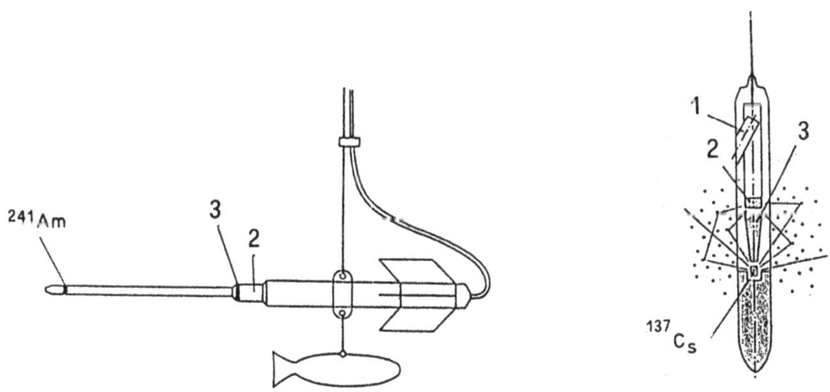

FIG. 2 - Gamma-ray scattering gauges for measuring suspended sediment concentration, A [after Florkowski, 1970] and for measuring bed density, B [after Caillot et al., 1984]. 1: depth probe; 2: NaI(Tl) gamma scintillation detector; 3: lead shild.

The sensitivity and accuracy of measurements increase when the X or γ-rays energy decreases. But it must be considered that the measurement depends on the elemental composition of the measured medium and for this reason low energy electromagnetic radiation gives an important variation on the counting with small variation of sediment concentration since then $\mu_s \gg \mu_w$. On the other hand the gauges with lower energy of γ-radiation will be more sensitive to variations of the chemical composition of sediments.

In practice the nuclear gauges have to be manufactured for each particular situation.

2.1. Calibration

The correct functioning of a nuclear gauge depends on calibrations undertaken both in the laboratory and in situ. In the first instance, calibrations are carried out in tanks using suitable salt solutions. A calibration curve is determined by relating the counting rate N to the varying concentration values. The validity of the curve is then verified in situ by sampling during flood events.

Fig. 3 shows an example of calibration curves for transmission gauges.

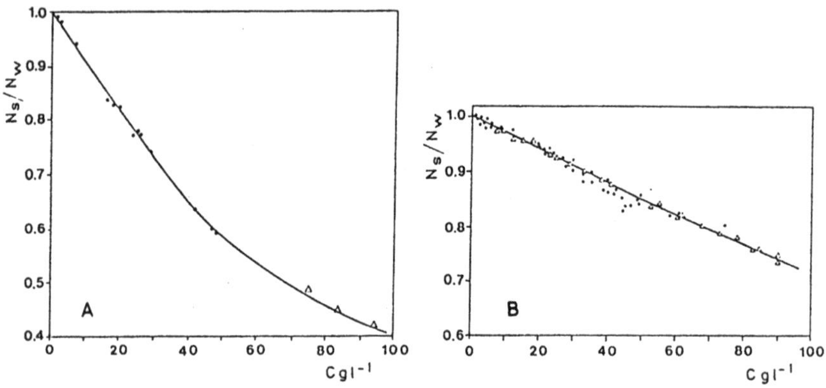

FIG. 3 - Calibration curves of Am-241 and Cs-137 gauges obtained with the use of silty clay and quartz-feldspathic sand. Triangles in Fig. A indicate samples values during flood events. N_s/N_w: ratio between observed count and clear-water count; A: Am-241 gauge; B: Cs-137 gauge [after Tazioli, 1980].

The relative sensitivity, S, of measurement is the ratio of the relative change in intensity, I, to the relative change in sediment concentration, C, as:

$$S = \frac{d\,I/I}{d\,C/C} \qquad (4)$$

The sensitivity of measurement is greater at lower energies because of the increased difference between μ_s and μ_w. For very low energies problems arise from the increased influence of variations in chemical composition of sediments. A reasonable compromise is provided by 241-Am sources, which have an energy of 60 KeV.

The accuracy of the measurements is extimated to be better than 1 gl^{-1}

for the ^{241}Am gauges and about 2 gl^{-1} for ^{137}Cs gauge, for counting times of 10^3s.

The major factors influencing occuracy are the variation of chemical composition of the sediments, the effect of temperature, the electronic drift and the variation of water salinity when working in estuaries.

Several years of field measurements in different sedimentological situations have however shown that the effects due to temperature, salinity and chemical composition of sediments are negligible if these parameters are taken into account with suitable recordings.

3. Field application

The measurement of sediment transport in streams and rivers carrying a heavy solids charge constitutes a major opportunity for the application of nuclear gauges. It is served both by portable gauges (Figs. 1 and 2) that are carried to site for intermittent but flexible measurements by trained personnel, and by fixed gauges established as permanent installations. The purpose of the latter is to allow continuous unattended recording of concentration variations with time (Figs. 4 and 5).

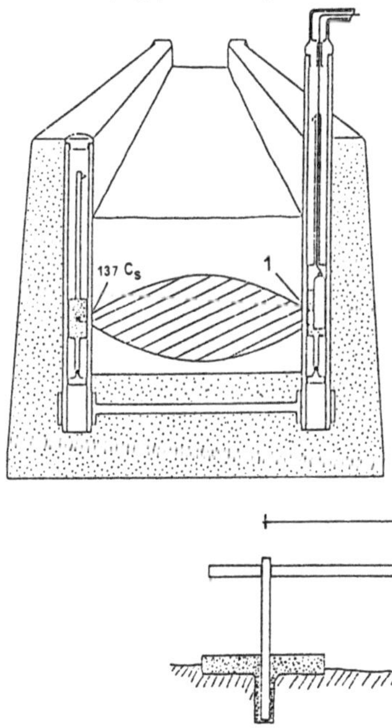

FIG. 4 - Stream gauging flume and sediment concentration gauge installed in a small watershed in Italy [after Frega et al., 1976]. 1: NaI(Tl) gamma scintillation detector.

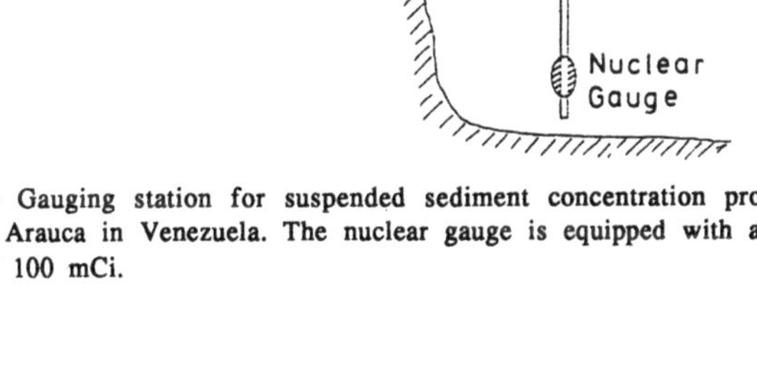

FIG. 5 - Gauging station for suspended sediment concentration projected for the river Arauca in Venezuela. The nuclear gauge is equipped with an ^{241}Am source of 100 mCi.

Field applications of nuclear gauge carried out during a period of 18 years in Italy for measuring suspended sediment concentration in torrential water courses have given interesting results. The measurement was made with fixed installations and in different hydrological situations [Tazioli, 1981].

4. Example of application

A gauging station is still functioning in Italy on the river Musone, near Ancona. The nuclear gauge is a permanent installation fixed by steel pipes on a concrete foundation in such a way that it measures the concentration in a solide of influence between 0.2 m from the river bottom to 0.50 m (Fig. 6).

The gauge is equipped with an americium-241 radioactive source of 100 mCi and a gamma scintillation detector (NaI/T1) which allows measurements of the concentration for values between 1 and 100 g l^{-1}, with an accuracy of about ± 1 gl^{-1} within the range of 1 to 50 gl^{-1}. The sensivity of detection is 50 counts per second (cps) for a variation of 1 gl^{-1}. The minimum water stage for accurate measurement is 0.5 metres from the river bottom.

The electronics for recording sediment concentration is made of a ratemater and a data-logger whose electronic memory allows measurements during a period of 3 months (Fig. 7). The same data logger is used for recording air and water temperature, rainfall and waterstage. The stored data are transferred each month in a magnetic disch with a portable computer.

The data, calculated from counts cumulated from 15 minutes periods during an entire flood, represent the average sediment concentration in the volume of water flowing through the volume of influence of the gauge. This volume can be estimated up to several hundred cubic metres of turbid water.

The stream flow velocities at the gauging station are quite uniform throughout the cross section, so that it can be inferred that the value of the sediment concentration has little lateral variation (Fig. 6). Controls carried out in the left and right branches, and on the vertical distribution of suspended sediment concentration, have shown that there is vertical and lateral homogeneity during flood events exceeding 30 m^3 s^{-1}.

The flood discharge is derived from rating curves drawn with data obtained from current meters and tracer measurements.

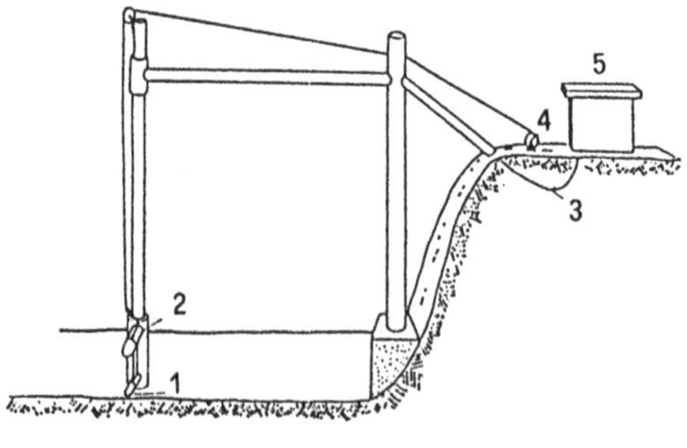

FIG. 6 - Gauging station on the river Musone in Italy. 1: ^{241}Am source (100 mCi); 2: gamma scintillation detector; 3: electric wires; 4: windless for moving vertically the gauge; 5: box for the electronic equipment.

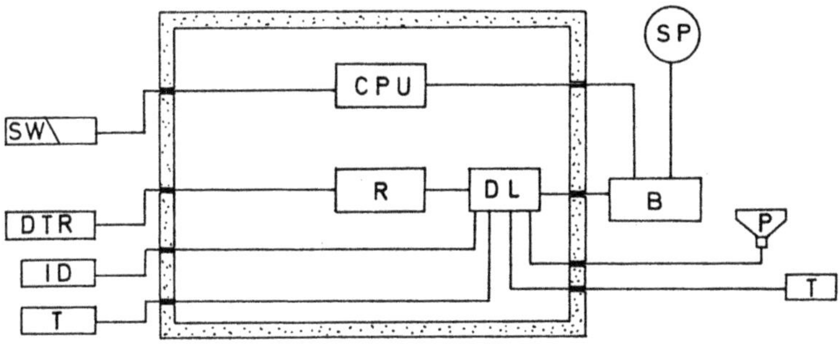

FIG. 7 - Block diagram of the turbidimeter. CUP: control Unit Power; R: ratemeter; DL: Data-logger; DTR: gamma-scintillation detector; WS: water stage; T: temperature; P: rainfall; B: lead batteries; SP: photovoltaic solar panel.

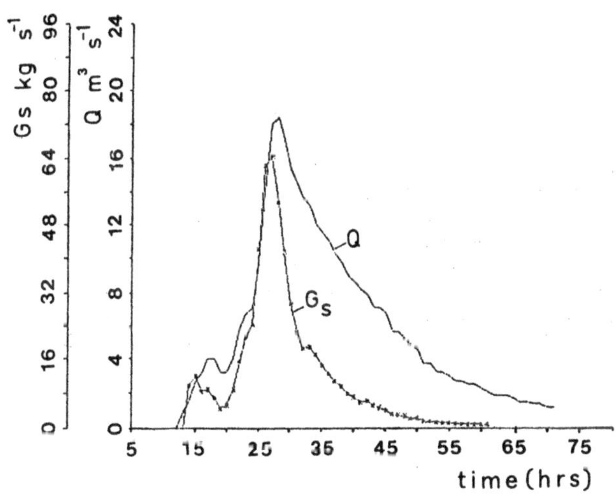

FIG. 8 - Example of a flood event recorded on the gauging station of river Musone.

Since runoff is torrential, and produces flash flooding, the gauge functions fully automatically by means of a switch which activates it when water depth exceeds 0.40 m.

The gauging station controls a watershed of 460 km^2 in surface. The maximum liquid discharge, Q, recorded is about 250 $m^3 s^{-1}$ with a suspended sediment concentration of 30 $g\ l^{-1}$.

An example of recording of flood event is shown in Fig. 8.

5. References

1. A. Caillot, G. Meyer, D. Chambellan & J.C. Tanguy, *A New nuclear gauge to measure the turbidity in muddy areas*, Coastal Engineering Conf., Huston (1984).
2. P. Ciet & G.S. Tazioli, *Metodologie radioisotopiche per la misura del trasporto solido in sospensione*, Geol. Appl. e Idrogeol., **XI**, parte I, (1976) 39-53.
3. M.J. Crickmore, G.S. Tazioli, P.G. Appleby & F. Oldfield, *Nuclear Gauging of Sediment Concentration*. In: The Use of Nuclear Techniques in Sediment Transport and sedimentation problems, Int. Hydrological Programme IHP-III, Project 5.2, SC-90/WS-49, UNESCO, (1990) 29-63.
4. G. Courtois, F. Anguenot C. & Magloire, *Turbidimétrie: Absorptiométrie X-g ou radioactivité naturelle?*, Proc. Symp. on Isotope Hydrology, IAEA, Vienna (1970) 555-577.
5. T. Florkowski, *Suspended load*. In: Guidebook on nuclear techn. in hydrology, Tech. Rep. Series n. 91, IAEA, Vienna (1968) 51-55.
6. T. Florkowski, *Portable radioisotope gauges for suspended sediments*, Proc. Symp. on Isotope Hydrology, IAEA, Vienna (1970) 545-554.
7. G.A. Frega, G. Lanzafame, V. Marone, G.S. Tazioli & G.G. Zuffa, *Il bacino sperimentale del T. Ilice (Calabria)*, Geol. Appl. e Idrogeol., **XI**, parte I, (1976) 65-95.
8. G.S. Tazioli, *The measurement of suspended sediment transport in natural streams using automatic radioisotope gauges*, J. of Hydrol., **47**, (1980) 173-185.
9. G.S. Tazioli, *Measurement of solid transport in natural rivers by nuclear techniques. Some examples of application in clayey basins*, (Invited paper),
   Int. Symp. on Erosion and Sediment transport measurement, Florence, 22-26 June 1981, IAHS-AISH, Publ. 133, (1981) 63-81.
10. G.S. Tazioli & A. Caillot, *Measurement of suspended sediment by nuclear techniques*. In: Guidebook on nuclear techniques in hydrology, Tech. Rep. Series 91, IAEA, Vienna (1983) 93-99.
11. G.S. Tazioli, P. Billi & P. Tacconi, *On Fluvial Dynamics in Italy*, in: The Lithosphere in Italy, Advances in Earth Science Research, Atti Conv. Lincei 80. Accademia Nazionale dei Lincei, Roma 5-6 May 1987, (1989) 467-501.

www.ingramcontent.com/pod-product-compliance
Lightning Source LLC
La Vergne TN
LVHW051053210926
848486LV00002B/96
* 9 7 8 9 8 1 0 2 1 8 5 4 6 *